D1087765

Springer Series on Environmental Management

Robert S. DeSanto, Series Editor

Global Fisheries

Perspectives for the 1980s

Edited by
Brian J. Rothschild

With 11 Figures

Springer-Verlag
New York Berlin Heidelberg Tokyo

Brian J. Rothschild
Center for Environmental and Estuarine Studies
University of Maryland
Solomons, Maryland 20688, U.S.A.

Library of Congress Cataloging in Publication Data
Main entry under title:
Global fisheries perspectives for the 1980s.
(Springer series on environmental management)
Contents: Extended fisheries jurisdiction and the new Law of the Sea / William T. Burke—Implementing the new Law of the Sea / Alberto Székely—Jeffersonian democracy and the fisheries revisited / J.L. McHugh—[etc.]
1. Fisheries. 2. Fishery management. 3. Fisheries—Research. 4. Fishery law and legislation. I. Rothschild, Brian J., 1934- . II. Series.
SH331.G57 1983 333.95′6 82-19460

Typeset by Publisher's Service, Bozeman, Montana.
Printed and bound by R.R. Donnelley & Sons, Harrisonburg, Virginia.
Printed in the United States of America.

9 8 7 6 5 4 3 2 1

ISBN 0-387-90772-6 Springer-Verlag New York Berlin Heidelberg Tokyo
ISBN 3-540-90772-6 Springer-Verlag Berlin Heidelberg New York Tokyo

Series Preface

This series is dedicated to serving the growing community of scholars and practitioners concerned with the principles and applications of environmental management. Each volume is a thorough treatment of a specific topic of importance for proper management practices. A fundamental objective of these books is to help the reader discern and implement man's stewardship of our environment and the world's renewable resources. For we must strive to understand the relationship between man and nature, act to bring harmony to it, and nurture an environment that is both stable and productive.

These objectives have often eluded us because the pursuit of other individual and societal goals has diverted us from a course of living in balance with the environment. At times, therefore, the environmental manager may have to exert restrictive control, which is usually best applied to man, not nature. Attempts to alter or harness nature have often failed or backfired, as exemplified by the results of imprudent use of herbicides, fertilizers, water, and other agents.

Each book in this series will shed light on the fundamental and applied aspects of environmental management. It is hoped that each will help solve a practical and serious environmental problem.

Robert S. DeSanto
East Lyme, Connecticut

Contents

Contributors

Lee G. Anderson, College of Marine Studies, University of Delaware, Newark, Delaware 19711, U.S.A.

Gary D. Brewer, School of Organization and Management, Yale University, New Haven, Connecticut 06520, U.S.A.

William T. Burke, School of Law, University of Washington, Seattle, Washington 98105, U.S.A.

David H. Cushing, 198 Yarmouth Road, Lowestoft, Suffolk, NR32 4AB, England

John A. Gulland, Marine Resources Service Fishery Resources and Evironment Division, Food and Agriculture Organization of the United Nations, 00100 Rome, Italy

James Joseph, Inter-American Tropical Tuna Commission, c/o Scripps Institute, La Jolla, California 92037, U.S.A.

G. L. Kesteven, 12 O'Brien's Road, Hurtsville, N.S.W. 2220, Australia

Peter A. Larkin, Institute of Animal Resource Ecology, The University of British Columbia, Vancouver, British Columbia, Canada V6T 1W5

J. L. McHugh, Marine Sciences Research Center, State University of New York, Stony Brook, New York 11794, U.S.A.

Brian J. Rothschild, Center for Environmental and Estuarine Studies, University of Maryland, Solomons, Maryland 20688, U.S.A.

Alberto Székely, Misión Permanente de México, Colonia Narvarte, Mexico 12, D. F., Mexico

Jean-Paul Troadec, ISTPM, Rue de l'Ile de Yeu, BP 1049, 44037 Nantes Cedex, France

1
Introduction

Brian J. Rothschild

The extension of fishery jurisdiction by most of the coastal States was the dominant event in global fisheries during the decade of the 1970s. The extensions changed the open-access regime to an extended jurisdiction fisheries management regime. Under the open-access regime, the coastal States had little management control over the stocks of fish. Fishing was accompanied by considerable economic waste, many stocks were overfished or depleted to historically low levels of abundance, fishing in the distant waters off coastal States diverted economic benefits away from those coastal States, and the capability and effectiveness of fishery management organizations became a matter of global concern.

The extensions of jurisdiction, however, provided the coastal States with the opportunity to control fishing and actually manage the stocks of fish in their newly claimed fishery zones. In principle, under extended jurisdiction, the coastal States had the authority to eliminate economic waste and to reduce excessive fishing pressure. In addition, the coastal States could eliminate extensive and intensive foreign fishing and thus channel benefits from their coastal resources into their own economies.

The extension of jurisdiction came at an especially propitious time. In the 1970s the annual increases in the global fish catches that had been experienced in earlier years had diminished considerably. The stabilization of the global catch required the notion that those stocks that comprised the catch had to be utilized with greater efficiency than they had in the past. This efficiency could only be insured through improved management, and the extended jurisdiction regime made effective and efficient management possible.

However, the anticipated benefits did not fully materialize. To be sure, distant water fishermen were driven from their traditional grounds off the coastal

States, or charged fees for the right to fish in the extended jurisdiction zones, but other than this, wherever active management was attempted, it did not appear to work well; many of the old problems of management under the open-access regime remained and new ones developed. Even in the absence of hard data on the economic performance of management, people haggled over boundaries, over objectives, over quotas, over the right to fish, over what optimum yield meant, over data, and even over whether fishery management was a worthwhile enterprise. In addition, basic fishery management data, the data that enable determination of the magnitude of the stock, deteriorated. Further, the applicability of standard management procedures, particularly to multiple-species fisheries, was challenged, and it became apparent to many that traditional approaches to enforcement of regulations was not cost effective.

Now, some several years after the advent of extended jurisdiction, the problems do not appear to have been resolved. What is the difficulty? Could it be that fisheries are just too complex, that fishery management is performing as well as it can, and we should therefore be satisfied with its performance. Likewise, could it be that fishery management as we have always known it and as we practice it now, while suitable for an open-access regime, is not particularly useful under extended jurisdiction? If so, what do we need to change? Are there concepts to be formulated, actions to be taken, and programs to be implemented in order to improve management and materially improve the economic efficiency and the societal benefits that accrue from harvesting the fishery resources of the world?

At present, there are no simple answers to these questions. Among the multitude of papers and books that have been written on fisheries under the extended jurisdiction regime it is difficult to find a systematic formulation of the management problem. One such formulation (FAO 1979, 1980) outlines a systematic analysis of the fishery management problem and calls for a broader and longer range view of the management problem than had heretofore been taken. For this book we have attempted to contribute to this broader and long-range view by assembling a series of personal perspectives on management strategy and policy formulation in global fisheries. The present book evolved from an earlier pre-extended jurisdiction study (Rothschild 1972). In developing the present collection of perspectives we took the opportunity to include, where possible, the contemporary views of the contributors to the early volume.

The first of these perspectives concerns the "Law of the Sea," the shaping force for the present extended jurisdiction regime. William T. Burke explores the evolution of the Law of the Sea during the past two decades. Burke outlines the claims and counterclaims regarding which States should control access to the stocks, the degree of exclusivity of coastal State authority, the obligations of the coastal States with respect to management, the species that should be managed by the coastal State, and the geographical extent of their jurisdiction. Burke traces the parallel stream of events involving the United Nations Conference on the Law of the Sea (UNCLOS) and the unilateral actions on the part of the coastal States to establish claims while the UNCLOS involved itself in the lengthy process of developing a draft treaty.

The emergent Law of the Sea has affected each coastal State. Alberto Székely outlines his perceptions of the Mexican experience in implementing their fishery jurisdiction. Székely describes how Mexico entered the Law of the Sea process as a moderate and how it took leadership in developing consensus views early in the Conference. Eventually Mexico became caught up in events and extended its fishery jurisdiction. Székely points out how extended jurisdiction left Mexico, on the one hand, with sudden control of its fishery resources and, on the other hand, with a need to determine the best way it could employ its human and budgetary resources to harvest the stocks, as well as to participate in difficult fishery negotiations with those States that fished off of the American coast before extended jurisdiction.

In the United States, fishery management was also greatly affected as a result of extended jurisdiction. J. L. McHugh discusses the performance of fishery management in selected fisheries of the northeast United States before and after extended jurisdiction. In this region, before extended jurisdiction the States managed only stocks that were found predominantly in the inshore areas (i.e., inland estuarine waters or 0 to 3 miles from the coast). Management performance was mixed and part of the difficulty related to the fact that the "Jeffersonian Democracy" contributed to dividing the management authority on single stocks among several States. After extended jurisdiction Regional Councils developed fishery management regulations. While the Councils were regional, there was a greater federal influence in that the U.S.Federal Government needed to approve management plans. McHugh outlines his perception of the yet unsolved difficulties of managing fisheries in the northeast United States.

Yet another case history of extended jurisdiction performance is given for the Northwest African fisheries by Jean-Paul Troadec. The area under consideration ranges along the coast of Africa, from Morocco to Guinea, one of the important fishery areas in the world. Catches in these waters in the late 1970s amounted to 2.7 million tons, but most of the fish were taken by distant water fishing countries, rather than by the coastal States. Extended jurisdiction has given the coastal States the authority to manage the stocks, but there are problems in developing the fisheries for the direct benefit of the coastal States. Troadec indicates that the development problems relate to the unwarranted use of the conventional fishery development model, oriented toward large-scale fishery operations, rather than a model that focuses on the unique properties of small-scale artisanal fisheries. He also sees the need to develop markets, not necessarily through conventional marketing approaches but through the development of infrastructure, which will enable fishery products to reach the markets.

While most coastal States had concern for the management of fish stocks that they perceived to be within their fishery zones, other problems began to emerge concerning the valuable tuna fisheries, single stocks of which existed in the fishery zones of numerous coastal States. Under the open-access regime tunas were taken primarily by distant water fishing nations. The Law of the Sea discussions stimulated anxiety in the relations between the distant water nations and the coastal States. The anxiety generated two lines of thought. The first was that jurisdiction over tuna management belonged to each coastal State in whose waters the tuna were found. The second was that because tuna were "highly migratory" or in any event because their range was intersected by many interstate boundaries as well as the boundary between the fishery zone of each state and the High Seas, tuna should be managed by an international organization which would have management competence over each stock throughout its range. James Joseph describes the various points of view along with the history of tuna management. He shows how the UNCLOS text postures the decade of the 1980s to provide for clearer definition of how tuna will be managed in the future. Joseph discusses the interpretation of the Law of the Sea text regarding tunas. If additional actions are to be taken with respect to managing the tuna, steps to implement the actions will be necessary and Joseph outlines a format for tuna management in the 1980s and develops a structure of a global tuna management body.

To deal with the variety of problems attendant to extended jurisdiction Brian J. Rothschild considers the evolution of the extended jurisdiction regime and observes that while it can be characterized by significant changes in authority, there have been few innovations to meet the new fishery management requirements. If improvements are to be made, they will most likely need to be made in terms of changing the setting under which the more traditional management operates. In changing the setting, various components of the management system need to be placed in the context of the extended jurisdiction regime. These include the involved individuals and their motivation, the structure of access, the quality and quantity of technical information, the role of governments, the role of international organizations, small-scale fisheries, and the traditional fishery management paradigm. Rothschild develops a perspective for fishery management in the 1980s based upon these considerations.

John A. Gulland addresses the imperfections in the fishery management system. He develops the insight that perfection in any setting as complex as fishery management will be difficult to attain. He analyzes the causes of the imperfections as they relate to fishery management system components: objectives, fishery system, data, analysis, decision making, implementation, and enforcement. He finds that "The aim of the fishery manager should therefore not attempt to determine and implement ideal management, but determine actions that are practicable and make the fishery better." At the same time he sees a need for a change in approach that will require "a wider recognition that whatever is being done at the moment will have to be changed." An important element of this change will be the need to take a broader view of fishery management.

Taking another strategic view of the problem, Garry D. Brewer describes the new requirements of fishery policy and management from the perspective of a policy expert. He observes that there are increasing pressures to intensify the harvest of the resources, and this intensification is likely to result in societal costs unless explicit management measures are developed. He categorizes these into preventative, active, and reactive management structures and shows how development of each can improve fishery management. Brewer perceives that the development of a new management architecture will require the direct involvement of the broad community of individuals who affect or are affected by fishery management. He sees educating the community on the benefits that could accrue from fishery management as laying at the foundation of the new fishery management architecture.

Lee G. Anderson considers criteria that will need to be developed to choose among alternatives in fishery policy development. Anderson focuses on management criteria not only as an aid to choosing among alternatives, but as a component of developing objectives. Further, because economic criteria are relatively quantifiable, they can often serve as points of departure to develop the more qualitative criteria that often constitute the essence of policy problems. Anderson places the policy development process in the context of economics. He stresses the nature of the process of policy development in setting management objectives.

Policy cannot be effective unless it can be implemented. The linkage between policy and implementation is thus critical. The nature of this linkage is often in itself a component of policy. The operation of this policy component resides to a large extent in the structure of the fishery management organization. Peter Larkin considers the fishery management organization in terms of how much management is needed in the context of the size of the management structure and how management should be organized. His analysis of the dynamics and statics of management agencies provide us with considerable and important insight as to constraints, incentives, and disincentives in large bureaucracies. These properties clearly affect creativity and efficiency and leave no doubt that policies need to be developed to improve the management environment. Larkin's observations are also critical to the design of new fishery organizations. Identification of past problems can assist in designing new systems that avoid the difficulties of the past.

The development of research and management organizations in developing countries is of particular concern to those involved in implementation and organization development. G. L. Kesteven shares some of his extensive experience in this field. He stresses the need for the developing countries to become self-sufficient in fishery research. He describes the nature of the research. In addition to focusing upon problems critical to the developing countries the research should be oriented toward developing a causal or predictive understanding of phenomena which can be used to develop as near real-time predictions as possible. The conduct of such research would then be of maximum use to decision making. Such an approach would, of course, reverse the usual trend of the investigation, thus concentrating on management per se first, and last on minute although not necessarily unimportant communication with the fishermen and the fishing industry. The development of Kesteven's philosophy will take considerable but worthwhile commitment, and for those who are willing to try, Kesteven provides in an Appendix a primer on running a fishery research organization.

Finally, David H. Cushing places the final layer on the palimpsest by providing an outlook for the next 10 years. He provides a background by elucidating the methodology of population dynamics which is central in contemporary fishery science and places in balance the events of the last 10 years with the events of the next 10 years. In the future Cushing sees fisheries addressing the specialized individual needs of the developing and developed countries especially with regard to the utilization of smaller and more efficient vessels. In addition, he foresees that fishery science will improve the techniques of cohort analysis, develop small experimental multispecies models, and continue to deal with the difficult problem of stock and recruitment. Also, management, not surprisingly, will become more complicated because there will be more choices for both fishermen and the managers. Cushing says, "the important point is that the coastal State has the power to act." To effect action, though, those involved need to clearly understand how the action will affect them and since fishery management is so complex, this will be a considerable challenge.

Thus, we have tried to provide a framework for thinking about fishery policy under the extended jurisdiction regime of the 1980s. The framework is a collection of perspectives on the state of fishery policy and management and what may be done about it. In any event, though, prescriptions can only be details without a concerted action on the part of the global fishery community in recognizing that there is a problem, carefully formulating it, and designing a strategy over the next decade or two for its resolution. Many thanks are owed to Janice Silverstein, who prepared the index.

References

FAO ACMRR Working Party on the Scientific Basis of Determining Management Measures. 1979. Interim Report of the ACMRR Working Party on the Scientific Basis of Determining Management Measures. Rome, 6-13 December 1979. FAO Fish. Circ. 718, 112 pp.

FAO ACMRR Working Party on the Scientific Basis of Determining Management Measures. 1980. Report of the ACMRR Working Party on the Scientific Basis of Determining Management Measures. Hong Kong, 10-15 December 1979. FAO Fish. Rept. 236, 149 pp.

Rothschild, B. J. (ed.). 1972. World Fisheries Policy. Multidisciplinary Views. Univ. Washington Press, Seattle. 202 pp.

2

Extended Fisheries Jurisdiction and the New Law of the Sea

William T. Burke

Introduction

International law concerning ocean fisheries is still in the process of radical transformation, but some important components of the new regime appear to be in place. The principal change, taking effect widely only since the mid-1970s, is the now commonly recognized extension of national jurisdiction to 200 nautical miles from the baseline. Accompanying this extension of national authority and in substantial measure the major impetus for it, the 3d United Nations Law of the Sea (UNCLOS) conference has elaborated jurisdictional and substantive principles for fisheries regulation in the Exclusive Economic Zone that is a major innovation of the conference.

The following discussion of extended fishery jurisdiction and international law examines major elements in the authority claimed by coastal States through national legislation or proposals considered in the UNCLOS negotiations that produced the Draft UNCLOS Convention. Opposition to these claims is similarly identified. For present purposes, the claims relate to the basic competence to extend coastal authority over fisheries beyond the Territorial Sea; the nature and scope of extended authority; the species subjected to coastal authority; and, finally, the outer limit of coastal jurisdiction.

William T. Burke is professor of law and of marine studies at the University of Washington in Seattle, where he teaches international law of the sea and participates in graduate programs of study in law and marine affairs, and in marine policy. He has served as an adviser to the United States Government agencies concerned with the Law of the Sea.

After brief identification of claims, a section on clarification of policy summarizes my preferences regarding the balance of competing interests at stake in relation to each category of claims and counterclaims identified. The perspective here is preferential, expressing, in abbreviated fashion, those considerations that appear to justify one policy outcome rather than another on each major issue addressed.

Discussion of trends in decision seeks to highlight the authoritative choices by decision makers on the various issues since the 1960 conference on the Law of the Sea. The major focus is upon state practice and the UNCLOS negotiations, with particular attention to the latter. Although the Draft UNCLOS Convention is, at this writing (in December 1981), still not in its final, adopted form, the provisions on fisheries in the Exclusive Economic Zone and the High Seas are generally considered to be in their final form. Accordingly, their interpretation and significance are prime ingredients of discussion.

A final section seeks to compare the policies recommended with the decisions expected after considering trends and to call brief attention to the problems that appear still to persist, where the discrepancy between goal and decision is especially noticeable.

Knowledgable participants in the decision process will hopefully seek to close the gap between preference and reality, but no effort is made here to recommend courses of action. It seems likely, however, that future efforts to cope with management of world fisheries must emphasize cooperative methods. The extension of coastal jurisdiction substantially accomplishes the nationalization of living marine resources but does so only partially and with less than complete effect. International cooperative measures will continue to be necessary for fisheries found within national zones as well as for those found mostly beyond. Coastal States in many instances cannot effectively manage fisheries wholly within their zone for lack of skill and resources. Perhaps, as time passes, the perspectives will emerge that are necessary to support more inclusive decision making.

Claims and Counterclaims

For centuries the basic claim by nations to exercise authority over marine fisheries insisted that access to them must be open to all beyond a narrow belt of national territory in the ocean. Under this traditional system claims to conserve and to manage marine fisheries essentially called for international agreement, because most fishery activities occurred beyond the Territorial Sea.

In the past two decades this pattern of national claims and responses has undergone a remarkable transformation. The major shift has been to claims to extend coastal authority outside national territory. Among the principal claims made, which will be examined in the discussion, are the following:

1. The competence to extend coastal authority over fisheries beyond the Territorial Sea
2. The nature and scope of extended authority, including effects on other uses
3. The fishery management objectives to be pursued

4. The species subject to extended jurisdiction
5. The outer limit of coastal State jurisdiction

First, the most fundamental claims over the past several decades regarding fisheries concern who should decide about their conservation and distribution among harvesters. Coastal States in increasing numbers over the years insisted through unilateral legislation that the coastal State could lawfully extend some degree of control over living resources beyond national territory. Other States, primarily those with significant distant water fishing interests, demanded either continued unrestricted access to coastal fisheries or an access subject only to mutually agreed restrictions.

Second, other conflicting claims centered about the nature and scope of coastal authority, however it might be established. Coastal States claimed exclusive competence to dispose of all, or nearly all, living resources within an area of extended jurisdiction and to prescribe and to enforce policy for managing and protecting such resources. Opposing claims rejected the notion of exclusive rights and advanced some lesser form of right, either preferential or even lesser. Some opposition claims would distinguish between States on the basis of their economic development, according greater coastal State rights to developing nations. The coastal State's authority to affect other uses of the ocean when exercising fishery jurisdiction gives rise to opposing claims to protect, *inter alia*, freedom of navigation.

Third, claims are also in conflict regarding the obligations of coastal States to observe certain constraints expressed as management objectives. A principal focus of management objectives has been upon the level of resource abundance that the fishery manager is to strive to restore or maintain. One claim is that the coastal State should be obliged to manage the fishery so as to maintain stock abundance at the level which will permit the maximum sustainable yield (MSY). Another and different objective is to manage the stock with the aim of satisfying social, economic, and political objectives although this may require reducing stock abundance below that which would provide the maximum sustainable catch. More significantly, seeking such objectives might mean that stock abundance must be maintained even higher than MSY would require and that no fishing be allowed despite the "surplus." There are demands also that management objectives, as constraints on coastal authority, take account of differences in species, especially marine mammals, and provide for the application of different policies to them.

A somewhat different set of interests and objectives comes into conflict in considering the scope of coastal authority to protect resources by measures that might affect freedom of navigation. Coastal State claims seek to protect fisheries against illegal harvesting, while flag States contend that no impediments can be introduced affecting navigation.

A fourth set of opposing claims concerns the scope of coastal authority in terms of the species included. The most comprehensive coastal claim embraces all living resources within the asserted jurisdiction but other States maintain that the life history of some species is such they they should be and are excluded from any single State authority. The latter claimants urge that international institutions be employed to undertake necessary management.

A fifth, and final, set of claims addresses the geographical extent of coastal State jurisdiction. This claim has varied through time beginning at a modest 12 miles and expanding over the years to 200 miles and sometimes beyond. Some States have advocated that no specific mileage limit be employed and that authority be distributed in accordance with the life history and migratory character of the animals.

The claims and counterclaims described above are to be seen both in national practice, manifested by legislation and executive pronouncements, and in the proposals advanced in multilateral conferences, including the Third United Nations LOS Conference.[1] The following section on clarification of policy discusses the policies at stake in resolving this welter of claims. The preferential statements record only my preferences.

Clarification of Policy

It is common knowledge by now that most coastal States assert and exercise jurisdiction over fisheries within a 200-mile exclusive zone. For the most part the following discussion centers about preferred policies in light of this situation. However the initial issue addressed, competence to exercise exclusive fishing authority, briefly considers the policy question as an original matter, without regard to the uniform extension of national jurisdiction. The remainder of the policy discussion either examines questions posed by such extension or questions pertinent to any management institution.

Competence to Exercise Exclusive Fishing Authority

It has been evident for decades that the single most pressing need of fishery management in the world ocean was for agreement on a single central authority to make enforceable decisions about discrete fisheries. It would do, and did, relatively little for effective management to adopt appropriate policies if there were no mechanism to make them effective.

An optimum arrangement would be a management entity negotiated among and established by those nations that were most concerned with and involved in the fisheries of interest. If such an entity could be established and provided with the necessary authority to prescribe and to apply policy, the result would be to permit those most affected by the decisions to have a direct role in making them and would assist substantially in securing compliance.

[1] The record of the above claims may be found in collections of national legislation and in the records of international negotiations. The principal documentation for state legislation is the *United Nations Legislative Series* and Nordquist *et al.*, *New Directions in the Law of the Sea*, a (currently) 11-volume compilation of documents involving the Law of the Sea. The prime source for proposals for the new law of the sea is in the records of the Third United Nations Conference on the Law of the Sea.

As will be noted further below, this approach was tried without sufficient access. For decades some of the most intensely fished areas of the world, in the northeast and northwest Atlantic, were subject to the jurisdiction of multilateral fishery commissions, established or adhered to by all interested States in addition to the coastal State in modest exclusive fishing zones. A prime difficulty, perhaps the most important, was that prescribed measures could not be made effective and some nations plainly did not comply with measures they agreed to employ. Elsewhere around the globe no comparable international agencies were established despite excessive fishing.[2]

If such international means of establishing control are ineffective or nonexistent, agreed extension of national jurisdiction is appropriate but such a preference is more complicated than might appear. The essential natural characteristics of the ocean, and of the living resources in it, make it unlikely in many important instances that a single coastal State can exercise the necessary authority effectively to control a fishery (Gulland, 1980). Fish are mobile and move laterally along coastlines, migrate from inshore, shallow water to offshore, deeper water (and vice versa), and in some instances range for hundreds and thousands of miles throughout entire expanses of ocean. In these circumstances a coastal State cannot by itself establish effective policies because exploitation may and does occur in regions beyond the reach of those policies. In such instances there should be an obligation on the States involved, whether it be adjacent coastal States or distant water fishing States whose vessels operate outside any national jurisdiction, to seek express agreement in the necessary management measures. Accommodation of interests by negotiation is more likely to leave these nations better off than to attempt to exercise conflicting exclusive controls.

The most difficult problem here is what to do in such circumstances if the States concerned seek agreement and fail. Since by definition no single State can wield effective authority, attempts to exercise control could be counterproductive or have minimal or no impact, depending on the particular situation. An obvious solution is to employ dispute settlement procedures to break the impasse. Should this fail also, the best the parties can do is to attempt to continue fishing but to do so with careful regard for coordination with foreign authority.

In a context of ineffective or unacceptable international institutions for management, the major problem is how to achieve agreed extension of national jurisdiction. Negotiations to that end are preferable because the effects of enlarged coastal jurisdiction will be widely felt and those affected should participate in the decision. Unilateral moves toward extended coastal control would be acceptable policy only as a last resort to install management where its absence caused general loss. Preemption of an internationally negotiated solution by unilateral State action should be avoided otherwise.

---

[2]For assessment of the state of fishery management prior to the extension of national jurisdiction see the following: Crutchfield and Lawson (1974), Marr (1976), Kasahara and Burke (1973), Tussing and Hiebert (1974), Saila and Norton, *Tuna: Status, Trends, and Alternative Management Arrangements* (1974). Each of the above was published as part of the Program of International Studies of Fishery Arrangements sponsored by Resources for the Future. See also Möcklinghoff (1973).

Nature of Extended Coastal State Authority

If some degree of coastal State authority over fisheries is desirable, or in any event unavoidable, its scope should be sufficiently broad that it permits the coastal State to manage effectively. If anything is apparent from the previous years of management history, it is that international agencies have not been able to manage well, primarily because they lacked the autonomy and adequate authority to be effective. Accordingly coastal States need to have exclusive authority to undertake all the functions of decision that are required for management. It is conceivable that one or another function could be more satisfactorily discharged by some other entity and the coastal State would be free to conclude such an arrangement.

The label attached to coastal authority that is sufficient for effective control over the resources is not important unless it has symbolic value. At the same time, however, no more authority need be claimed or conferred than will enable the coastal State to deal with the management problem of fisheries, including the disposition of them. The region needing management need not be incorporated into the State as part of national territory, nor does authority need to be claimed over any other activity except as the latter may interfere with the coastal State's capacity to protect fisheries from unlawful encroachment or exploitation.

It is a delicate problem to accommodate coastal authority over fisheries with other activities in the marine environment. In general, conflicts between fisheries and navigation, or sea bed exploration and exploitation, or disposal of wastes are not a major difficulty. Insofar as the coastal State also has exclusive authority to control the other activity[3] (which is, or will usually be, the case for other resources or for dumping of wastes) the choice of priorities is solely its own decision. However this is not the case with navigation, which is generally considered to be protected as part of the freedom of the sea and therefore not within coastal State resource jurisdiction.

Fishing rights and navigation could come into conflict because the coastal State may feel a need to adopt protective measures to insure that vessels in transit of its jurisdictional area do not violate fisheries laws and regulations. Flag States would be interested in assuring their vessels the full protection of freedom of navigation in these circumstances and would be especially concerned that no precedent was set that might threaten navigation more generally.

The conflicting interests at stake can best be accommodated by restricting coastal authority over navigation to the maximum degree possible and by recognizing such authority only when it is essential for effective management, particularly enforcement, of fisheries that are unusually beneficial to a coastal State or group thereof. The general community interest in freedom of navigation would thus be largely preserved against encroachment but this policy would give way in

[3]Under current customary law, most States already claim or could claim authority over other resources. However most developed States have not yet established Exclusive Economic Zones and therefore do not have laws and regulations for dealing with this issue. For discussion of the problem and a survey of national legislation see Burke (1981).

situations where the coastal State's need for adequate protective measures was great.[4]

Management Objectives: The Clarification and Specification of Conservation and Allocation Goals

For present purposes the two main elements of policy pertaining to management objectives include, first, the desirability of clarity in expression or communication of such objectives and, second, preferences for the substantive policies concerning conservation and allocation. In this context the key issue to be addressed in allocation is access by foreign vessels to fisheries under national jurisdiction.

It can make a great deal of difference whether or not fishery management objectives are spelled out in an authoritative way. In the first place, there are numerous objectives that might be sought, each of them, or each set of them, serving quite different interests and general purposes (Christy, 1977; Gulland, 1977; FAO Advisory Committee on Marine Resource Research, 1980, pp. 24-25). Such objectives may be in direct conflict and therefore some choice is required at least at the level of implementation and perhaps also at the earlier stage when political choices must be made among different interests and benefits. Accordingly at this stage clear formulation of objectives helps to reduce conflict, to promote accommodation of divergent interests, and to determine priorities.

Another, very important, purpose served by clear expression of objectives is to seek to insure the accountability of officials charged with management responsibilities. Unless the manager is charged with pursuit of identifiable aims, it will be difficult and perhaps impossible to assess the performance of the management entity. It is understandable for this reason that some management officials may be unenthusiastic about clear specification of objectives. From a national perspective it is especially important that such assessment is feasible so that external agencies can determine whether support or assistance is desirable and serves useful purposes.

From a different perspective, unambiguous statement of objectives serves also to support officials who must decide among competing interests and groups. Capacity and willingness to make hard choices will be favorably affected by provision of clear guidelines for choosing.

Third, explicit objectives are needed to guide all the elements of the decision process that must come together for effective management. This point has two dimensions, one internal to the coastal State and the other external. Presumably there is within any coastal State a more or less identifiable management process that involves the performance of various decision-making functions. Carrying out these functions satisfactorily, i.e., in accordance with criteria of effective management, requires that all functional components understand the objectives of management. The following discussion explains this briefly in terms of various

---

[4] For general discussion of the policy problem in light of the Draft UNCLOS Convention, see Brown (1977), Oxman (1977), and Extavour (1979).

functions commonly performed including intelligence, promotion, prescription, invocation, application, appraisal, and termination.[5]

The function of intelligence is to gather data and information, to process it through analysis and interpretation and to disseminate it (McDougal *et al.*, 1973). Clarity about objectives is important for guiding the performance of this function by identifying the critical needs for data and information and by informing officials adequately to select research priorities. A simple illustration may suggest the importance of understanding objectives. Thus, when it is recognized that the principal management objective is to attain certain social goals rather than seek strictly biological aims, managers can allocate resources to development of satisfactory data pertinent to such goals. Obviously there is a difference between research focusing on economic and cultural factors and that directed at fishing mortality.

The promotion function refers to the task of bringing policies to the attention of those charged with their consideration and adoption. In the present context this function is especially important in connection with shared stocks where other nations or entities are necessarily involved in management. If a coastal State wishes to protect its interests satisfactorily it must be prepared to present a cogent set of policies designed to achieve a recognizable objective. If management objectives are vague and ambiguous or erratically changeable, it may be difficult to elicit accommodation and coordination by other decision makers.

The prescription function in this context refers to the promulgation of regulations (detailed policies) (McDougal and Reisman, 1980). It hardly needs to be said that choice among a range of management measures is facilitated when the prescriber has a clear idea of what aim is to be achieved. An individual measure may serve more than one goal, of course, but this does not obviate the need to aim for some objective among the many conceivable. Further, the scope, timing, and detailed content of a regulation may be dependent on the precise objective in mind.

Invocation is the process of setting another decision function into motion and particularly involves the provisional characterization of conduct deviating from established regulations. In situations involving scarce resources, priority among objectives may be determinative in the decision to activate specific regulations.

Application refers to the specific enforcement of general regulations. Various types of enforcement measures are available and choice among them may be influenced by the principal objective being sought.

Appraisal is the function of assessing whether the management process attains the objectives set for it. Obviously operational indices must be set for the general objectives sought if this function is to be performed. It is implied that objectives must be formulated with sufficient specificity to allow evaluation of progress toward them. A critical part of the appraisal function is in providing for the accountability of the managers.

Termination refers to the ending of prior policies and regulations and the institution of new arrangements. Adoption of new objectives may well require a

[5]For discussion of these decision functions see McDougal *et al.* (1981).

change in regulatory measures as well as in the performance of other decision functions.

The external elements of the decision process involve all of the above functions operating in other national agencies or institutions or in international entities. As noted elsewhere, more than one State may need to be involved in management because stocks are shared or because they do not fall wholly within any combination of fishery zones. In such circumstances it is especially important that the management objectives in the minds of one State be known to other entities. The total decision process involved can range from bilateral relationships, through multilateral, regional, and global.

A final point concerns where objectives should be stated. It may not be vitally important, and may not even always be possible, that precise objectives be spelled out in legislation. At some stage, however, the specific aims of government intervention should be recorded so they can be communicated to those who are concerned, including other management entities and international institutions.

Turning from procedural aspects of establishing fishery management objectives to their substance, the second set of policies of interest are those that concern conservation and allocation of living resources. With respect to the coastal State's authority to adopt conservation measures, an initial question is whether there should be an international standard or requirement that the allowable catch cannot exceed some maximum number or fall below a minimum. These possibilities pose quite different questions. The questions involved in setting an upper limit on harvesting relate to conservation in the sense of maintaining some level of stock abundance. The questions raised by establishing a lower level do not concern conservation but are pertinent to whether foreign fishermen will be allowed access. The former issue, the maximum allowable catch, furthermore concerns not only exploited (target) species, but also those affected by harvesting, i.e., species caught incidentally (including those deliberately taken "incidentally" as well as those otherwise "intentionally" caught) and species affected by the removal of food sources or predators.

With respect to the question of a required minimum level of allowable catch, the more specific issue is whether a coastal State should be allowed to limit harvesting to whatever level will satisfy its interests even if the effect is to withhold fish from any foreign harvest in its zone. To answer this question in the unqualified affirmative would appear to require acceptance either that *any* coastal interest is superior to *any* interest of other nations in the resources of the zone involved or that in specific instances the coastal interest outweighs the general interest in using a resource as a source of food or for other gain. Desirable policy would concede the latter choice but reject the former.

The issue of setting a standard for an upper limit on catch translates into considering whether it is desirable policy to maintain living resources at some particular level of abundance above zero. It is taken for granted that community policy ought to require avoiding not only the extinction of a species but also threats to reduce it to the point that reversing a downward trend in abundance is no longer possible. Accordingly exploitation should not be permitted, directly or indirectly, to reduce population levels to the point that fishing mortality coupled with natural fluctuations might threaten the survival of a species. What this level might be in particular instances would of course depend on scientific information.

For the taking of target species, the above standard presents no particular difficulty. In relation to marine finfish (as distinguished from mammals and diadromous species) the standard would only rarely have any relevance. There is only one known instance of a commercial marine finfish population being considered as theatened.[6] With respect to marine mammals, however, there is no doubt that harvesting could exert sufficient pressure either directly, or as incidental catch, or because of food deprivation, to threaten extinction. Such goal has a meaning in this context although implementation could be very difficult in specific instances, particularly in connection with changes in the food web.

In the case of diadromous fish, there is no question that extinction could result from deliberate or inadvertent overharvesting (as well as from habitat destruction and degradation), and all this should be avoided even if no other nations expect to join in harvesting than the coastal State.

The same standard might present difficult problems of implementation in connection with incidentally caught species or those associated with or dependent on the target species. To take account of other affected species requires a knowledge of the ecosystem that mostly does not exist. While with incidentally caught species it is at least possible to know what mortality is inflicted and thus permit assessments of impact, even this is very difficult when the affected species is not caught at all in quantity but is a potential predator or prey of the target or, worse, more distantly associated.

Lack of fully adequate theory and gaps in knowledge need not necessarily halt decision making involving the standard mentioned. One device is to require only that choices be supported by the best available scientific information, which permits officials to proceed even though the information is not necessarily of high quality. A variant of this which could act as a brake on use of inadequate information is to allow the use of the best available information so long as it substantially and credibly supports the conclusions reached. This might help rule ou out the use of information that is poor and unreliable but still the best available.

The policy proposed departs substantially from the notion that the maximum catch from a fishery should not exceed the maximum sustainable yield. Such a standard certainly would assure that target species are well protected from excessive exploitation threatening irreversible effect, but it is rejected as appropriate here either as a ceiling or a floor on harvesting. There are a number of reasons for discarding MSY as an objective of conservation, a major one being that sustainable catches can be taken at a lower level of abundance than will yield the maximum sustainable. Such a policy may be considered desirable by a coastal State and it need not offend against the basic protective policy already spelled out.

The more difficult policy problem is whether the coastal State should be allowed to set a total allowable catch (TAC) which clearly will not reduce abun-

[6]The totoaba of the Gulf of California is listed as endangered under the U.S. Endangered Species Act (U.S. Code of Fed. Reg. vol. 50, §17.11). It seems unlikely that this status arose solely from commercial exploitation.

dance to that which would produce the maximum sustainable yield.[7] At such a point catches could be increased on a sustained basis. It is at this point that the maximum sustainable yield concept becomes more controversial. The argument thus is advanced that a coastal State should not be permitted in this fashion to set the TAC lower than the MSY figure because this would mean that the stock in question was not being fully harvested. I have pointed out that this would not offend desirable conservation policy but some believe it would not be consonant with desirable full utilization of the fishery. If the catch is less than the maximum sustainable because the level of abundance is maintained at too high a level, the effect may be to keep resources in the water that could be taken consistent with continued harvesting at a higher level.

Another way of putting this issue is to ask whether there is a "surplus" when fishing is stopped short of the MSY catch. In terms of weight of available fish the answer is yes. In terms of the social and economic context in which the coastal State figures its interests, the answer can well be no. Why would a coastal State set an allowable catch at this level? The reason is that a lower level of abundance, which would exist if harvests were increased, increases the cost of fishing and thus offers a smaller net economic return. In this sense there is no "surplus"; to the contrary the net economic return would be reduced if catches were higher. Another way of putting this point is that excluding foreigners permits coastal fishermen to realize a greater catch per unit of effort and, accordingly, to achieve a greater net benefit from a given expenditure of effort.

The critical question regarding allocation is whether there should be constraints on coastal State exclusive management authority concerning its competence to implement its policies by excluding or significantly restricting foreign access to a fishery. This question is not whether the coastal State should be authorized to limit foreign access in order to avoid overfishing for a biological reason, but whether there should be constraints on coastal State authority to limit access for social, cultural, economic, and political reasons. The answer is that the coastal State should, within limits, be permitted to satisfy its own interests. At the same time, a coastal State should not be permitted to exclude foreign fishing for just any reason that may appeal to it at the moment. Living marine resources are significant not only to coastal States but to other communities as well and these wider interests should also be protected and accommodated.[8] Certainly many coastal nations have an intense interest in adjacent fisheries and, whether always sensible or not, are now generally considered entitled to a prior-

[7]It is recognized that there is a certain degree of unreality in this proposition. Animal populations fluctuate, sometimes wildly, and are subject to influences that are far more determinative of abundance than fishing. The MSY figure is considered an average over time and does not represent an actual projection of what can be done on an annual basis (see Gulland, 1977, pp. 1-2).

[8]The question posed here is whether States generally should agree (by whatever method) on limiting their exclusive authority over living resources that are otherwise within their jurisdiction. In this sense these resources are no different from any other resources that fall within the exclusive authority of the States. It is commonplace that, for example, minerals, oil, and timber are shared among States pursuant to agreement (see Arsanjani, 1981).

ity in benefitting from them. Thus the key question is what limits there should be on depriving foreigners for the benefit of coastal nations.

A principal distinction is between developed and developing nations.[9] A nation which is considered developed should not be permitted to withhold access to the stock on terms that are reasonable. The developed State is in a better position to deal with the problem of reduced income for domestic fishermen that may result from this policy. In the converse situation, on the other hand, developing States should be permitted to withhold access to a surplus over domestic harvesting capacity if it perceives an economic or social advantage in doing so.

The situation involving two developing States, one being coastal, can probably best be handled by leaving the parties as they are. This means that the coastal State should be permitted to give priority to its own interests and that may entail allowing access by a developed State in preference to a developing. The key factor is the coastal State's perception of its interests.

Extent of Authority Conferred: Geographical Limits and Species Coverage

If extended coastal State jurisdiction is to be employed to improve fishing conservation and management, its delimitation ought to be in terms that will facilitate and promote effective decision making. The difficulties here revolve around the territorial limits of the State *vis-à-vis* other States and the nature of living marine resources. With respect to the former it must be recognized that except for some isolated island nations, the lateral boundaries of States prevent extension and exercise of exclusive coastal control over stocks that move across the boundary or are distributed on both sides of it.[10] Neither State concerned may be able to exercise exclusive management responsibility over a unit of stock in the absence of agreement. This difficulty is expected to be continuing and critical and beyond solution by unilateral measures.

The external limit to jurisdiction relates to the migratory nature of fish. Many fish species move from deep to shallow waters and vice versa and others migrate over very large ocean expanses (Gulland, 1980). With respect to fish associated with coastal regions it would be most appropriate to define the limit of coastal authority in terms of the range of the habitat of the fish. This might be

[9]A recent analysis distinguishes these in terms of per capita product and uses $2245 per capita product as the dividing line. In this scheme there are 113 developing and 53 developed nations (Block, 1981, p. 5). The number of those engaged in fishing is much lower.

For discussion of fishery exploitation and trade in terms of a similar, but undefined, distinction see Holt and Vanderbilt (1980).

FAO reports its statistics by "countries" designated as developed and developing, but the former term includes also "cities, land areas, provinces, districts, enclaves, exclaves and other parts of territories or combinations of countries . . ." (FAO, 1980, p. 27).

[10]Of course even for the isolated island nations the fish move beyond their jurisdiction and may require management by international agreement.

combined with a specific mileage limit in order to create a stable or continuing management situation.

Another approach is to define the outer limit of coastal State jurisdiction as a specific distance, sharply separating the area of coastal jurisdiction from the area beyond. This approach fixes coastal jurisdiction and adds an element of certainty, but does so at the price, compared to a functional definition, of depriving numerous stocks of a single, coherent management authority. While international agreement on regulatory measures for the High Seas/fisheries zone may prove to be adequate, the suspicion is that this route will work inefficiently, awkwardly, and sometimes not at all.

Neither a species-oriented nor a fixed limit boundary is able to cope with the problems of highly migratory or anadromous species, both of which are available to harvesting gear far beyond any conceivable boundary. With respect to the highly migratory species, there can be no serious question that management would be most effectively achieved by means of agreement on an agency endowed with the necessary authority and resources to do so for the entire exploited species. These stocks move into national jurisdiction, and sometimes are predominantly accessible there, and then move outside national jurisdiction where they are also accessible. Effective management in either place alone is not possible. While there may be conceivable alternatives, an international institution seems preferable if those concerned can establish one that can do an effective job. It seems likely that the agency would be most effective if its task were limited to conservation problems rather than allocation. Preferably an international agency for conservation would be delegated the authority to prescribe standards and limit catches on the basis of research conducted by its own staff. Allocation of benefits among nations is an extremely difficult assignment and may need to be left to individual negotiations between coastal States, as a group or individually, and other fishing nations.[11]

The problem with anadromous fish is at least as complicated as others. On the one hand anadromous fish depend for their survival and continued productivity on adoption of appropriate policies by the host nation or nations, the one(s) in whose waters the fish begin life and later, on return, spawn and die. On the other hand these fish live and get their sustenance over much of their life span in the open ocean and thus derive essential support from the common pasture (Fredin *et al.*, 1977). In these circumstances should the management responsibility be allocated to the host State or should it be delegated to all nations who take a share of the resources, including those who take it on the High Seas?

The common preference is that the State of origin, whose rivers the fish emerge from and return to, ought to be considered competent to make all management decisions, including allocation of access and benefits. While High Seas fishing States have some modest historical claim on which to base demand for access to salmon and other anadromous species during their ocean sojourn, this

[11]These few sentences obviously do not attempt to address the complex problem of tuna management, including allocation. For full treatment see Joseph and Greenough (1979).

method of harvesting is inefficient and cannot be conducted with due regard for escapement goals of any particular run. Coastal States can, on the other hand, fine tune regulations to meet escapement goals (as well as others) and should have responsibility for safeguarding the resource and its habitat.

Trends In Decision

Claims to Competence to Establish an Exclusive Fishery Zone

Historically the Law of the Sea has evolved through the nation-state system according to which claims to authority are advanced for specific purposes and other States respond by accepting or rejecting the authority asserted. In more recent times, this pattern of law creation has been partially displaced by efforts to establish coastal exclusive fishing authority by explicit agreement. After World War II, States began to claim greatly extended fishery authority and an effort was made to reach explicit agreement on this and other elements of the Law of the Sea.

It is well known that the first (1958) and second (1960) UNCLOS conferences were unable to agree on an extension of the Territorial Sea or on an exclusive fishery zone in the water column. Nonetheless these conferences left no doubt that a 3-mile Territorial Sea had little international support while a wider area of exclusive coastal control and preferential rights over fisheries met with widespread approval. For living resources on the bottom, the sedentary species, the 1958 conference was able to establish that coastal State sovereign rights extended throughout the continental shelf as defined in the Convention on that area.

The failure of the conferences to reach agreement on the limits issue, coupled with increased fishing pressure exerted by the continued growth in distant water fishing fleets, meant that problems of fishery conservation and of allocation of catches could not be resolved except by further international agreement or by unilateral extensions of coastal authority.

While international negotiations were pursued by many States during the 1960s, a substantial number also adopted larger fishery zones through unilateral action either to establish such a zone or to extend their Territorial Seas. From 1958 to 1968 the number of exclusive fishery zones quadrupled while the number of Territorial Seas beyond the "traditional" 3 miles increased from much less than half claiming over 3 miles to about two-thirds claiming in excess of 3 miles. In 1966 the United States changed its historic policy of adhering to fisheries jurisdiction solely within a 3-mile Territorial Sea and added a 9-mile zone of federal fishery jurisdiction. At the time this change was being considered in the U.S. Senate, the Legal Adviser of the Department of State pointed out that "there has been a definite trend toward the establishment of a 12 mile fisheries rule in international practice in recent years and stated that the establishment of a 12 mile fishery zone under the legislation would not be contrary to international law (*Congressional Record*, 1966).

During the period following the 1958 Conference until the beginning of the third UNCLOS conference in 1974, two major centers of controversy dominated the trend toward wider fishing limits: the successive claims by Iceland that lead to conflict with the United Kingdom and others and the claims by Chile, Ecuador, and Peru that occasioned conflict with the United States.

The most significant conflict in terms of confirming the trend in decision involved Iceland's attempts to extend its fisheries limit to 12 and later to 50 miles. Initially Iceland and the United Kingdom jousted over the 12-mile claim and this was settled by agreement in 1961 when the United Kingdom conditionally accepted 12 miles and also got what it considered assurance of judicial settlement of any future disputes over fishery limits. Iceland again extended its fishery limit in 1972, this time to 50 miles. When the parties failed to settle this matter through negotiation, the United Kingdom took the issue to the International Court of Justice (ICJ), whose jurisdiction Iceland refused to accept at any stage of the case. Although the Court refused to decide upon the permissibility of the 50-mile zone in general international law, instead limiting its decision to declaring that Iceland's zone was invalid as against the United Kingdom, it did confirm that the concept of an exclusive fishery zone of some width was accepted in international law. According to the Court this principle was established both by the "general consensus revealed" in the 1958 and 1960 UNCLOS conferences and by State practice based thereon (United Kingdom *v.* Iceland, 1974; for a helpful and incisive analysis of this decision see Khan, 1975).

The other center of conflict over fishery limits pitted the States of Chile, Ecuador, and Peru (the CEP nations) against the United States whose flag was flown by the High-Seas tuna-seining fleet in the eastern tropical Pacific, the vessels of which were most affected by the 200 mile claims of the CEP nations (Hollick, 1981). Over the period from 1947 to the 1970s the United States opposed the 200-mile claims advanced by these States and through special protective legislation attempted to maintain this opposition while shielding its flag fishing vessels from the financial burdens and losses that arose from coastal State enforcement of these claims. Over the years, therefore, the controversies off South America did not come to a head as did that in the North Atlantic and the coastal States continued to maintain their claims. In refusing to use naval force, the United States obviously placed greater weight on maintaining harmonious hemispheric relations than on its global policy of opposing extended jurisdiction in the ocean. Over time this weighting of policy interests contributed to the success of the trend toward extended jurisdiction and specifically toward acceptance of the 200-mile limit.

By the eve of the Third United Nations LOS Conference it was, therefore, well established in international law that coastal States could exercise jurisdiction over fisheries in a zone beyond its Territorial Sea. Although States had been unable to agree on a precise expression of this principle in the two earlier UNCLOS conferences, nations had acted unilaterally with sufficient uniformity that the concept was "crystallized as customary law," as the ICJ put it.

An area of exclusive fishery jurisdiction was thus widely endorsed long before the 1973 UNCLOS Conference convened and there was no need for a Conference struggle to achieve its acceptance. As will be noted below, however, the geographical extent and the exact scope of coastal authority over the zone were not similarly accepted and remained for negotiation along with other details of a new regime for fisheries.

Scope of Authority: Nature of Coastal State Rights

The earliest extended jurisdiction claims of Chile, Ecuador, and Peru were very clear in expressing complete authority and control at least over the living resources within 200 nautical miles. Later claims by other nations to narrower zones, during the late 1950s and 1960s, also asserted full power to regulate and to dispose of fisheries. In the first United States legislation on this subject, establishing in 1966 a 9-mile contiguous zone for fisheries, the United States claimed the same rights and authority over fisheries within the extended zone as it exercised within its Territorial Sea (U.S. Congress, 1970). The Submerged Lands Act of 1953 left no doubt that Congress believed the living resources of the Territorial Sea were wholly subject to disposition by the United States.[12] Although the 1958 and 1960 UNCLOS Conferences did not reach agreement on an (exclusive fishing zone), it may be recalled that the Canadian-American proposal at the 1960 Conference barely failed to secure two-thirds approval and it provided for the same rights to fisheries in a 6-mile fisheries zone as pertained in a 6-mile Territorial Sea.

During the Third United Nations LOS Conference States advanced a great many fisheries proposals, upwards of 60-odd, in the preparatory stage and at the second session in Caracas (United Nations, 1974b; FAO Committee on Fisheries, 1974). The contrasts of greatest interest are in those of coastal fishing States (most of them developing countries) in relation to those of the major distant water fishing States. The United States views are also noteworthy because of its position astraddle very valuable distant water fisheries, on the one hand, and politically significant and regionally important coastal fisheries, on the other. Although the United States struggled to reconcile these interests at the first two UNCLOS conferences, the task at this time was even more difficult and, as it turned out, impossible to achieve except for its anadromous fisheries.

The vast majority of fishery proposals in the pre-Caracas period and at Caracas were directed simply at acquiring full coastal authority over living resources out to 200 nautical miles (Johnson, 1975; Koers, 1974; Kury, 1975; Taft, 1975). Although economic zone proposals differed in a great many respects, they seldom differed on this specific issue of full coastal authority. This congruence of view did not come as a surprise, of course, as the preconference consultations and expressions of collective opinion left little doubt about the expectations of large numbers of developing coastal States. Two of the prominent expressions were the report of the Yaounde Seminar and the Declaration of Santo Domingo,

---

[12]43 U.S. Code §1301 defines the United States boundary to be 3 miles (except in the Gulf of Mexico); §1301 defines natural resources to include fish, shrimp, oysters, clams, etc.; §1311 declares that title to and owernship of the natural resources within United States boundaries vest in the States.

both of which were circulated as Sea Bed Committee documents in July1972. The former declared the right of African States to establish "an Economic Zone over which they will have an exclusive jurisdiction for the purpose of control, regulation and national exploitation of the living resources of the sea and their reservation for the primary benefit of their peoples and their respective economies. . . ." (United Nations, 1972, pp. 73, 74). The Declaration of Santo Domingo (United Nations, 1972, p. 70) had a more complex course to steer between or among the strongly held and conflicting views of Latin American nations regarding the nature and scope of offshore jurisdiction (Székely, 1976, pp. 266-270). The nations signing the Declaration subscribed to an extension of lesser authority than claims to added national territory and denominated an area the "patrimonial sea" within which the coastal State had "sovereign rights" to the renewable and nonrenewable resources. The area of combined territorial and patrimonial seas could not exceed 200 nautical miles. Despite the difference in the label both pronouncements envisaged full coastal State authority over fisheries within the zone, although the Yaounde Report looked to participation by agreement by the landlocked and near landlocked African States. The Yaounde Report also did not mention a specific mileage limit for the outer limit of the zone.

As the major distant water fishing States, the Soviet Union and Japanese proposals indicate a considerable apprehension about coastal State jurisdiction. The former's draft articles (United Nations, 1972, p. 158) were carefully formulated both to insulate USSR fisheries from encroachment by the coastal States off whose coasts the Soviets mostly fished and to retain its control over fishing by Soviet vessels beyond a 12-nautical mile fishery zone or Territorial Sea. Article 1 limited to developing coastal States the right to reserve part of the allowable catch in High Seas areas adjacent to a 12-mile zone. This would effectively exclude the United States, Canada, and European States from such action and preserve Soviet catches in the areas of the northeast Pacific, northwest Atlantic, and northeast Atlantic where it was most significantly involved. At the same time, however, Soviet interests in Asian salmon were preserved by allowing any coastal State to reserve a portion of the allowable catch of anadromous fish spawning in its rivers. Coastal State regulatory authority beyond a 12-mile fishing limit would be deflected by requiring that any regulations be established either through an existing international organization or by agreement with the States fishing in the area. In sum, the Soviet Union's proposal recognized a certain priority to adjacent fisheries by developing coastal States but acknowledged no independent regulatory authority in any coastal State.

The Japanese proposal (United Nations, 1972, p. 188) sought also to protect its fishery interests, partly through distinguishing between developing and developed States but mainly by requiring that implementation of any preferential rights to an allocation of resources in High Seas areas beyond a 12-mile limit required agreement of the States concerned. Thus the substance of preferential rights would have had to be negotiated and agreed with the distant water fishing States. Additionally the Japanese proposal excluded highly migratory, including anadromous, stocks from any "special Status" for conservation and from preferential rights of catch. Further protection against disruption would follow from the provision that for developed States any preferential rights applied only to "an allocation of resources necessary for the maintenance of its locally conducted small-scale coastal fishery." This would insulate distant water States, such as Japan, who harvest high-volume, low-unit-value stocks.

The United States' proposal (United Nations, 1972, p. 175) was certainly not less self-serving than others. It would have provided for preferential rights to the catch by coastal fishermen of coastal species to the limits of their migratory range, excluded highly migratory species from any coastal-State authority, and recognized the preferential right of the spawning State to anadromous resources. At the same time the United States draft called for maximum utilization of coastal resources and required the grant of foreign access to the resources that the coastal State could not harvest. These provisions were an attempt to balance the pressures of the United States coastal fishermen and the tuna, shrimp, and salmon fishermen. This formula would quiet restive coastal fishermen in New England and Alaska, maintain access by the California tuna fleet to the South and Central American areas of the eastern tropical Pacific, and placate the salmon industry of the Pacific Northwest and Alaska.

In the outcome as measured by the 1981 Draft Convention (United Nations, 1981), the interests of coastal fishermen are given greatest protection. In general the Draft UNCLOS Treaty establishes broad and exclusive coastal State authority over fisheries within a zone of 200 nautical miles measured from the baseline for the Territorial Sea. This quality of "exclusiveness" in relation to authority over economic resources of the economic zone, including fisheries, is emphasized by repetitive mention. Part V of the Convention is entitled "Exclusive Economic Zone" and this phrase or label is repeated in every general article concerning the zone (Articles 55-60). The Convention seeks to make certain that the degree of coastal authority over the economic zone derives wholly from the Convention and not from other sources. Article 55 defines the zone as "an area beyond and adjacent to the territorial seas, subject to the *specific legal regime* established in this Part" (emphasis is added).

The coastal State's rights in the zone are declared to be "sovereign" for certain particular purposes, namely exploring and exploiting, conserving and managing the natural resources, whether living or nonliving, of the sea bed and subsoil and the superjacent waters. This means that in the matters mentioned the coastal State has the final authority to choose one way or another, whatever its obligations may be to consult or to take account of specific considerations, views, or evidence. The sovereign rights of the coastal State exclude either decision making by another entity or a claim to participate in decision making with respect to the subject matter of living resources. Thus, other nations, international organizations, business associations, or other private groups cannot lawfully claim to share equally with the coastal State in choices relating to or involving exploration, exploitation, conservation, and management of the living resources of the zone. A coastal State may, in the exercise of its sovereign rights, accord a role, perhaps a decisive role, to another entity but it is not obliged to do so by the provisions of the Draft Convention.[13]

As noted elsewhere in this discussion, the sovereign rights of the coastal State are not unqualified. It is obliged to observe certain standards, to have regard to certain considerations in making decisions and to engage in unspecified cooperative behavior with other entities, but these obligations do not displace its final decision-making authority. These instances and their effects will be discussed in a later section.

Perhaps it might be noted, in passing, that question has been raised concerning whether or not the draft treaty rights over living marine fisheries can properly be labeled as "ownership" and it is urged by one observer that to do so is incorrect because the rights are specifically limited by the treaty (Binnie, 1980). In major, and perhaps all, respects this position appears to raise a false issue because it appears to assume that "ownership" of property somehow signifies unrestricted authority of disposal or the authority freely to decide on the most important elements of use. This is simply incorrect—property owners are commonly burdened by a variety of restrictions on the employment of their land or buildings or other property and this is not considered to abolish the rights of "property." The point is immaterial, however, since the policy questions always are who may exercise authority under what conditions. The focus should be on these substantive policy questions, not on whether the possessor of rights is the "owner" and, therefore, can be presumed to have full control or must necessarily have such control.

National legislation extending jurisdiction over fisheries in many instances either closely resembles the Draft UNCLOS Convention formulation of sovereign rights over living resources or claims exclusive rights directly. Assertion of such rights over living resources out to 200 nautical miles has occasioned no protest in recent years and it seems unlikely that there will be any challenge to such coastal authority. There if a firm, almost universally shared, expectation that the coastal State has full authority over all living resources in the zone. Very few States would need to amend asserted authority over their fishing zone to accommodate the UNCLOS Convention provision on this point.

Turning to the problem of accommodating coastal State rights over fisheries with other uses of the fisheries or economic zone, the principal question arises in connection with freedom of navigation. To what extent, if at all, may a coastal State seek protection of living resources (as by prevention or discouragement of illegal fishing) by measures which have some impact on navigation? Such measures might include attempts to prohibit access by transitting fishing vessels, special sealanes, gear stowage, and reporting procedures, to mention some alternatives.

[13]Article 64 refers to establishing international organizations for highly migratory species but does not provide that such agencies shall have any final decision-making authority (see discussion in footnote 30).

State practice does not shed much light on what might be permissible under customary law. Apart from the general adoption of 200-nautical mile zones for control of fishing, few States have enacted specific measures to deal with possible evasion of its regulations by vessels ostensibly in passage. Some have general legislation that would permit regulations that affect navigation,[14] but very few have specific legislation. The Maldives is the most prominent exception in that it specifically prohibits passage of fishing vessels without its consent (United Nations, 1980).

The more immediate question about customary law is whether or not the general (nearly universal) adoption of 200-mile zones carries with it acceptance also of measures which have some effect upon navigation by fishing vessels. It hardly seems radical to argue that the evolution of extended fishery jurisdiction should be accompanied by the authority necessary to protect the living resources subject to that jurisdiction. At the same time the established right to freedom of navigation must also be protected against undue interference.

Under customary law freedom of navigation has never been regarded as absolute even though its protection has been widely recognized as an important and vital interest of all States.[15] The traditional means of reconciling coastal and flag State interests has employed the standard of reasonableness, balancing the interests at stake and judging the permissibility of restrictions in terms of that standard (McDougal and Burke, 1962, p. 765). Over the centuries and including the most recent times, limitations on freedom of navigation have been accepted as new exclusive interests have come to be recognized. A prime recent illustration, which occasions not even a raised eyebrow today, arose from the extension of coastal authority over the Continental Shelf. It was impossible to make adequate provision for shelf exploration and exploitation under coastal State control unless freedom of navigation admitted of some modification (McDougal and Burke, 1962, pp. 692-712). Similarly the 1958 Fishing and Conservation Convention and the later modest extension of exclusive fishing rights required acceptance of some slight encroachment on absolute freedom of navigation (McDougal and Burke, 1962, pp. 979-998).

The problem with 200-mile fishery zones is thus simply a larger instance of the familiar difficulty of accommodating enlarged coastal State exclusive interests with general community interests. A reasonableness standard does not call for absolute freedom for navigation nor do coastal State sovereign rights require or justify negation of that freedom. In the circumstances coastal State protective measures affecting navigation could be considered reasonable where they are necessary for effective management and enforcement, hold unusual benefit for a

[14]Economic zone laws providing for this include India, Seychelles, Pakistan, and Guyana.

[15]In discussing States' attitudes toward freedom of navigation on the High Seas McDougal and Burke (1962, p. 768) observe: "It is not to be inferred that this widespread acceptance of the general doctrine prescribing freedom of access for navigation absolutely prohibits any activity or authority which may interfere with such freedom. To the contrary, states commonly make certain claims to exclusive use and authority, interfering with inclusive use, which are recognized by the community to be consistent with international law, and specifically, with provision for substantial freedom of access." See also Extavour (1979, pp. 235-237).

particular coastal State or States, and impose modest or slight burdens on navigation. Such measures would not be directly aimed at regulating navigation as such, but designed to protect resources subject to coastal State jurisdiction. It is conceivable that in some circumstances, limited probably only to certain key geographical and other circumstances, a sealane for fishing vessels would be a permissible development when reasonableness is assessed in terms of the factors mentioned.

The question under the Draft Treaty is whether a sealanes requirement is compatible with Article 58 which declares that "In the exclusive economic zone, all States, whether coastal or landlocked, enjoy, subject to the relevant provisions of this Convention, the freedoms referred to in Article 87 of navigation and overflight. . . ." Can the coastal State, consistent with this provision, require passing fishing vessels to use designated sealanes? There are two points to notice immediately. One is that the freedom of navigation mentioned is that found in Article 87, which is interpreted by some to mean this freedom is the High Seas freedom of navigation as traditionally understood, i.e., before exclusive fishing zones were widely accepted (Oxman, 1977, pp. 263-264; Richardson, 1980). If this were all that were relevant, the question might be answered immediately—no coastal State has authority to require vessels passing on the High Seas to employ any specific route for such passage. If it were desirable to use the sealanes concept to strengthen coastal authority to protect its interests in living resources, it would be necessary to seek international agreement to that end.

The second point, however, is that Article 58 does not confer a pure or undiluted right to freedom of navigation in the zone. The enjoyment of this freedom is "subject to the relevant provisions of this Convention." The relevant provisions include the coastal State's "sovereign rights for the purpose of exploring and exploiting, conserving the managing" living resources (Brown, 1977; Extavour, 1977, pp. 236-237). This broad grant of authority is elaborated upon in Article 61 and following to give the coastal State complete power to dispose of the fisheries in its zone, subject to certain general standards and to requirements to cooperate with adjoining States and other States fishing in the region. Articles 61 and 62, in particular, provide the coastal State with ample general authority to protect the living resources of its zone from unauthorized foreign fishing and list specific competences to regulate all fishing activities by foreign vessels. The question that is left open, or not explicitly answered by these provisions, is how far the coastal State may go in affecting passing vessels as distinct from those expected to fish in the zone.

It does not seem unreasonable that in some circumstances a coastal State should be allowed to reduce an onerous regulatory and enforcement burden by measures that affect passing vessels. The question is, how much of an effect is reasonable? Assuming some impact on passing vessels does not unduly compromise the right of freedom of navigation, what are factors relevant to determining reasonableness?

Important factors surely must be the size of the zone relative to land mass and the value of resources therein to the national economy. In some parts of the ocean the ratio of water to land area of the State is high and there are enormous obstacles to maintaining adequate monitoring and surveillance of fishing activities involving valuable resources. Designating sealanes for passing fishing vessels might in some circumstances make the difference between effective and ineffective enforcement and this, in turn, could have a significant impact on coastal State revenues that are especially important to that State. Where the advantage to the coastal State is so high, the added burden on passing vessels might not be unreasonable. On the other hand, it is also in the large ocean areas that the freedom to select the exact route may have special importance because of the vast distances to be overcome and the hazards of long-distance transit. The skippers of fishing vessels may need to change course and normal navigation routes in order to avoid bad weather or to take advantage of new information regarding desirable fishing areas. Perhaps these conflicting considerations could be accommodated by contingent designations of required sealanes or by an approach that would allow for departure from a sealane under specified conditions. If the designation can be made flexible without eliminating its usefulness, there would be a stronger argument for the reasonableness of a sealane requirement.

On the other hand, where sealanes requirements add significant time or hazard to fishing vessels passage, the impact on freedom of navigation would be impermissible and flag States would have good reason to object.

On balance under the Draft UNCLOS Treaty, the requirement of due regard by the coastal State for freedom of navigation imposes a heavy burden that can probably be discharged only in select and somewhat unusual circumstances. Restrictions on movement of any vessel should be limited to the exceptional situation. The main factors of importance in applying the "due regard" standard appear to be the coastal State's difficulty in securing adequate enforcement and the contribution that the fishery makes to the national economy. If the coastal State confronts unusual enforcement problems and the result is costly to a State that has significant dependence on revenues from fisheries, then the case for some modest interference with navigation would be persuasive.

Management Objectives: The Specification of Conservation and Allocation Goals

In the traditional international law of the sea, the issue of management objectives affecting foreign fishing arose in the process of establishing agreed measures for fishing beyond national territory. The latter area was in nearly all instances so confined that coastal State regulations were of insignificant interest and effect. Virtually all international agreements mentioning objectives shared by the participants declared that the purpose was to maintain a level of abundance that would permit taking of the maximum sustainable yield. At the 1958 UNCLOS Conference, agreement was reached on the phrase "optimum sustainable yield" as the goal of conservation measures "so as to secure a maximum supply of food and other marine products." This was widely understood as referring to maximum sustainable yield.

Although the 1958 Conference produced a convention on fisheries conservation, with the conservation objective noted, the more significant negotiations centered on access to fisheries in a proposed exclusive fishing zone; however, the Conference failed to resolve this issue by the narrow margin of one vote. It is significant to note, too, that at the 1960 Conference the near agreement on a 6-mile Territorial Sea plus 6-mile exclusive fisheries zone was significantly promoted by an amendment proposed by Cuba, Venezuela, and Mexico that provided for coastal State preferential rights beyond the fisheries zone. In short the really critical issue revolved about who could take scarce resources rather than how to regulate fishing in order to sustain yields.

After the 1958 Conference the access issue continued to dominate the question of fisheries jurisdiction but the focus turned from international negotiation to unilateral claims to national jurisdiction. With the advent of widespread extended national jurisdiction, the objectives found in national legislation were of great importance to foreign fishermen. The following discussion looks briefly at these objectives in national fishing and economic zone legislation from many nations and then considers the provisions of the Draft UNCLOS Treaty.

Surveying legislative objectives for fishery management[16] in a 200-mile zone leads to the following conclusions: (1) a surprisingly large number of enactments contain no statement of objective at all[17]; (2) virtually all legislation covered in this survey that does formulate objectives does so very broadly and offers little specific guidance[18]; (3) where objectives can be identified the preponderant emphasis is upon general economic and social goals rather than upon biological; (4) objectives that are spelled out are sometimes so numerous and varied that it continues to be difficult to determine what guidance they offer; (5) the objectives are frequently conflicting. It may be significant that the major FAO study on coastal legislation refrains from discussing the objectives adopted by States, noting that "in most cases, anyway, these management objectives, if formulated at all, are somewhat loosely defined" (Moore, 1981).

It may be entirely misleading of course that no objective is expressed in national legislation inasmuch as officials do seek in practice to attain certain general goals. For some developing nations the goal of "fishery development" is

---

[16] Fisheries and economic zone legislation was examined for the following nations: Australia, Bahamas, Bangladesh, Barbadoes, Burma, Canada, Comoros I., Costa Rica, Dominican Rep., Fiji, Guyana, Guinea, The Gambia, Iceland, Kenya, Japan, Maldives, Mauritania, Mexico, Naru, Norway, New Zealand, Pakistan, Papua New Guinea, Philippine I., Portugal, Poland, Sao Tome and Principe, Senegal, Seychelles, Sierre Leone, Solomon I., Sri Lanka, Spain, Suriname, United Kingdom, United States, USSR, Uruguay, Venezuela, Vietnam, Western Samoa, and Yemen (PDR).

[17] Barbadoes, Burma, Comoros I., Dominican Republic, Fiji, Guyana, Mauritius, Pakistan, Philippines, Portugal, Seychelles, Spain, Solomon I., Sri Lanka, Suriname, Uruguay, Vietnam, W. Samoa. It is not known whether all pertinent laws and regulations were examined.

[18] This includes the United States, which spells out objectives more comprehensively and in more detail than anywhere else. Nonetheless this legislation is also confusing on this point. For discussion of the original legislation adopted in 1976 see Christy (1977).

obviously a prime concern (FAO, 1980a). Insofar, however, as this is taken as an expression of a coherent goal, it should be noted that it embraces a considerable range of outcomes of varying benefits to a society as a whole and to different groups therein. Thus fishery development might involve increased production for export but this may involve wholly different species and fishing interests than increased production for the local market. Similarly development efforts aimed at increasing efficiency in the harvesting sector are in conflict with those to increase employment. Likewise, encouraging foreign fishermen to increase capital investment and to employ new technology may lead to development at the price of discouraging local groups. Accordingly, while "fishery development" can be cited as an objective of management, it needs further specification to serve as a useful guide.

The level of generality in expression of objectives may also be seen in the frequent use of "optimum" to qualify such other terms as "utilization" (Australia), "yield" (United States), and "sustainable yield" (Papua New Guinea). In essence, legislation with such terms serves the function of delegating adoption of more specific objectives to subsidiary bodies. Similarly objectives expressed as "ensure rational use" (Poland), "proper conservation and management" (Japan), or "maximizing net benefits" (Maldives) do not mandate any identifiable immediate goal but rather direct or allow lower levels in the decision process to adopt goals and to change them as circumstances demand.

When legislation is framed so generally the outside observer without ready access to the operations of the management process cannot comment knowledgeably regarding objectives sought in practice. Certainly delegation of the task of specifying goals is a sensible procedure in many instances, such as where many different locales and fishing situations must be dealt with or where final decisions have a substantial element of local influence in them. Be this as it may, the necessity for clear identification of objective remains. There also remains the suspicion that in numerous, perhaps many, instances, objectives are ultimately not specified or are identified only in most general terms. Given the high importance and utility in clarification of goals, such failures in the management process are regrettable to the extent they occur.

Protection of resources is a major aim of a good deal of legislation, although seldom the sole one. Prevention of depletion,[19] avoiding excessive or overexploitation,[20] avoiding overfishing,[21] and protection and maintenance of stocks of fish[22] are the principal formulas employed for this purpose. In almost every instance the reference of these formulations appears limited to targets of exploitation. Only the United States legislation, which is the most detailed and comprehensive, makes provision for nontarget species. Ecosystem management seems to be virtually unheard of at the national legislative level.

[19]Bangladesh, Costa Rica, Fiji, Guinea, Mauritania, Senegal.

[20]Australia, Iceland, Mexico, Venezuela.

[21]United States.

[22]USSR, Sierre Leone.

Allocation of stocks as between national and foreign fishermen is not always mentioned specifically in fishery or economic zone legislation, but it appears to be safe to conclude that all assertions of exclusive or sovereign rights to fisheries contemplate exclusion of foreign fishing if so desired. The Icelandic law appears to be the most explicit in limiting catches within 200 miles to Icelandic fishermen but other legislation, including that of the United States, clearly is aimed at phasing out of foreign fishing effort. Numerous laws specify that foreign fishing requires a permit or must be preceded by an international agreement. Many laws contemplate that foreign fishing will be allowed only if there is a surplus over the domestic catch and some, such as Australia and the United States, would reduce foreign catches on the basis of expectations about future domestic harvests (Moore, 1981).

Whatever their specific provisions on foreign fishing, national laws uniformly assume that within the zone of coastal authority foreign fishing is wholly subject to the control of that State and may be allowed or denied as the coastal State law may decide. In most instances in practice, foreign fishing continues to be allowed although reductions in catch and effort are also common.

The Draft UNCLOS Convention is considerably more forthright in identifying objectives for fisheries management than most national legislation. The Draft addresses both of the major international problems of fishery management, conservation (Article 61), and allocation to users (Articles 61 and 62) and confers exclusive authority on coastal States to make decisions on such issues (Articles 56, 61, 62) in the area out to 200 nautical miles. Provisions on dispute settlement (Part XV) establish that these decisions are subject to the discretion of the coastal State, which cannot be obliged to submit disputes concerning them to dispute settlement as provided in the treaty. The ultimate thrust of the Draft Treaty on living resources is to permit the coastal State either itself to take the entire allowable catch and exclude foreigners from the 200-mile zone or, if it decides a surplus exists, to grant access on terms and conditions that are satisfactory to it and do not arbitrarily restrict foreign fishing for the surplus.

In the Draft Treaty the major stated objectives of management are:

1. To ensure that living resources in the zone are not endangered by overexploitation [Article 61 (2)].
2. To maintain associated or dependent species above levels at which their reproduction may become seriously threatened [Article 61(4)].
3. To provide for harvesting the entire allowable catch of living resources within the zone; the portion in excess of the domestic harvesting capacity shall be made available by agreement to foreign fishing subject to coastal regulations.

A major unstated objective, but a direct effect of the several provisions, is to allow the coastal State to manage fisheries, within the above "limits" or "standards," to derive the optimum benefit therefrom, as it defines optimum benefit. In doing so it is not burdened by any obligation to make resources available to other States except as it sees fit to do so. As discussed further below, the only coastal State obligation in this regard is to establish an allowable catch, whose dimensions are solely within its discretion to determine, and, if it cannot harvest this catch, to grant access to other States on such terms and conditions as it is willing to accept and are not unreasonable.

The UNCLOS Treaty formulation of the conservation goal of avoiding over-exploitation is not precise by any means and the obligation arising from it is unclear. Article 61(2) provides that the coastal State must "ensure" that "the maintenance of the living resources in the economic zone is not endangered by over-exploitation." The expression "maintenance of living resources" needs clarification in at least two respects and the reference of the concept "endangered" is also not obvious.

With respect to the former phase, the term "maintenance" does not specify the level of the resource to be maintained and held safe from overexploitation. The second ambiguity centers upon the term "living resources," specifically whether it refers to a stock, a population, or a species. The difference in reference may be largely theoretical and not practical because it is in any event difficult to identify component stocks and populations in a fishery and to determine what is harvested from them.

The concept "endangered" could be interpreted as implying a threat to survival, i.e., in danger of extinction, but it is doubtful if this is very meaningful or helpful. In the first place extinction of a true marine finfish as a result of overfishing is so remote a possibility that it is not a reasonable objective of conservation. So far as is known no marine finfish species has been reduced to extinction through overexploitation in the past several thousand years, despite intense exploitation of some for hundreds of years.[23] Moreover if one is interested in continued exploitation of marine fish, apprehension about endangering a resource arises much earlier than any realistic worry over extinction.

Another interpretation is that endangering in this context means reductions in abundance that amount to commercial extinction. A further possibility is a reference to reductions of such magnitude that a species is likely to become endangered unless protective action is taken. Even a seriously reduced stock might support a marginal fishery, but the added mortality might tip the balance toward irreversibility and this should be prevented.

Whatever else an appropriate interpretation may be, we are able to exclude one alternative. Article 61(3) states that the measures to avoid endangering are also to "maintain or restore populations of harvested species at levels which can produce the maximum sustainable yield, as qualified by relevant environmental and economic factors. . . ." Accordingly it is appropriate to establish such management objectives as environmental and economic factors, among others,[24] may suggest and to pursue such quantities of harvested fish as are appropriate to these objectives, including yield figures in excess of MSY if this is in the coastal State's interest. Temporarily taking a yield greater than MSY does not constitute

---

[23] See note 6.

[24] The term "social" does not appear in the Draft UNCLOS Treaty but this does not seem important. There is no doubt of the coastal State's discretion to set the allowable catch and this certainly includes the competence to take social factors into account.

overfishing in the sense of necessarily threatening irreversible effects; hence such fishing does not endanger maintenance of living resources in the Exclusive Economic Zone.

The second conservation objective of the Draft Treaty, concerning "the effects on species associated with or dependent upon harvested species" is quite ambitious. Article 61(4) provides that the coastal State, in establishing conservation and management measures "shall take into consideration the effects on species associated with or dependent upon harvested species with a view to maintaining or restoring populations of such associated or dependent species above levels at which their reproduction may become seriously threatened." One observer notes that the terms "associated or dependent species" have as yet no common usage and are vague (Dawson, 1980).

For present purposes this objective includes two different situations: (1) effects on associated species, which is here interpreted to include incidental catches or by-catches and (2) effects on dependent species such as predator-prey or more distant food or other biological relationships. These effects are mentioned also in Article 61(3) which requires that in determining conservation and management measures the coastal State should take into account the interdependence of stocks.

Each of these impacts of fishing may relate to other finfish or to mammals. Insofar as the incidental catch or an otherwise affected dependent species is composed of marine fish, the limitation imposed by this objective on coastal State measures seems minimal. If it is difficult (and perhaps impossible in any practical sense) to threaten continued reproduction by targeting on a species, this is even less likely to be accomplished, in most instances,[25] by incidental catch or by reducing food levels or the like. Indeed there may be reason to doubt that fishery science is sufficiently precise and well developed that such a nice determination could even be made.[26] On the other hand the very existence of uncertainty about impacts on associated or dependent species of marine fish might argue that the coastal State should be cautious in adopting particular conservation measures and employ a more restrictive regime than would otherwise be the case.

The problem of incidental catch (or other impact) on marine mammals is different for these populations can be seriously reduced by fishing activities either directly as incidental catch or, perhaps, indirectly by deprivation of food. Because their rate of reproduction is low, a marine mammal species can be reduced to extinction.

Turning to the allocation objective of management, some considerable difficulties were encountered in the UNCLOS negotiations. In the initial preparatory meetings, as States began to come to grips directly with the question of coastal authority in an extended zone, there was a noticeable concern to make provision for full utilization of exploited stocks. It was acknowledged that many coastal

---

[25] If a stock is naturally small or has already been severely reduced by fishing or natural mortality or both, the by-catch may take a larger proportion of the stock than would be taken of a target species by the directed fishery.

[26] An earlier preliminary version of the ACMRR Report (Dawson, 1980) made this point in connection with ecosystem management. See FAO (1979, p. 20).

States were unprepared to make full use of available resources and that access to them should be maintained on behalf of foreign fishermen. The United States, for example, when it tabled revised fisheries articles in 1972 provided for coastal State preferential rights to coastal resources beyond the Territorial Seas but, in recognition of its own distant water interests, sought to provide also for maximum (not optimum) utilization thereof by requiring the surplus over coastal harvesting to be allocated to foreign fishing. The Soviet Union, Canadian, and Japanese proposals at this time made similar provision. On the other hand, the proposals tabled by Kenya and Australia and New Zealand would give the coastal State virtually complete discretion to decide whether any foreign access would be allowed. It may be assumed that the latter proposals more closely reflected the views of coastal States.[27]

In the end there can be no doubt that coastal State views prevailed on the question of exclusive authority over access of foreign vessels to fish within the Exclusive Economic Zone (EEZ). Under the Draft UNCLOS Treaty the coastal State has exclusive authority to establish an allowable catch and there are no requirements that set a floor on the catch. Consistent with the treaty the allowable catch could be very low in relation to the magnitude of the resource.[28] A coastal State may, if it wishes, decide that the total allowable catch in its EEZ is exactly equal to the domestic harvesting capacity. Under Article 61 the coastal State is obliged to establish an allowable catch but the only stated requirements regarding its level are that (1) maintenance of the living resource is not endangered by overexploitation and (2) that the living resource be maintained or restored at levels which will produce the maximum sustained yield as qualified by environmental and economic factors. The former of these seeks to prevent an excessive catch, while the latter permits the coastal State to reduce stock abundance below the level that can produce a maximum sustainable catch. Neither of these requirements significantly qualifies the level of allowable catch except to avoid setting one that is so high it may endanger the resource. In short the coastal State may set an allowable catch that equals its domestic harvesting capacity and therefore excludes all foreign access.

Under the Draft Treaty the coastal State may be obligated to permit foreign access if it sets an allowable catch that exceeds its domestic harvesting capacity as it determines that capacity. However this obligation is qualified substantially by the provision that such access is subject to the terms, conditions, and regulations established by the coastal State.

That a coastal State can decide to exclude foreign fishing in whole or in part despite the availability of fish follows reasonably from Articles 61 and 62. The

[27]The principal negotiations on the fisheries articles, except for anadromous species, were in the Evensen Group, whose final (sixth) revision of its product emphasized the rights of coastal State as appears in the Draft UNCLOS Convention. There has been only minor change in the basic thrust and terminology of the articles resulting from this negotiation.

[28]For example in the United States its responsible management body recommended an allowable catch of 15,000 metric tons of tanner crab as the optimum yield when it was agreed that the MSY figure exceeded 100,000 metric tons.

latter declares that the obligation of the coastal State is "to promote the objective of optimum utilization . . . without prejudice to article 61." There are two particular points to emphasize about this: (1) the aim is defined as "optimum" utilization, not full and not maximum, and (2) such objective is constrained by the coastal State's rights under Article 61. The preeminent right of the coastal State under the latter article is to determine the allowable catch. This decision is exclusively within coastal State authority which, according to Article 297(3)(a), is a discretionary power whose exercise the coastal State cannot be obliged to submit to dispute settlement under the Convention.

Article 62 envisages foreign fishing where the coastal State does not have the capacity to harvest the allowable catch, but the coastal State has sole and unchallengeable authority to determine its own capacity. Article 297(3)(a) categorizes this power also as discretionary.

According to the Draft Treaty the coastal State does have some obligation to declare an allowable catch in its zone, whatever such catch might be.[29] If the coastal State does not establish a TAC and, when requested to do so, arbitrarily refuses to make the determination, Article 297(b) provides that at the request of a disputant State, the issue must be submitted to a conciliation procedure specified in the Draft Treaty.

The net effect of these provisions is to underscore the complete authority of the coastal State to decide who can take how much fish under what conditions in its Exclusive Economic Zone. It is only when the coastal State decides upon an allowable catch and upon its capacity to catch it that any foreign fleet might be given access to the living resources in the EEZ. If the coastal State decides the TAC is larger than its capacity to take it, then foreign fishing may be permitted.

The question then involves the conditions under which foreign access is permitted. Article 62 makes it clear that it is the coastal State which decides upon the conditions of access. Foreign access is permissible pursuant to an agreement or other arrangement and pursuant to terms, conditions and regulations established by the coastal State. In practical terms access is dependent upon negotiations between the coastal and foreign fishing State and the outcome revolves around political relations between the two as well as other conditions of the situation.

In determining who gets access, the landlocked States and those with special geographical characteristics are singled out for special mention in Article 62(2)

[29]It must be said, however, that the nature of this obligation is not as clear as it might seem because of the unsatisfactory drafting of Article 61. The latter does not provide in so many words that the coastal State must declare the TAC for a stock or all stocks. Article 61 simply states that the coastal State shall "determine" the allowable catch, which could be accomplished without any declaration or pronouncement. Second, the obligation relates not to any single specified stock but to the "living resources" of the zone. Presumably this refers to exploited stocks but it can also, and should also, be construed to include incidental catches. However, as stated, it does not require differentiation by stock or species or anything but refers to an overall allowable catch. Such an interpretation, however, would serve little or no purpose because (a) such a calculation would only rarely, if ever, be made, and (b) useful determinations would relate to particular exploited and affected stocks. Few, if any, States would even have the capability to make a determination of allowable catch of "living resources."

and (3). It is doubtful, however, that this special mention translates into any priority of treatment by way of preferred access. It is true that Article 62 calls for the coastal State to have "particular regard" to Articles 69 and 70 and requires the coastal State to "take into account all relevant factors" including the provisions of Articles 69 and 70 which deal with these two categories of nations. Nonetheless access by these nations or by any other is by agreement only and on terms and conditions acceptable to the coastal State.

Articles 69 and 70 do not alter the fundamental balance of authority favorable to the coastal State even though both articles refer to the right of landlocked States and of States with special geographical characteristics to participate in the exploitation of the same subregion or region. This is because participation must be in conformity with Articles 61 and 62, which envisage agreement in accordance with terms and conditions negotiated with the coastal State. Accordingly the "right" to participate is so substantially modified by conditions subsequent that are outside the control of those States that the "right" is more apparent than real.

These conclusions are not substantially modified by consideration of other specific terms of Articles 69 and 70. Paragraphs 1 of both these articles contain provisions of such vagueness and generality that it is extremely difficult to spell out any effective right of participation. It does not help appreciably that the "right to participate" is qualified by the phrase "on an equitable basis," since what is "equitable" depends on a great many variables. The right, it is clear, also does not apply to the "surplus" as such but only to an "appropriate part" of such as may exist in the economic zones of "the same subregion or region." The right to participate is further eroded by the injunction to "take into account the relevant economic and geographical circumstances of all the States concerned . . . ." In light of all these qualifications and conditions, insistence upon an effective right of access by either category of States seems unlikely. It is entirely conceivable that coastal States may wish to permit access as provided for in Articles 69 and 70, but the forces driving the accommodations that may be made stem from the political and social constraints and opportunities that prevail rather than from the legal technicalities of the Draft Treaty.

Scope of Authority: Species Coverage

In the traditional international law of the sea, with narrow zones of coastal fishery authority and freedom of fishing everywhere else, the only occasion for distinctive treatment among living marine resources came in the shaping of international arrangements for their conservation and allocation. Numerous species agreements were concluded for marine mammals (whales, seals), highly migratory species (tuna), and anadromous species (salmon). With extended jurisdiction, fishery interest groups (including fishermen and also those concerned to protect certain animals) urged that distinction be made either in national legislation or new international agreement or both. The species singled out for special treatment include highly migratory species, anadromous (and catadromous) stocks, and marine mammals. After initial reference to national legislation relating to these, I shall discuss how each is treated in the Draft UNCLOS Convention.

With slight exception, national laws assert authority over all living marine resources found within the area of claimed national jurisdiction. A few (Australia, Fiji, New Zealand, Papua New Guinea, Portugal) do not include sedentary species in their general fishery legislation, but in all probability include these animals in laws dealing with their continental shelf. The United States excludes marine mammals from its major fishery law for the 200-mile zone (and Australia excepts whales) but asserts jurisdiction in its Marine Mammal Protection Act. Only the United States and the Bahamas appear to exclude an important commercial fish stock; these States do not claim jurisdiction over tuna (but do claim jurisdiction over other highly migratory species). Japan's law on this is interesting because it claims jurisdiction over all fish within 200 miles but does not prohibit the catch of highly migratory species by foreigners.

With these very slight exceptions States appear to be all-embracing in their claims over fish within 200 nautical miles. This position also is embodied in the Draft UNCLOS Convention although the Treaty provisions are more detailed and deserve more detailed explanation.

Highly Migratory Species. The Draft UNCLOS Treaty has a separate article (Article 64) on some highly migratory species (HMS) (most tuna species and several other species which are listed in an Annex), but the predominant view is that these species are treated exactly the same as other living resources in the sense that they fall within exclusive coastal authority in the economic zone. The following discussion first examines some bases for the view that tuna are not subject to exclusive coastal authority in the zone and also considers possible interpretations of the obligations in Article 64.

Arguments of varying plausibility are advanced to support the contention that coastal State authority does not include the highly migratory species mentioned in Annex I. One is that the separate provision of Article 63 on High Seas/economic zone stocks, which would include highly migratory species, establishes that HMS are not within coastal authority in the zone. By its very existence this separate provision serves, it is urged, to take HMS out of the general reference to living resources in Article 56 and Articles 61 and 62 (which contain the basic general articles on fisheries). Unfortunately nothing in Article 63 suggests that fisheries there mentioned are not within coastal authority. Furthermore while Article 64 on tuna does not explicitly declare that coastal States have control of tuna within their zone, it does state that the provision of Article 64(1) is "*in addition to* other provisions of this part" (emphasis added). It is argued that this form of cross reference is unique and should be understood as meaning "instead of" or "exclusive of." It is true that this cross reference does not appear elsewhere in the Treaty, but this does not suffice by itself to reverse the otherwise apparent meaning of "in addition to" which obviously assumes the applicability of the other provisions of Part V.

Also relevant is that Annex I species include cetaceans, although these are clearly subject to coastal State authority subject to the limitation of Article 65.

This makes it even more difficult to read Article 64 as excluding tuna from any coastal State authority.

A weightier basis for the contention about Article 64 is that it makes different provision for Annex I species than for other species that migrate great distances in the ocean or that inhabit both the zone and the High Seas. Article 63 applies to the "same stock or stocks of associated species" that "occur both within the exclusive economic zone and in an area beyond and adjacent to the zone." The coastal State and "States fishing for such stocks in the adjacent area" are obliged to seek "to agree upon the measures necessary for the conservation of these stocks in the adjacent area." In contrast the coastal State's duty under Article 64 is to "cooperate" with other States whose nationals fish in the region for Annex I species both for conservation and for optimum utilization "throughout the region, both within and beyond the exclusive economic zone."

Although there are, thus, some differences in the coastal State's duties under the two articles, the substance of the coastal State's rights appears to be the same under both. Nothing in the language quoted goes to alter the all-inclusiveness of Articles 56, 61, and 62, especially in light of Article 64(2)'s provision that Article 64(1) applies "in addition to" other provisions of Part V. Article 63 and Article 64 require the coastal State, respectively, to seek to agree or to cooperate with certain other States, but either obligation is fully consistent with, and can indeed be argued to rest upon, the full authority of the coastal State over living resources in its zone. It is true that the Article 63 duty to seek to agree applies to "measures necessary for conservation" outside the zone whereas Article 64 calls for cooperation regarding activities within and beyond the zone. However, the duty to cooperate is not inconsistent with "sovereign rights" or with recognition that final decisions on the cooperative action that results, or the unilateral action required if cooperation breaks down, is in coastal State hands.

If the coastal State has such authority over Annex I species, as appears to be the case, what is the substance of its obligation under Article 64 to cooperate with other States?

A plausible interpretation is that the coastal State must cooperate with the other States fishing in the region regarding conservation and utilization, but if other States do not cooperate or agreement is not reached the coastal State has the exclusive authority to make final decisions in its zone on the substantive conservation measures and on the sharing of catches within its zone. Outside the zone the initiation of measures also is supposed to result from cooperation between coastal and fishing States, but if the latter refuse to cooperate its right to fish (which is subject to the coastal State's rights and interests, according to Article 116) may be forfeited and the coastal State may initiate conservation and utilization measures.

A requirement of negotiation for conservation and optimum utilization within and without the zone makes sense because it fits the nature of the resources in question, i.e., because they are highly mobile and can be caught within and without the zone, the coastal State cannot conserve them acting alone nor can other fishing States do this on their own. To the extent amounts caught within the zone are dependent on harvesting outside it is also true that allocation of access and levying of fees cannot be considered isolated tasks within the zone and the coastal State cannot determine these independently (Joseph and Greenough, 1979, Chap. 6).

In the end, Article 64 of the Draft UNCLOS Treaty appears to preserve the full authority of the coastal State to make the final choices in the zone, but the injunction that it cooperate with other States (and vice versa) is recognition that these decisions cannot simply reflect the exclusive interests of the coastal State. To the extent that highly migratory species are available outside coastal authority, as by definition they are in one or another degree, the questions of how much of the stock can be taken, where in the zone, and by whom cannot really be resolved without international cooperation. The ultimate authority of the coastal State in its zone works in such a context to pressure those concerned to reach agreement that balances the interests involved so that all are "satisfied." While the noncoastal fishing States do not have the weapon of exclusion to force agreement on more favorable terms, their actions can impact seriously on future conditions to the disadvantage of the coastal State; hence such States are not without strength as they attempt to cooperate with coastal States. The balance in pressure between the coastal State and other States depends upon the accessibility of HMS to those concerned. The more fishing must be conducted within the zone, the more influence the coastal officials may exert. Contrariwise, the greater the success normally enjoyed by those fishing outside the zone, the greater the pressure on the coastal State to conclude an arrangement satisfactory to the other States.

One issue that has been much discussed pertains to the form of cooperation under Article 64 and particularly the use of an international agency. In examining Article 64 it is striking that it does not require the establishment of an international organization but rather allows the States concerned, coastal and distant water, to cooperate directly. Even if one could read a requirement for such an agency into Article 64, the States involved are left free to determine its membership, structure, voting provisions, operating practices, procedural rules, and functions. Article 64 is silent on all these elements.

Fishing for highly migratory species beyond the exclusive conomic zone is also dealt with in the Draft Treaty in Articles 87 and 116-120. Article 87 provides that freedom of the seas includes freedom of fishing subject to Articles 116-120. In turn Article 116 declares that "All states have the right for their nationals to engage in fishing on the high seas subject to (a) their treaty obligations; (b) the rights and duties as well as the interests of coastal States provided for, *inter alia*, in Article 63, paragraph 2, and Articles 64 to 67; and (3) the provisions of this section." These provisions introduce an entirely new condition affecting fishing on the High Seas, one that had not previously been accepted. That condition is that freedom of fishing on the High Seas is subject to coastal State rights, duties, and interests as defined in Articles 63(2) and 64-67. This appears to say that High-Seas fishing for tuna is subject to coastal State rights in the Exclusive Economic Zone. Such rights include the establishment of the total allowable catch, the determination of management measures, decisions on its own harvesting capacity, and the conditions of an allocation, if any, to other nations. All of these rights pertain to the same stock of tuna that is caught on the High Seas.

The coastal State, and the other States fishing in the region, also have duties and they include the duty to seek to agree on conservation measures [Article 63(2)] and the duty to cooperate to ensure conservation and to promote optimum utilization (Article 64). The Treaty thus directs all the States concerned (coastal and distant water) to negotiate to establish conservation measures (Articles 118 and 119) and to promote optimum utilization (Article 64) involving High-Seas fishing. The rights of coastal States would come into play in the negotiating process, just as would the rights of distant water States.

The principal effect of Article 116 perhaps arises in connection with the obligation of States to cooperate by negotiating in good faith. If the States involved do genuinely seek agreement and are unable to reach it despite good faith efforts, the Draft Treaty appears to leave them as they are, subject to whatever dispute settlement provisions are available (which would be none regarding the coastal State's decisions in its zone, but some alternatives are available regarding High-Seas fishing). On the other hand, if a distant water fishing nation did not cooperate (outright refusal to negotiate) Article 116 would indicate that its right to fish on the High Seas was no longer operative and that coastal States could unilaterally decide on measures applicable in the area. Such an eventuality would seem very remote. As noted earlier the thrust of the draft treaty is in support of negotiated arrangements for tuna and other highly migratory species and this probably reflects the views of all those concerned.

Anadromous Species. The draft convention provides that "States in whose rivers anadromous stocks originate shall have the primary interest in and responsibility for such stocks." This is clear enough where the waters of only one coastal State are involved because the fish spawn there and leave for the ocean from such waters. The draft is not clear, however, where more than one State's waters are involved, as where the fish spawn in one State's river or lake but leave for the ocean via another State's. Such situation is not infrequent. In the Pacific Northwest, the United States and Canada share at least seven major salmon streams where one or the other is the spawning State and the other is the exit State. The treaty seems to assume that the exit State is the State of origin because other provisions refer to the latter's authority to regulate in its economic zone and beyond. The spawning State has no economic zone and hence cannot be in a position to insure conservation by regulation in that area.

Under the treaty the State of origin has full authority over anadromous species throughout their migratory range except where they enter another State's economic zone or other national waters. The State of origin may prescribe regulations that salmon fishing is to be conducted only within the 200-mile limit except when "this would result in economic dislocation for a State other than the State of origin." Too, while the State of origin may establish TACs for its stocks, it can do so only "after consultations with other States" fishing these stocks beyond any zone and in another State's zone. For fishing beyond the economic zone, the "States concerned shall maintain consultations with a view to achieving agreement on terms and conditions of such fishing giving due regard to conservation requirements and needs of the State of origin in respect of these stocks." The State of origin has the further obligation that it "shall cooperate in minimizing economic dislocation in such other States fishing these stocks . . . ."

Although the Draft Treaty provides for continued High-Seas fishing where its cessation would result in economic dislocation, the treaty does appear to require prior agreement between the fishing State and the State of origin on the terms and conditions of such activity. The text contemplates that regulations, terms, and conditions are to be promulgated by the State of origin after consultations with a view to agreement "giving due regard to the conservation requirements and needs of the State of origin in respect of these stocks." The preeminent position of the latter State is further indicated because it is obligated to minimize economic dislocation [Article 66(3)(b)], strongly suggesting that the State of origin has final authority over the levels of effort and catch. Subparagraph (c) seems particularly significant in this context. It specifies that High-Seas fishing States participate "by agreement with the State of origin in measures to renew anadromous stocks, particularly by expenditures for that purpose" and, as a consequence, "shall be given special consideration by the State of origin in the harvesting of stocks originating in its rivers." This appears clearly to anticipate prior agreement on conservation and management measures and on the sharing of costs for conservation, and to establish these as preconditions to special consideration by the State of origin. Such special consideration is the allowance of fishing beyond the economic zone, which greatly complicates the problem of management and is contrary to the normal pattern of fishing envisaged by Article 66.

Sedentary Species. As noted above, the legislation of most States treats sedentary species the same as any other found within their economic or fishery zone. The United States is a noticeable exception, however, because it claims exclusive authority over such species wherever they may be found beyond the fishery zone.

In its treatment of sedentary species the United States law both resembles and differs from the Draft Treaty. It resembles the Draft Treaty in asserting authority over sedentary species beyond 200 nautical miles. Under the treaty coastal State sovereign rights to the shelf and its resources may extend beyond 200 nautical miles to as far as 350 nautical miles or 100 nautical miles from the 2500-m isobath.

The treaty provisions on the economic zone do not apply to sedentary species as defined in the treaty. Accordingly no obligations regarding yields or surplus are applicable to coastal State regulations for these species and the coastal State may deal with these resources entirely as it pleases. In contrast, sedentary species in United States law are treated the same as any fish subject to the MFCMA and therefore the various yield determinations and foreign fishing quotas are applicable. Thus United States fishery management law recognizes greater foreign rights to these resources than it is required to by the Draft Treaty.

Marine Mammals. All marine mammals are subject to the sovereign rights of the coastal State spelled out in Article 56 and the rights and obligations of Articles 61 and 62 but Articles 64 and 65 add certain responsibilities and clarify the scope of coastal authority.

The clarification in Article 65 applies to all marine mammals and makes it clear that within its jurisdiction the coastal State "may prohibit, limit or regulate the exploitation of marine mammals more strictly than provided in this Part." This means that the coastal State need not permit any exploitation of marine mammals if it so wishes. It may decide that the allowable catch is zero, no matter what the abundance of the animals may be. On the other hand, the coastal State continues to be obliged to observe convention standards that conservation and management measures should take into consideration that incidentally affected species should be maintained above levels at which their reproduction may become seriously threatened.

Article 65 calls for cooperation in conservation of marine mammals and singles out cetaceans by calling for States to "work through the appropriate international organizations for their conservation, management and study."

Article 64 also calls for coastal and other States to cooperate regarding cetaceans. Annex I contains a list of highly migratory species of which cetaceans are one category. For such species coastal and other fishing groups are to cooperate directly or though appropriate international organizations to conserve and to promote the objective of optimum utilization. The more detailed requirements of such cooperation are spelled out in prior discussion.

Areal Extent of Coastal Authority

During the 15 years between the opening of the First and Third United Nations LOS Conferences, expectations about the areal limits on coastal State authority to manage fisheries altered radically. In 1958 and 1960 the serious debate centered about the conditions to attach to an extension of coastal fisheries authority to 12 miles and about whether this alone met coastal State interests. Over the next few years States extended either their fishery limit or their Territorial Sea to 12 miles. By 1966, 49 coastal States had one or another 12-mile limit and this increased to 60 in 1968 and 66 in 1972. What was not entirely clear was whether 12 miles was the maximum fisheries limit a coastal State could adopt as against other States or whether a claimed wider limit must be accepted.

This question was answered in the mid-1970s as a combined and interrelated consequence of the UNCLOS negotiations and uniform national claims, but before this occurred the question was presented to the International Court of Justice when Iceland sought to impose a 50-mile exclusive fishery limit in 1972. The United Kingdom and the Federal Republic of Germany both objected and when negotiations failed to resolve the issue they brought the matter to the Court. The Court's judgment in 1974 did not decide the issues of whether Iceland could unilaterally impose a 50-mile fisheries limit under customary law but rather went off on the ground that Iceland could not oppose such a limit specifically to the United Kingdom. In several dissenting and concurring opinions six of the judges indicated their view [illegible] mile limit could not be imposed on other nations because international law [illegible] require its acceptance. A joint separate opinion by five judges (in which ano[illegible] concurred) expressed the view that international law did not limit coastal [illegible] jurisdiction to 12 miles, but in a delicately inadequate explanation these judges [illegible] not also go on to say that other States must accept a limit wider than 12 miles.

The record of the 1960s demonstrates that most States were agreeable to at least a 12-mile limit in practice, with a small if appreciable number claiming an even wider limit than 12 miles, but by 1972 it was becoming clearer that a much larger fishing zone would be widely acceptable as a product of the UNCLOS negotiations involving numerous other issues. From the beginning discussions of the Enlarged Sea Bed Committee in 1971 preparatory to a new UNCLOS conference, the concept of a 200-nautical mile Exclusive Economic Zone for resources, including fisheries, enjoyed wide support, and by the time of the first substantive session of the conference in Caracas in 1974 the general concept was virtually uncontested, although there was still the task of negotiating the more specific content of the zone. The extent of the EEZ, however, was already agreed to be 200 nautical miles, as indicated by the fact that the only two alternatives incorporated among main trends emanating from the Caracas session called for a zone not to exceed 200 nautical miles (United Nations, 1974). All subsequent negotiating texts provided for a 200-nautical mile Exclusive Economic Zone.

So far as fisheries are concerned, however, events outside the UNCLOS Conference, inspired by the consensus revealed within, came to have decisive significance. By 1975 it became more and more evident that despite general agreement in the UNCLOS Conference on expanding coastal State jurisdiction it would probably require prolonged negotiations on other issues before a comprehensive UNCLOS Treaty could be concluded. At the same time, however, coastal States continued to feel pressure to take unilateral action to resolve difficulties arising from foreign fishing activities. The continued recalcitrance of some distant water fishing nations, apparently oblivious to the political consequences, in the end proved to be the trigger that unleashed a massive dose of 200-mile fishery and economic zones.[30] In 1975 the U.S. Senate passed legislation establishing a 200-mile exclusive fishery zone and in early 1976 the House and the Senate agreed on the form of such legislation. The effect of this was to spur other unilateral actions to claim 200-mile zones. Soon, the USSR, Japan, the United Kingdom, and other European Economic Community (EEC) nations had all extended fisheries jurisdiction to 200 nautical miles. By the beginning of 1977 there could be no serious doubt that coastal States could exercise exclusive fishery management out to 200 nautical miles. At the present it appears to be the general expectation that this is the maximum extent of national fishery jurisdiction.

By the end of its first substantive session in Caracas it was already clear that the 200-mile economic zone enjoyed nearly universal support, whatever reservations distant water fishing States held. The Draft Convention records this agree-

[30]A recent compilation shows 70 such zones established between January 1976 and February 1981 (U.S. Department of State, 1981, pp. 2-7).

ment in Article 57. Coastal authority over sedentary and anadromous species is recognized beyond 200 nautical miles in Articles 76 and 66.

Appraisal of Goals and Decisions

The following offers appraisal of a few major points in comparison of goals and decisions. It is not intended to be a comprehensive or detailed assessment.

Competence to Extend Exclusive Fishing Authority

The preference for a negotiated extension of national jurisdiction has, surprisingly, been largely met by developments in customary law and in negotiations in the Third UNCLOS Conference. Although the current widespread adoption of 200-mile zones began relatively early in the UNCLOS negotiations, the foundation of this development essentially rested in the general consensus that was made apparent in the negotiations leading up to that conference. The evolution of the 200-nautical mile fisheries zone did not occur simply as a series of reciprocal national claims from which one could infer that such zones were not in conflict with international law. This was a case of unilateral actions being taken after a consensus was established by negotiations which, because of their breadth and complexity, did not reach a timely conclusion. Once it became apparent, as it did by 1975, that States generally were agreed on coastal State control of fisheries out to 200 nautical miles, then the avalanche of 200-mile claims began.

The degree of consensus on fisheries is not matched by agreement on other elements of coastal authority in a 200-mile zone. Although 90 States now assert fisheries control out to 200 miles, only a handful of them agree that this area should be considered as national territory, with the relatively greater authority such a claim makes regarding navigation and other uses. A substantial minority of this 90 (32, including most of the industrially developed world) refrain from claims regarding research and pollution and in all probability there is serious disagreement with some of the EEZ legislation provisions that are aimed at navigation. Fewer than half of the coastal States have acted to create a 200-mile EEZ, even though this concept also is incorporated in the Draft UNCLOS Treaty. The one element of extended jurisdiction that enjoys universal acceptance is coastal State control of fisheries.

The major difficulty with the inexorable trend toward national control of fisheries out to 200 miles is that it virtually insures that most fishery management around the globe will be within the exclusive authority of nations lacking the skilled personnel, the funding, the knowledge, and the institutions that are needed for effective management. The capacity to manage fisheries is distributed thinly and unevenly around the globe and does not match the distribution of fishery resources or stage of economic development.

It does not follow that all coastal States should be endowed with or furnished with the people, funding, facilities, and institutions necessary for managing satisfactorily whatever marine resources are subject to each one's jurisdiction. In some instances, at least, the resources available within coastal jurisdiction are not worth the investment required to establish an appropriate management mechanism and process. In other instances, probably a majority, the coastal State can only manage effectively in conjunction with other coastal States or through an international institution. On some occasions, it may well be cost effective that the necessary investments be made to establish an effective management regime.

Judgments about the appropriate course of action depend on numerous factors. FAO assistance should only be extended where it would actually make a difference in producing identifiable net gains for the coastal State. As noted in earlier discussion, coastal States must identify management objectives with such specificity that evaluation of gain/loss is feasible. Individual States must assess their position carefully and assess alternatives with a view to a cost-effective means of acting so that management needs are met and satisfactory benefits realized.

Nature of Extended Coastal State Authority

Decisions categorized either as customary or conventional law are wholly congruent with the policy goal of providing the coastal State with the authority needed for effective management. National legislation and the Draft UNCLOS Treaty both confer the necessary competence to carry out the tasks essential to management. Coastal State authority is exclusive and without significant qualification. The next section offers comment about standards that the coastal State must observe in the exercise of its authority.

Decisions regarding the accommodation of exclusive fishery management authority with freedom of navigation are yet to occur with sufficient frequency to mark a trend. The Draft UNCLOS Treaty does not make precise provision on this point and its interpretation is still controverted. Nonetheless its provisions do provide a basis for working out an accommodation which would permit both sets of interests to be satisfied. Only a few States appear to require protection measures that would affect navigation and any other instances of interference should be rejected.

Management Objectives: The Specification of Conservation and Allocation Goals

Policy and decision about objectives diverge sharply at several important points, including insufficient attention to objectives in much national legislation, a welcome emphasis on economic and social goals in the Draft UNCLOS Treaty but with an accompanying unfortunately narrow formulation of these goals solely in terms of the coastal State's interests, and a generally acceptable formulation of a conservation goal which is, however, embedded in a management system that makes it unusually difficult to achieve.

As noted earlier, national legislation frequently makes no mention of fishery management objectives at all or expresses them too generally or in a confusing fashion. As a result it is difficult if not impossible in many instances to assess how management is proceeding in relation to goals. This is accompanied by all the complications noted in previous discussion.

Insofar as the Draft UNCLOS Treaty objectives are concerned there is a notable shift from the previous pattern of international agreements that with only slight exception emphasized securing the maximum sustainable yield (MSY). Although in practice nations undoubtedly sought within this rubric to maximize their own political, economic, and social goals, the agreed formula stressed an almost nonexistent biological goal. The treaty changes this by making it clear that conservation measures may depart from MSY in order to satisfy other interests. The difficulty with this approach is not in accepting a departure from MSY, for this development was too long in coming, but that the coastal State is given substantially complete discretion to manage the fisheries for its own exclusive interests, however narrowly and selfishly conceived they might be.

This major shift in perspectives about fisheries is a retrogressive development in some ways. Although this approach permits developing nations to secure benefits from living resources within 200 nautical miles, it also enables these States and developed States to ignore the interests of other developing nations who might benefit from gaining access to the resources. Formerly, fisheries management was necessarily an effort to accommodate the interests of various user groups, with a tilt to adjacent coastal States in some instances, but the allocation of discretionary authority to coastal States permits it to make decisions affecting others without weighing their interests, however justifiable their accommodation may be as seen by an objective observer. An allocation of authority that allows arbitrary treatment of foreigners is not desirable policy.

Another difficulty with the Draft UNCLOS Treaty arises from the combination of the two previous points. The Draft UNCLOS Treaty affirms that the coastal State may allocate access to fisheries on terms acceptable to it, thus encouraging that State to seek maximum benefits from exploitation. Yet in the aggregate few coastal States have the capacity to adopt adequate conservation and management measures. The result may well be, at least in the short run, an unhealthy emphasis on exploitation without adequate attention to proper limitation and regulation of effort.

Scope of Authority: Species Coverage

A major problem in the Draft UNCLOS Treaty concerns the regime for highly migratory species (HMS). The inclusion of HMS within coastal authority in the EEZ as are all other living resources may considerably complicate the difficulty of reaching agreement on their management, although it must be admitted that there might have been at least as much difficulty if they had not been included. The advantage of the Draft Treaty is that it encourages negotiations among coastal and distant water States and does not seek to set out any rigid formulas for the institutional arrangements that might be made. Perhaps the treaty provision for ultimate coastal control of HMS in the EEZ will operate to promote a realistic consensus.

Areal Extent of Coastal Authority

Provision for 200-mile zones of fishery control is not less arbitrary than a 12-mile Territorial Sea, but at least it is extensive enough to embrace many stocks that need management. It also, by definition, divides some stocks, or fails to embrace all of a stock, and requires further agreement to be reached regarding their management. This problem may become particularly acute in the Pacific in connection with island States and tuna fishing and may require some particularly creative solutions.

References

Arsanjani, M. H. 1981. International Regulation of Internal Resources. Chap. III.

Birnie, P. 1980. Scientific fisheries management (book review). Mar. Policy 4: 333.

Block, H. 1981. The Planetary Product in 1980: A Creative Pause?

Brown, E. D. 1977. The exclusive economic zone; criteria and machinery for the resolution of international conflicts between different users of the EEZ. Marit. Pol. 4:323.

Burke, W. J. 1981. National Legislation on Ocean Authority Zones and the Contemporary Law of the Sea. Ocean Dev. & Int'l L. 3:289.

Christy, R. T., Jr. 1977. The Fishery Conservation and Management Act of 1976: Management objectives and the distribution of benefits and costs. Wash. Law Rev. 52:657.

Congressional Record. 1966. June 20. Vol. 122, p. 13608.

Crutchfield, J. A., and R. Lawson. 1974. West African Marine Fisheries: Alternatives for Management.

Dawson, C. L. 1980. Glossary of terms and concepts used in fisheries management. *In*: FAO Advisory Committee on Marine Resources Research, Report of Working Party on the Scientific Basis of Determining Management Measures. FAO Fish. Rept. No. 236, p. 117.

Extavour, W. C. 1979. The Exclusive Economic Zone. pp. 233-242.

FAO. 1979. Report of ACMRR Working Party on the Scientific Basis of Determining Management Measures. FAO Fish. Circ. No. 718.

FAO. 1980a. Report on the CIDA/FAO/CECAF Workshop on Fishery Development Planning and Management. Rome, 6-17 February 1978. FAO Doc. No. FAO/TF/INT 180(b) (CAN).

FAO. 1980b. Yearbook of Fishery Statistics. vol. 48.

FAO Advisory Committee on Marine Resource Research. 1980. Report on the Scientific Basis of Determining Management Measures. FAO Fish. Rept. No. 236.

FAO Committee on Fisheries. 1974. Statements and Proposals Made at the Third United Nationsl Conference on the Law of the Sea, Caracas, 20 June-29 August 1974. FAO Doc. No. COFI/74/Inf., 12 October.

Fredin, R. A., R. L. Major, R. G. Bakkala, and E. K. Tanonaka. 1977. Pacific Salmon and the High Seas Salmon Fisheries of Japan. Northwest and Alaska Fisheries Center, Seattle, WA.

Gulland, J. A. 1977. Goals and Objectives of Fishery Management. FAO Fish. Tech. Pap. No. 166.

Gulland, J. A. 1980, Some Problems of the Management of Shared Stocks. FAO Fish. Tech. Pap. No. 206.

Hollick, A. L. 1981. U.S. Foreign Policy and the Law of the Sea. pp. 67-95, 160-165.

Holt, S. J., and C. Vanderbilt. 1980. Marine fisheries. *In*: Ocean Yearbook. vol. 2, p. 9.

Johnson, B. 1975. A review of fisheries proposals made at the Caracas session of LOS III. Ocean Mgmt. 2:285.

Joseph, J., and J. W. Greenough. 1979. International Management of Tuna, Porpoise, and Billfish.

Kahn, R. 1975. The fisheries jurisdiction case—A critique. Indian J. Int. Law 15:1.

Kasahara, H., and W. T. Burke. 1973. North Pacific Fishery Management.

Koers, A. W. 1974. Fishery proposals in the United Nation Sea Bed Committee: An evaluation. J. Mar. Law Comm. 5:183.

Kury, C. 1975. The fisheries proposals; An assessment. San Diego Law Rev. 12:644.

Marr, J. C. 1976. Fishery and Resource Management in Southeast Asia.

McDougal, M. S., and W. T. Burke. 1962. The Public Order of the Oceans.

McDougal, M. S., and W. M. Reisman. 1980. The Prescribing Function in the World Constitutive Process. Yale Studies in World Public Order, Yale Univ. Press, New Haven, vol. 6, p. 249.

McDougal, M. S., H. O. Lasswell, and W. M. Reisman. 1973. The intelligence function and world public order. Temple Law Quart. 46:365.

McDougal, M. S., H. O. Lasswell, and W. M. Reisman. 1981. The world constitutive process of decision. *In*: M. S. McDougal and W. M. Reisman (Eds.), International Law Essays. pp. 268-286.

Möcklinghoff, O. O. 1973. Management and Development of Fisheries in the North Atlantic. FAO Doc. No. FI:FMD/73/S-45.

Moore, G. 1981. Legislation on Coastal State Requirements for Foreign Fishing. vol. 6. FAO Legislative Study No. 21.

Nordquist, M., S. H. Lay, and K. R. Simmondo. 1976-1981. New Directions in the Law of the Sea. Currently 11 vols.

Oxman, B. H. 1977. The Third United Nations Conference on the Law of the Sea: The 1976 New York sessions. Amer. J. Int. Law 71:247.

Richardson, E. L. 1980. Power, Mobility, and the Law of the Sea. Foreign Aff. 58:902, 916.

Saila, S. B., and V. J. Nelson. 1974. Tuna: Status, Trends, and Alternative Management Arrangements.

Székely, A. 1976. Latin America and the Development of the Law of the Sea.

Taft, G. 1975. The Third U.N. Law of the Sea Conference: Major unresolved fisheries issues. Columbia J. Transnatl. Law 14:112.

Tussing, A. R., and R. A. Hiebert. 1974. Fisheries of the Indian Ocean.

United Kingdom *v*. Iceland. 1974. Int. Court Just. Rept. 4.

United Nations. 1972. Report of the Committee on the Peaceful Uses of the Sea-Bed and Ocean Floor Beyond the Limits of National Jurisdiction. GAOR, vol. 27. U.N. Doc. A/8721, suppl. 21, pp. 73-74.

United Nations. 1974a. U.N. Doc. No. A/Conf 62/c.2/SP.1, 15 October, pp. 58-59.

United Nations. 1974b. Comparative Table of Proposals Related Directly to Living Resources. U.N. Doc. No. A/Conf. 62/C.II/L.1, 21 February (submitted by the United States).

United Nations. 1980. Law No. 32/76, 5 Dec. 1976, National Legislation and Treaties Relating to the Law of the Sea. *In*: U.N. Legislative Series. U.N. Doc. No. ST/LEG/SER.B/19.

United Nations. 1981. U.N. Doc. A/Conf. 62/L.78, 28 August.

U.S. Code. 1970. The Bartlett Act, Title 16, Sections 1081-1094.

U.S. Department of State. 1981. National Claims to Maritime Jurisdiction. 4th Rev.

3

Implementing the New Law of the Sea: The Mexican Experience

Alberto Székely

Introduction

Mexico's foreign policy traditionally has been characterized by its persevering reliance on a handful of general and fundamental principles. Its behavior abroad has been guided by them on all fronts and in all forums. Championing nonintervention, self-determination, a generous policy for the indiscriminate granting of asylum, the peaceful settlement of international disputes, and the permanent sovereignty of people over their natural resources has only been the result of deep committment to the strict adherance to the binding rules of international law.

Perhaps in no other field of international activity has this policy been so jealously followed as in the country's maritime practice. This may be no more than a natural reaction of a nation whose international experience with its territory in general has been, as it has been especially during the past century, a victim of repeated mutilating assaults to its integrity,–assaults which were, in turn, a result of an expansionist policy totally without respect to international law.

This perservering attitude, as applied to marine matters, has not always worked to the benefit of Mexico. Other countries which have acted more

Alberto Székely, the alternate Representative of Mexico to the United Nations in Geneva, is currently on leave from his post at the University of Mexico. He has been a legal advisor to now Foreign Minister, Jorge Castañeda. He has been a representative of his country to the Third United Nations Conference on the Law of the Sea and has been directly involved in all recent fisheries negotiations involving Mexico. He is the author of *Latin America and the Development of the Law of the Sea* (Oceana Publications, New York) and *Mexico y el Derecho Internacional del Mar* (University of Mexico).

liberally and unilaterally, sometimes in contradiction to generally agreed rules of international conduct and specifically in the establishment of extended national marine jurisdiction zones, have been more successful *vis-à-vis* other interested states in affirming and implementing their claimed rights over living resources. These days, such radical states as Peru and Ecuador, claiming sovereignty over a 200-mile territorial sea which is not recognized in international law, seem to have a more cordial and workable fisheries relationship with countries which historically have fished off their shores or are interested in doing so, than a law-abiding coastal state such as Mexico whose fisheries policy is earnestly implemented solely on the basis of what has been universally agreed to.

The purpose of this chapter is to study the Mexican experience in putting into practice the new legal order that has emerged from the decade-long international negotiations at the Third United Nations Conference on the Law of the Sea. Special regard is given to the field of Mexico's international fisheries relations, negotiations in which the major fishing powers participated for fear of what they perceived as a creeping jurisdiction tendency over the oceans by some states. Such study will lead to the daring insinuation that the lack of willingness on the part of some of those powers to stick to the new internationally agreed rules of the game may have precisely the effect of speeding the materialization in real life of the doctrine of creeping jurisdiction, even by those who have been seriously intent on remaining within the confines of international law by pursuing moderate practices.

The Elements of Mexican Practice in Traditional Law of the Sea

The opportunity of a claim to a marine zone of national jurisdiction, coupled with the extent of the claim, are excellent indicators as to the readiness of a coastal state to abide by the prescriptions of the international law in force on the matter. It is against this criterion that the assertions in the Introduction of this chapter will be tested.

The first time Mexico put forward a maritime claim was in 1902, a time when it could still be alleged that the traditional 3-mile rule, as maximum breadth for the territorial sea, was in force. Such was the distance legislated that year in the Law of National Properties (*Diario Oficial,* 1902). This was certainly not a very difficult decision to make, as the first evidences that emerged at the turn of the century of the progressive decay of such traditional rule were confined to observations by the academic community. It was still very much a time when customary rules took immemorial lengths of time both to be born and to be changed or derogated. By 1917, however, when the new Constitution was being drafted at the end of the revolution, the legislature was apparently already conscious that a significant change was in the making, as a result of state practice. When the time came to include a provision on the breadth of the territorial sea, extreme care was taken to avoid adopting a text that would lack the required flexibility to accomodate itself to the forseen change in the customary rules of international law on the matter. At the same time, the legislature took care not to adopt a claim that would eventually become contrary to whatever new rule would eventually take shape. Thus, an intelligent formula was incorporated in Article 27 of the Constitution, according to which the territorial sea was declared under national sovereignty, "with the breadth and in the terms set by international law" (*Diario Oficial,* 1917). The formula has proved to be quite useful, as the subsequent changes in international law have not rendered it necessary to amend the Constitution.

The 1930 Hague Conference for the Codification of International Law, sponsored by the League of Nations, included in its agenda an item that would allow the participants to come to grips with what by then had become evident, namely, that the 3-mile rule had been abandoned by such a number of states that it was impossible to speak of its universal customary validity. The failure of the Conference to even come to a vote on the question of the maximum allowed breadth for the territorial sea made it difficult for states to put forward their claims with a minimum degree of confidence that, if one did not stick to the 3-mile rule, other states would not object. However, the range of the proposals advanced by the participants at the Hague (between 3 and 12 miles) constituted some kind of an indicator to the direction in which the practice of states was moving. Any claim in between those two figures was a risk.

For the Mexican legislature, such risk was a calculated one, when in drafting the new Law of National Properties, in 1935 (*Diario Oficial*, 1935), it was decided to specify the width of the territorial sea at 9 miles. Even when this claim met with the objections of some of the largest maritime powers, Mexico was able to proceed with it for more than three decades, precisely the ones during which the international law on the matter passed through a period of turmoil before reaching a final accomodation. It can be asserted that during this time Mexico remained in a moderate position as compared with other states, specially those in its own Latin American region.

It was precisely when its southern neighbors initiated a trend toward extended national marine jurisdiction, as a reaction to the 1945 Truman Continental Shelf Proclamation (Truman, 1945), that Mexico had a chance to demonstrate its willingness to remain a moderate element.

The fact that the said Proclamation was issued on the same day as another one, on the "Policy of the United States of America with Respect to Coastal Fisheries in Certain Areas of the High Seas," (Truman, 1945b) apparently lent itself to some confusion and misinterpretation, in the sense that the United States claim over the Continental Shelf included not only the soil and subsoil, but the superjacent waters as well. Mexico was the first state to follow suit on the claim, and also the first to fall victim of such erroneous appraisal of its extent.

On 6 December 1945, a bill was introduced to Mexico's Congress by the President to amend the Constitution in order to claim both the Shelf and the waters above it, an act which was clearly contrary to international law. The bill went through all stages of the legislative process, being approved by both chambers of Congress and by the required two-thirds of the local legislative assemblies. It was precisely at that time that the Foreign Ministry became aware of the mistake that was about to take place, as it was then understood that the bill meant an illegal unilateral appropriation of a part of the High Seas. The President abstained from publishing the approved bill in the official gazette and, thus, it never came into force (Székely, 1979), a fact which maintained Mexico within the confines of legality.

The latter episode was perhaps a lasting lesson for Mexico, since from then on its behavior on these matters became even more cautious. No legislation was adopted on the law of the sea until the international community had pronounced itself through well-settled general guidelines and rules as to the limits of unilateral legislation in this field. Thus, it became apparent that the better strategy, at least as compared with the one undertaken by the sister republics of South America, was to actively participate in the international process of codification and development of international law, to ensure the expected pronouncement of the community of nations.

That is probably the best description of the attitude that accompanied the Mexican Delegation to the First United Nations Conference on the Law of the Sea, in Geneva, in 1958, when the participation of the country translated into the advancement of one of the proposals that nearly saved the day of that most important international event. This active participation signaled the central role that Mexico was to play in the future in all global law of the sea negotiations. This role put Mexico in the leadership of the movement favoring a 12-mile Territorial Sea, a movement which crystalized not in the Conference but shortly thereafter through customary international law. The Mexican-Indian proposal in Geneva, favoring the described breadth, received 35 votes in favor, 35 against and 12 abstentions in the First Committee of the Conference and, under its Rules of Procedure, was regarded as defeated (UNCLOS, 1900). At the Plenary Session, Mexico made a new attempt, cosponsoring the same proposal with seven other delegations. This proposal was defeated by 39 votes in favor, 38 against and eight abstentions (Garcia Robles, 1961).

Nonetheless, it later became evident that the Mexican position was an unequivocal signal as to the direction in which the international community was inevitably moving on this most fundamental subject of the law of the sea. However, despite its strong commitment to such position, Mexico abstained from incorporating it into its national legal system, until it was collectively enshrined by state practice in international law. This happened shortly after the Second United Nations Conference on the Law of the Sea, which was held in 1960 in Geneva, and where an additional effort to reach an agreement on the matter had also failed. By the mid-1960s, it became easy to find clear evidence that the majority of nations had finally embraced the 12-mile rule. It was then that Mexico felt confident and comfortable about proceeding to change its national legislation accordingly. However, such an undertaking faced some grave problems of an international nature, the careful handling of which would determine its international legitimacy and effectiveness.

Being rich in living resources, the seas adjacent to the national coasts had for a long time been attractive to powerful foreign fishing fleets, especially those of the United States and Japan. The extension of the Mexican Territorial Sea would thus mean tampering with created interests, whether legitimate or not. It was not difficult for Mexico to understand that in order for its new claim to be successful, it would be necessary to deal intelligently with such interests. Thus, it proceeded in two well-planned stages. First, it took the moderate step of creating only a 3-mile exclusive fishing zone (*Diario Ofical,* 1967), additional to its 9-mile Territorial Sea, thus keeping itself within the spirit of its 1958 proposal at the Plenary of the Conference, which provided just that, that is, that when a state did not claim a full 12-mile Territorial Sea, it would be entitled to an exclusive fishing zone up to that limit. This in itself would serve two other well-thought out objectives. On the one hand, it would afford additional time for the 12-mile customary rule to consolidate itself through further state practice, a matter which was envisioned as highly likely, given the increasing number of states achieving their independence precisely during those years. On the other, it would facilitate dealing smoothly with foreign fishermen, as the claim would not involve the national sovereignty and, thus , any problems and even a lack of agreement would not give rise to the same degree of nationalistic response in the country. The latter proved to be true when the second step was undertaken, which consisted in the negotiating of exemplary phaseout agreements with both the United States and Japan (Székely, 1979, p. 10), which by 1973 voided all fishing claims by the fleets of both countries. This was a major diplomatic success, which set a precedent in international fisheries relations. It was possible, obviously, only because those two states had already resigned themselves to the proved derogation of the traditional 3-mile Territorial Sea (Public Law 89-658, 14 October, 1966).

The latter developments allowed Mexico to move progressively toward the broadening of its Territorial Sea by 1969, up to the 12-mile limit (*Diario Oficial,* 1969), safeguarding of course the provisions of the fisheries agreements with the United States and Japan, which were still in force. This was also the culmination of a long process, which formed an integral part of Mexico's perception of the limits of its behavior under the framework of international law. This perception, once again, sharply contrasted with the superficially more attractive one adopted by other Latin American countries, in its turn based more on a policy of confrontation, whose effectiveness was sustained with the inevitable support of the popular nationalistic feelings to which it gave birth.

Mexican Contribution to the Emergence of a New Law of the Sea

Such a moderate practice left Mexico in an ideal position to play a significant role when the international community decided once again to negotiate a new regime for the oceans. This was true specially *vis-á-vis* the highly antagonistic positions held by the different groups of countries at those negotiations. Its record in the field could not have been more propitious for playing a moderating and conciliatory role. Mexico envisioned such a challenge and undertook to face it fully, supported by its own high interest in securing a new regime benefitial to it, as well as by its own priveleged situation both geographically and economically in the sea.

It is extremely important to understand that, up to that moment, that is, when the negotiations for the new Law of the Sea commenced at the turn of the decade, the Mexican perception of the potential benefits that it stood to gain in the negotiations for a new international regime evolved around a narrow conception of the economic meaning of the oceans, that is, the concern for achieving exclusivity over the living resources of the sea. This conception was widely shared by most developing countries. In many ways, even when the pretext for the restarting of Law of the Sea negotiations at the global level was originally given by the interest and need to legislate a regime for the sea bed and ocean floor, and the subsoil therof, beyond the limits of national jurisdiction, which stemmed from the Maltese proposal in the U.N. General Assembly in 1967 (UN Document A6695 and 1967 *Yearbook of the United Nations*, p. 41). Most of those countries embarked on the negotiating process with one fundamental object in mind, namely, securing exclusive access to adjacent fisheries resources. It was during that process that the perception was gradually broadened, to finally embrace the impressive subject matter which today constitutes the new Law of the Sea. This was perhaps the result of an increasing conscience, among the members of the international community, regarding the concept of the unity of the oceans and was visible in the difficult bargaining that took place in the negotiation of the list of issues and items for the Third United Nations Conference on the Law of the Sea.[1]

Aside from the international sea-bed question, at the beginning it was mostly as though a new international fisheries regime was the main subject of the negotiation. In this precise aspect Mexico, being as it is endowed with a particularly large wealth in living marine resources, had much to struggle for.

It was then not surprising that the Mexican Delegation was the first to submit a written proposal, at the Sea-Bed Committee created to prepare the Conference, in a first attempt to bring together the conflicting positions that had taken shape on the question of a 200-mile zone (UN Document A/AC. 138/SC. II/SR. 11).

During the preparatory work of the Committee from 1970 to 1973, it was easy for the Mexican Delegation to identify the elements separating, on the one side, the big maritime powers and, on the other, the increasingly large number of developing countries which, on the experience and inspiration of certain Latin American countries, strove for extended national maritime jurisdiction. And it was in those elements that Mexico, together with a handful of other delegations, such as Kenya in Africa and Colombia and Venezuela in Latin America, envisioned the possibility of a compromise.

The Mexican proposal embodied the conciliatory element, one which recognized the resource-oriented aspirations of the developing countries while, at

---

[1]The original list was adopted by the Committee on the Peaceful Uses of the Sea-Bed and the Ocean Floor Beyond the Limits of National Jurisdiction, on 16 August 1972 (*Official Records*. U.N. General Assembly, Supplement No. 21, A/8721, p. 5).

the same time, safeguarding the security and strategic concerns of the maritime powers. It thus brought out the idea of a "special jurisdiction zone," where the coastal state would conserve, administer, and exclusively utilize its living resources, prevent pollution, and observe other applicable rules of international law. This latter item was a clear reference to the maintenance of such rights of the international community as the traditional freedoms of navigation, of overflight, and for laying submarine cables and pipelines. However, in reality it remained a fisheries zone. Mexico had still a bit to mature regarding its position on the 200-mile zone. But for the moment it was contributing with an eclectic approach to the matter.

It did not take long for Mexico to expand its view on the matter. As early as the beginning of 1972, President Echeverria embraced the "patrimonial sea" concept (Székeley, 1979, p. 134), originated in a speech in 1970 by the Chilean jurist Edmundo Vargas Carreño (1973), which lent itself to an all-resource approach but preserved its eclectic element, with its built-in *quid pro quo*. This was first carried out through the all-out support given by the Mexican Delegation to the Work of the Santo Domingo Specialized Caribbean Conference on Problems of the Sea, held in June of 1972, which went beyond the preferential or even exclusive rights over the 200-mile zone, which had so far been advocated by Mexico, to claim for rights of "sovereignty" and not only over living resources but over the nonliving resources as well (UN Document A/AC 138/80). This Conference produced a strong coalition of Latin American countries, which once and for all disassociated itself from the "territorialist" South American group, that is, those which advocated a 200-mile Territorial Sea (Székely, 1976) and which eventually became a strong conciliatory force in the Conference, the leadership of which Mexico was to embrace and hold till the end of the global negotiations. This force neutralized the strength of the two radical positions, offering a basis for negotiation and giving room, into which the extremists could move, to a central unifying criterion. This force was substantially strengthened with the emergence of a basically identical position in another geographical area, through the Kenyan proposal for an Exclusive Economic Zone, a proposal which put the two regions in touch on the matter (UN Document A/AC. 138/SC. II/L. 40).

The Santo Domingo position was formally introduced by Mexico, Colombia, and Venezuela to the Sea-Bed Committee (UN Document A/AC. 138/SC. II/L. 21), in 1973. Shortly thereafter, the Kenyan position was equally introduced by the African countries. The great degree of coincidence between the two documents made it easy for the Latin American patrimonialists to adhere to the concept of an Exclusive Economic Zone, thus giving birth to a more expanded coalition. A reading of both proposals, today, gives an unequivocal idea of their measure of success, as they have been literally adopted by the international community as a whole.

The process to ensure such success was a long and difficult one, from the beginning of the Conference in 1973, in New York, throughout its substantive sessions. The general concept of the Exclusive Economic Zone stated emerging clearly during the 1974 Caracas session, and an implicit consensus was already detectable in Geneva in 1975. Mexico took the leadership as Chairman of the Coastal State Group, a body whose merit consisted in embracing not only developing countries from the three regions, but some important developed ones as well. Even more important than that, the Group was able to attract some of the leaders of the radical territorialist group, such as Peru and Ecuador. In the end, the whole negotiation for the 200-mile zone was conducted between this Group and the one representing the largest maritime powers, with the United States as its main spokesman. An outstanding virtue of the Coastal States Group was that it allowed for the formation of smaller negotiating groups, through which they were represented by a handful of countries to whom they gave their trust. The most important of these was the one that received the name of its main sponsors, Castañeda of Mexico and Vindenes of Norway, which invited the maritime powers to negotiate informally through an ingenious system of working papers which, little by little, brought them closer to a consensus, a consensus which was submitted to a continuous referendum by the rest of the membership of the Coastal States Group. During the Work of this small group, the Mexican Delegation proposed the elements which gradually gave juridical contents to the institution of the 200-mile Exclusive Economic Zone, through negotiating and conciliatory formulas that solved the hard problems regarding the legal nature of the zone (as a zone which is neither part of the Territorial Sea nor of the High Seas), the utilization of the living resources at the optimum level (within the limits of the maximum sustainable yield and through the system of surpluses), the residual rights which had to be recognized either to the coastal state or to the international community, and many others. Many of those formulas are identified with the name of the leader of the Mexican Delegation and remain as testimony of its contribution to the new law of the sea. The subsequent texts which appeared at the end of the sessions of the Conference (A/Conf. 62/WP. 8 and Rev. 1, 10 and Rev. 1 to 3, and A/Conf. 62/L. 78.) incorporated to the letter the agreements of the said negotiating group, since its successful results were invariably regarded as the best indication of the will of the rest of the Conference, despite the totally unofficial nature of its work. At the end, it was from there that the final consensus on the Exclusive Economic Zone was taken, on the most central question of the new Law of the Sea.

Establishing the Exclusive Economic Zone

The Political Reasons

By the time the Exclusive Economic Zone (EEZ) became a reality in the Third United Nations Conference on the Law of the Sea, the Mexican administration which had so devotedly struggled for it was rapidly coming to an end. This administration, headed by President Luis Echeverria, had distinguished itself in the international level for its strong and intense commitment to a new international economic order, of which the EEZ was regarded as an integral part.

Even when it was obviously never stated officially, it was quite understandable to expect the outgoing President to wish to see the establishment of the Mexican Exclusive Economic Zone during his administration. This was regarded as possible and desirable, given the fact that all necessary ingredients were considered attained, specially the legal ones.

On the other hand, it was becoming increasingly difficult to sustain the situation then present, one which concurrently included the strong advocacy, at the international level, of legal institutions to drive away foreign fishing in the adjacent seas as well as the permanence of such activity in real life. The two were evidently contradictory, and few were the remaining reasons to preserve such a *status quo*. In many ways, the national rhetoric surrounding the matter made it untenable not to proceed to put into effect what was already reagrded as accomplished internationally.

The Legal Justification

There were many arguments for considering that the moment was ripe for proceeding to the unilateral act of creating the 200-mile zone, but the legal ones had to be dealt with carefully.

The timing of the claim, if it was to be consistent with Mexico's traditional practice in the Law of the Sea, had to coincide with the existence of an international rule validating it. In 1976, the international community had been able to reach a basic agreement on the general concept of the Extended Economic Zone (EEZ), but it had not reached the stage, and has not so far, for that matter, of incorporating such agreement in an international treaty with binding force. Moreover, a great deal of detailed matters pertaining the EEZ remained to be negotiated, especially in the field of protection of the environment and scientific research. The consensus that had emerged revolved solely on the right of the coastal state to establish such a zone of national marine jurisdiction, its breadth, a list of the powers vested in it over the zone and its resources (sovereign rights, exclusive and concurrent jurisdictions, etc.), and the regime for the management and conservation of the living resources therein.

Also, the question of the existence of a consensus was far from settled, except as applied exclusivly to the very general concept, and even here it was far from a universal one.

The only documentary evidence as to the existence of the general consensus was the Informal Single Negotiating Text (A/Conf. 62/WP. 8) whose very title is the best description of the distance between it and a treaty in force. For Mexico, this document had the value of consolidating the basic and implicit consensus on the general concept of the Zone and of being clear evidence that the institution of the EEZ, as a part of international law, had emerged. Although Mexico was fully conscious that the said document could not, by itself, afford the legal basis for a unilateral act, its practical viability had a greater weight than the formality or, rather, informality, surrounding it. Indeed, the greatest strength of the Informal Single Negotiating Text was that it evidenced not how far the international community would go on the matter, but how far it had gone already. In other words, despite potential lack of agreement on more specific details regarding the scope and extent of the exercise of certain rights and jurisdictions, a basic right had already been recognized to all coastal states to establish an EEZ, and it would have been totally unrealistic to expect that, because of possible failures in reaching further detailed agreement, the coastal states would have relinquished that basic right. Briefly, Mexico identified this as an irreversible process. There was no way back on the general issue, no matter how controversial its implementation could become. It was inconceivable to Mexico that the great majority of states which had for so long and so arduously worked in order to see the acceptance of the zone, would ever resign themselves to not being able to legally create a 200-mile zone such as the one that had in general terms been negotiated. Even the counterparts in the negotiations had embarked, in many instances, in the process of creating a zone for their own under such agreed general guidelines. The sum of these considerations led Mexico to the conclusion that if it certainly was a well-calculated one. After the 1975 Geneva Session of the Conference, a new Law of the Sea, whether conventional or customary, was utterly unthinkable. It all then depended on the scope and extent of the desired claim. If it went too far into the details of the claimed rights and jurisdictions, the country would more likely than not run into international opposition.

When the national legislation was finally published in February of 1976 (*Diario Oficial,* 1976a), creating the Mexican Exclusive Economic Zone, the only country to submit a protest was the United States, arguing that the unilateral act was contrary to the spirit of the Conference, as it was likely to obstruct its work by stimulating a series of similar claims. This, as it will be seen shortly, came ill out of the mouth of a country that was to take no time in signifying itself as the main threat to the "spirit" of the Conference and as its main obstacle to success.

International Implications

The only way in which it could have been argued that Mexico had acted against the spirit of the Conference was that its claim contained elements contrary to what had been agreed internationally. The Mexican claim, as a matter of fact, followed the agreed texts on the Zone to the letter, as textually as it could be, and concerned itself exclusively with the very general concept of the zone, and not with the detailed aspects of the several rights and jurisdictions involved. This concept never varied in the subsequent negotiations. As to the argument that the Mexican claim would obstruct the work of the Conference, time disproved it, since the negotiations progressed steadily on the subject and the claim did not stimulate a series of similar unilateral actions. A large number of states did proceed eventually with the creation of their own 200-mile zones, but not as a result of the Mexican unilateral act. It could be said that they all did it, including Mexico, as a reaction to the progress in the international negotiations.

There is, however, an even stronger element to prove the good intentions of Mexico, in the sense of ensuring that its unilateral claim would fully conform with the will of the international community.

The bill that was introduced by the President to Congress (*Diario Oficial,* 1975), to establish the Mexican EEZ contained a provision regarding a *vacatio legis* for its entry into force, designed precisely to give the Conference an additional opportunity to arrive at a more formal and, if possible, written agreement on the matter. Its entry into force was set for June 1976, that is, a full month after the end of the next session of the Conference.

The willingness of Mexico to conform with its traditional practice in the Law of the Sea was the object of a still further effort to afford time for the exercise of its claim in the framework of a healthy international environment. Even when the said *vacatio legis* was completed, Mexico practically postponed the implementation of its law establishing the zone, up to 31 July, through an ingenious formula incorporated in the decree through which the outer limit of the EEZ was set (*Diario Oficial,* 1976b). Under the pretext that sufficient time was required for the destinataries of the claim to become aware of the limits of the claimed zone, and in order to enhance the chances of its effectiveness by ensuring that the zone would be respected, the implementation of the claimed sovereign rights and jurisdictions was extended to the said date. In reality, however, what was intended with such action was to give extra time to the Mexican negotiators, who had been hard at work trying to conclude agreements with both the United States and Cuba, regarding both their access to Mexican fisheries surpluses in the new zone and the delimitation of the respective boundaries.

Problems of Implementation

It soon became clear that the Cuban and American agreements were the main initial problems the country would have to face in implementing its recent claim. This stemmed from the compromise in the international negotiaition which had given way to an agreement on the Exclusive Economic Zone, according to which a state that lacked the capacity to fish the maximum allowable catch of its living resources would be obliged to give access to foreigners to the resulting surpluses. This, of course, would entail the granting of fishing permits, for which the foreign fleets would have to pay, fees which would mean a real economic benefit for the coastal state.

The act of establishing the EEZ would, then, mean a total review of the country's fisheries policy. If a real exclusivity were desired over the resources of the zone, it would be necessary to eliminate all foreign fishing in it. For this, the only possible recipe would be to increase the national effort, in order to make the surpluses disappear. This, in turn, would require the launching of a major program for the optimum and rational utilization of the newly acquired living resources. Mexico was forced to look at itself, and to make an evaluation as to the meaning of these resources for its economy and their place in its development programs and strategy.

Mexico's Marine Wealth and New Fisheries Policy

Before embarking in an intensified effort to utilize the living resources of the EEZ, it was obviously necessary to know the nature and dimensions of the wealth to be exploited, in the framework of a conceptual development plan. This brought to the surface the degree of importance attached by the country to this field of activity.

Mexico has 10,000 km of coastline on four seas: the Gulf of California, the Pacific Ocean, the Gulf of Mexico and the Caribbean. It finds itself among the largest coastal states in the world, and the dimensions of its Exclusive Economic Zone, about 2000 km^2, which are a bit more than the surface of its land territory, are in an proportion to the size of the wealth it contains, a phenomenon which is rarely true in other countries.

Only a few other states among the about 130 bordering the seas are in a position to claim a similar or larger marine wealth. These are probably only eight coastal states, namely, the United States, Australia, Indonesia, New Zealand, the Soviet Union, Japan, and Brazil (*Neptune*, 1975).

The Scripps Institute of Oceanography (1976) was among the first to recognize the great potential wealth of the country as a result of the establishment of the Exclusive Economic Zone. What many failed to see in the turmoil of the national enthusiasm on the matter, especially as enhanced by government rhetoric and the mass media, was precisely the mere "potential" of such wealth, as this would only translate in concrete benefits for the country if a major political decision was taken to make its actual full utilized scale (within ecologic limits), an issue of the highest national priority.

Immediately before the establishment of the 200-mile zone, marine fisheries in Mexico were barely taken seriously. Only in a very few years previously had it received attention. Still, in 1976 only seven species constituted 73% of the national catch, of which 54% was shrimp (Carranza Frazer, 1978). Of this total, about half was devoted to human consumption. Also, approximately 81% of the 340,000 tons of the total catch was exported (*El Nacional* (Mexico City), 2 January 1982). It was, thus, a very incipient industry.

It is widely agreed that the only way to quantify a species is by fishing it up to the moment when the statistics show that the maximum allowable catch has been reached, and that in order to maintain a maximum sustainable yield and avoid overexploitation that catch is in fact the rational limit. A country that concentrates its fishing efforts in only a handful of species, and merely because they are lucrative in the international market, cannot properly evaluate its entire fishing stock. Fishing has to be diversified.

In 1975, the National Fisheries Institute estimated that there were in Mexico 504 species of commercial interest. However, only 100 of them had been subject to some degree of exploitation and, according to Dr. Alejandro Villamar (1976), only 20 of them constituted the backbone of the national fisheries industry. Still, only in the case of five of those 20, was there acceptable knowledge as to their real abundance.

The place of this activity in the economy was meager, as fisheries only represented 0.16% of the National Product, and the per capita consumption was a ridiculous 4.5 kg. This is hardly acceptable in a country surrounded by such marine wealth, and especially in a country where such a large proportion of its population suffers from hunger and malnutrition.

The new administration, which took over at the end of 1976, at the very beginning adopted a more agressive attitude. First, it reestablished the Department of Fisheries as part of the Federal Government. Then it adopted its 1977-1982 National Program for Fisheries Development, with the aim of increasing the national catch to 2.4 million tons (Departamento de Pesca, 1977). At the end of 1981, the official figures put the annual catch at 1.8 million tons, that is, six times that when EEZ was established (*Anuario Estadistico del Dapartamento de Pesca,* 1981). Still, there is undue concentration of effort on few species; that proportion of human consumption has not changed and the national market has not been substantially expanded. However, the first steps have been taken, and a stronger political will to give this field the priority it deserves will result in the real benefits expected from such wealth for such poor country.

Dealing with Foreign Fishing

The Agreement with the United States

The U.S. shrimp fleet was the primary foreign fishing fleet in the waters which, before the establishment of the zone, were considered a part of the High Seas. There interests had to be heavily affected if implementation of Mexico's sovereign rights was to be put into effect. The first obstacle was to convince the United States Government that Mexico's catch capacity for shrimp was large enough to justify excluding the U.S. shrimp fleet from Mexico's fish zone. For this it was necessary to negotiate and, in so doing, to come to a common understanding of Mexico's rights.

Differences in Positions and in Interpretations. The United States had expressed its objections to Mexico's unilateral claim. However, these stemmed only from the all-embracing resource-oriented nature of the claim, that is, the fact that it included not only the living resources in the water column but the nonliving resources in the soil and subsoil. The latter resources, in the United States view, were not yet a part of the package agreed in principle at the Third United Nations Conference on the Law of the Sea. That is why, when establishing its own 200-mile zone, the United States limited its unilateral claim to the fishing and conservation of its living resources (Public Law 94-265). This had been done even before the Mexican claim, and thus the U.S. Government had already found it necessary to embark in a long process of negotiating with many countries for fisheries surpluses, both within its zone for foreign fleets as well as for its own fleet in foreign zones. However, in any such negotiations, such as the one with Mexico, the United States insisted on a disclaimer clause that would ensure its nonrecognition policy to any claim that went beyond its own. As pointed out, however, the differences in position did not prevent it from negotiating with Mexico on this matter.

The Shrimp Problem. At the beginning of the negotiations, Mexico had sought to limit United States access to the shrimp fishery as much as possible. The talks proved to be one of the first instances of the great difficulties involved in bilaterally implementing the international agreement reached at the Conference. These consisted mostly in the hard bargaining involved in determining the size of a surplus for a given species.

Mexico had decided, from the very start, that any surpluses would be negotiated only with two of the interested countries, namely, the United States and Cuba. This left Japan out because of its indiscriminate fishing practices.

Both sides, obviously, exaggerated the numbers upwards or downwards in accordance with their own interests. Agreeing on the historical statistics to show both the abundance of the resource and the catch capacity of the Mexican fleet was no easy matter. All this was largely complicated by one internal issue in Mexico, which delayed the negotiations. According to the 1972 Fisheries Development Law, shrimp is one of the species whose exploitation is reserved for the social sector of the industry, that is, the cooperatives, which excludes the private sector. This latter interest, with great investments in fisheries, had indirectly benefitted from shrimping by cooperatives by renting boats to them. However, the value of the shrimp had always generated their interest in gaining direct access to it. Thus, they found it incomprehensible, and totally unacceptable, that the Mexican Government would be willing to give that access to foreigners, no matter how large or reduced the agreed surplus while at the same time excluding an important sector of its own nationals. Needless to say, the nationalistic cards in their hands were many and powerful, and they set out to play them. For them, the agreement with the United States could only be acceptable if the Mexican Government allowed the private sector to start the exploitation of the resource by themselves. Moreover, they were in a position to argue that their participation in the fishery would help Mexico do away with the surplus all the sooner. The government, at the the highest levels, sought to justify the agreements as arising from an obligation under international law but resisted till the end the attempts of the private sector.

In the negotiations the United States disputed not only Mexico's figures but its allegations that it would soon increase its capacity up to the maximum allowable catch. It was perhaps this lack of credence in Mexico's ability to move toward that goal that finally led the United States to accept a Mexican proposal, according to which a phase-out system would be adopted, through a period of 3 years, by which time the American shrimp fleet would finally fade from the fishery. Each year the United States fleet would be given a smaller quota and, by 31 December 1979, a date when Mexico estimated it would be in a position to catch up to the limit, it would stop operations in the Mexican Exclusive Economic Zone (see, *Treaties and Other International Agreements on Fisheries, Oceanographic Resources, and Wildlife Involving the United States* (U.S. Government Printing Office: Washington), 1977, p. 997). The above assertion was subsequently confirmed each of the several times that the United States sought to extend the agreement in order to save its access to the fishery. This happened even after the said date, when in fact the Mexican fleet had the capacity to catch the resource without leaving any surpluses.

Fishing of Other Species in the Gulf. In the same agreement, Mexico agreed to allow the fishing by the American fleet of a token amount of other species' surpluses, snapper and grouper, of up to 450 tons in the Gulf of Mexico and, as in the case of the shrimp, outside of the Territorial Sea.

The fishing of these species by the United States was pursued with little interest. It formally stopped when the Fisheries Agreement of 1976 was denounced by Mexico in 1980, once it was decided that there were no bases for a mutual fisheries relationship, given the tuna controversy (which will be reviewed in the next section) and given the fact that the other Fishery Agreement concluded by both countries in 1977, for fishing of certain surpluses by Mexican vessels in the U.S. Fisheries Management and Conservation 200-mile Zone, had been breached by the United States, by not granting the quotas provided by it for one of the species involved. The denunciation took effect on 31 December 1981 and, with it, all activities by American fishing vessels within the Mexican Exclusive Economic Zone ceased.

Concession for the Small California Fleet. It must be pointed out that another provision of the 1976 Agreement concerned the small fleet belonging to American families from California, which had traditionally fished off the Baja California coast and which depended for its survival on continuing to do so. The Mexican's generous willingness to allow them to continue operating, not in the EEZ but in the Territorial Sea itself, was a true concession which also fell victim of the tuna controversy and of the lack of willingness by the United States to reciprocate with Mexico by respecting the 1977 Agreement.

Leaving the Tuna Question Pending. At the time of the negotiations for the 1976 Fisheries Agreement, the only other fishery being undertaken by United States boats was that of tuna in the Mexican part of the eastern Pacific Ocean.

At the Law of the Sea Conference, an agreement had already been made in responding to the highly migratory nature of tuna species by subjecting them to a special sort of legal regime, both for their conservation and for their exploitation. Since a great proportion of the tuna in the region had been caught in waters which were part of the High Seas but which, by virtue of the progress registered at the Conference, had become part of exclusive economic zones, the rules of the game were different as applied to their exploitation and it was thus necessary to renegotiate them. Because of these developments, Mexico had, at the 1975 annual meeting of the regional organization which had been competent in the conservation of the resource, the Inter-American Tropical Tuna Commission (1 UST 230; TIAS 2044), announced its intention to call for a special conference in order to adjust the regime of conservation and exploitation of eastern

Pacific tunas to the new legal realities. It also convinced the Commissioners to entrust the Director of Investigations of the Commission, Dr. James Joseph to draft a paper with the possible alternatives available for that purpose.

Thus, in the 1976 Agreement all these matters were incorporated in its provisions, and it was established that, in the meantime, the United States tuna vessels would be required to obtain a certificate of access to the Mexican EEZ, for which they would have to pay a symbolic fee of $20.00 U.S. Mexico, in turn, would give those certificates automatically upon the request of the United States Government through a list of the vessels involved. This provision involved partially a recognition on the part of the United States, in the sense that tuna would be subject to a different regime than the one applicable to all other species. However, a strong disclaimer was equally embodied in the Agreement, to preserve the positions of each country on the matter which, by then, were perceived as being quite different. The whole issue would then be settled finally at the Conference which Mexico planned to call, and which the United States was willing to attend.

The Agreement with Cuba

The Cuban Agreement represented very little problem for Mexico from the very start. Aside from the internal matter relating to the reserved status of shrimp, for the cooperatives, the Agreement was successfully negotiated, giving the Cuban fleet the other portion of the Mexican shrimp surplus, and in a different geographical area than the one given to the United States fleet, as well as the other part of the snapper and grouper surpluses, plus a small quota of associated benthic species.

This is today the only bilateral agreement of fisheries in force in Mexico, and it may constitute a good example of international cooperation in this field.

Attempts by Japan

As was already said earlier, the Japanese Government tried first to persuade Mexico not to extend jurisdiction before the completion of the Conference. Once Mexico extended its jurisdiction the Japanese tried to gain access to Mexican surpluses.

For Mexico, thanks to the power left to the coastal state in the agreements reached at the Conference to freely choose with which countries to negotiate its surpluses, it was hard enough to try to obtain an agreement with its neighbors. The indiscriminate fishing practices of the Japanese fleet, not well known for their conservationist orientation, did not make it any more attractive for Mexico to negotiate with that country.

Finally, the geographical location of Japan did not involve any strategic concerns for Mexico, at least not enough to constitute a significant pressure to negotiate. From the start Mexico made it clear that it was not in its interest to establish a fisheries relationship with Japan in its Exclusive Economic Zone.

A New Regime for Eastern Pacific Tuna

Background

Up to 1976, the highly migratory species of the eastern Pacific Ocean had been regulated by the Inter-American Tropical Tuna Commission (IATTC). These species have to be regulated internationlly since by their migratory nature, they cross maritime borders established by states. In the eastern Pacific, the tunas may travel from the border between Mexico and the United States to the border between Chile and Peru. Any actions unilaterally undertaken by a coastal state while the resource migrates through its marine jurisdiction may affect its abundance on the waters of other states and, more than that, its own conservation. Thus, it is everybody's concern what each coastal state does while the resource travels in its marine zones. Therefore, the conservation and exploitation of these species have to be subjected to regional action and cooperation.

Such was the purpose of the 1948 Convention which established the Commission, orignially between Costa Rica and the United States and eventually joined by Panama in 1953, Mexico in 1964, and Nicaragua in 1973. Ecuador had joined it as well but denounced it in 1967. Other states from outside the region, but with interests in the fishery, adhered to the Convention as well, such as Canada in 1968, Japan in 1970, and France in 1973.

Originally, the Commission was established only to study and research all matters related to the conservation of these species. This was the practice of the Commission, the one that gave birth to a real conservation regime, at this regional level and through the actions undertaken by its annual meetings, where the Commissioners adopted by an additional meeting of government representatives (Székely, 1979, pp. 159-174).

In 1962 the Commission created the Commission's Yellowfin Regulated Area (CYRA), which delimited the portion of the eastern Pacific within which it exercises its conservation functions. This area ranges from San Francisco to Valparaiso, comprising a zone of more than 5000 square nautical miles and which, in some sections extends more than 1000 miles from the coastline. Given the observed depletion in the abundance of yellowfin tuna, in 1966 the Commission accepted a 1961 proposal by the Director of Investigations through which a global catch quota was set for that species. Since then, the practice was established that, on the recommendation of the Director of Investigations through which a global catch quota was set for that species. Since then, the practice was established that, on the recommendation of the Director, made on the best available scientific evidence, a maximum catch quota would be established for yellowfin tuna for the next season. The catch was carried out on a first come, first served basis, which implied that all states were free to catch up to their capacity as long as the quota was not reached. Once the quota was about to be exhausted, the Director would close the season and allow only a last trip, during which each vessel would be allowed to catch its carrying capacity.

From 1969, special allocations were authorized for the closed season in favor of small or newly built vessels from developing member states. These allocations were aimed at Mexico, in order to allow its fleet to compete with the bigger United States fleet. Ever since, Mexico obtained its allocation every year, but only after painful bargaining with the United States, a country which had the power to veto Mexican proposals since IATTC decisions were adopted unanimously. In 1975, other developing countries in the region began to obtain similar special allocations.

While this system lasted it was highly successful, and it constituted a singular precedent and example in the world of regional fisheries cooperation. For instance, the global quota for 1966 was 79,300 tons, while for the 1977 season the conservation efforts allowed for a quota of 210,000 tons. This applied only to the one species subjected to a conservation regime, yellowfin tuna since the other important species in the fishery, skipjack tuna, has been regarded as too abundant to require any conservations measures yet.

Effects of the New Law of the Sea Negotiations on the IATTC Regime

With the emergence of a consensus on the right of the coastal state to establish a 200-mile Exclusive Economic Zone at the United Nations Law of the Sea Conference, a good part of the CYRA became integrated within the zone. This immediately meant that the resource could not continue under the regime of freedom of fishing on which the IATTC system was built. Since an important volume of the highly migratory species concentrates itself within the 200 miles, it would have been useless to maintain the system only in the High Seas part of the CYRA.

According to the different texts negotiated at the Conference (see, A/Conf. 62/WP. 8 and Rev. 1, 10 and Rev. 1 to 3, and A/Conf. 62/L. 78.), the coastal state has sovereign rights for the purpose of exploring and exploiting, conserving and managing the natural resources, whether living or nonliving, in the EEZ. Recognizing the special nature of the highly migratory species, the texts, in their Article 64, provide (first paragraph) that their conservation and exploitation should be a matter of regional cooperation. It also provides, though, that these resources are subject to the same regime applicable to other living resources in the EEZ (second paragraph), which means that the coastal state exercises the same sovereign rights over them as over all other species. This was, at least, the understanding of the delegations at the Conference, specially those with abundance of tuna within their zones. The regional cooperation is to be carried out through existing competent organizations or through those established for that purpose. Mexico thought from the start that the IATTC would not be an adequate regional organization to administrate the new regime, since it was based on a totally different legal regime. Thus, it was necessary to create a new organization, based on the new rules of the game. Mexico therefore took the initiative for the convening of a conference to discuss a new organization.

Mexico's First Initiative and the Costa Rican Link

This Conference was cosponsored by Mexcio and Costa Rica and was convened in San Jose in September of 1977.

Plenipotentiaries participated from all IATTC members, as well as from the remaining coastal states in the region. At the 1976 Meeting of IATTC, the Director of Investigations had distributed the document requested by Mexico, describing the alternative regimes available to the participants (Joseph and Greenough, 1979). In San Jose, Mexico and Costa Rica submitted a proposal based largely on one of those alternatives, which contained the following basic elements: A new organization would be established, to supercede the IATTC. A maximum global quota would be adopted for species which required a conservation regime (yellowfin). Of this quota, coastal states would be recognized a guaranteed quota, equal to the concentration of the resource within its zone. The concentration would be calculated on the basis of the statistical catches by the international fleet in the last 5 years. Whenever the coastal state did not have the capacity to catch its guaranteed quota, the surplus would be available to other states, in exchange for the payment of a fee. There would be a single permit to fish in the entire area of application of the new treaty, issued by the organization, and each vessel would pay a participation fee for each ton of fish caught, equal to 5% of its commercial value. The sum of the paid fees would cover the organization's expenses to the coastal states in proportion to the concentration of the resources in their respective zones, and the rest to all participating states in proportion to their respective catches. The season would be closed by the Director of Investigations, and the coastal states would be allowed to complete their guaranteed quotas during the closed season. However, the proposal called for the abandonment of the practice of the last free trip, since it involved reducing considerably the closed season and greater difficulty for the coastal states to enjoy their guaranteed quotas.

Principles and Positions in the Negotiations

The reaction of the United States was quite negative. It rejected most of the Mexican-Costa Rican proposal, especially the concept of the concentration of the resource, and proposed formulas which in reality were aimed at preserving the *status quo.* This was understandable, since the position of that country was shifting toward the concept that highly migratory species were not subject to the sovereign rights of the coastal state in the EEZ.

A Second Plenipotentiary Conference in Mexico City produced a joint document, sponsored by Costa Rica, Mexico, and the United States which listed the different alternative formulas representing their different points of view on the most important issues. This effort was followed by more than a dozen other meetings in San Jose, Mexico City, and Washington over the next 2½ years. Mexico, given the failure of the Plenipotentiaries Conferences, decided to denounce the IATTC Convention. The denunciation took effect on 8 November 1978. Costa Rica did not take long in following suit. Still, Mexico participated and sponsored more and more negotiating efforts, to no avail. In fact, for a long time, even when it had already left the IATTC, it provisionally adopted its annual resolutions, in the hopes that an agreement could be reached. When it was no longer possible to postpone the adoption of unilateral action, that is, when it was observed that the lack of regional measures was detrimental to the conservation of the resources, Mexico adopted a decree subjecting highly migratory species in its EEZ to the normal system of surpluses applicable to all other species (*Diario Oficial,* 1980).

Attempts To Attract South America

Mexico could not have found itself in a more difficult position on the tuna issue, when it thought that other coastal states in the region would find common interests sufficient bases for the adoption of common positions.

From the beginning of the San Jose negotiations, Mexico invited all coastal states, from Guatemala to Chile, even when they were not members of IATTC, in order to have a truly regional regime. Their participation was, from the start, with little enthusiam. Ecuador and Peru, and subsequently Colombia, found any proposal unacceptable as long as their radical territorialist position was not respected.

When it was realized that they would not help in the negotiations *vis-à-vis* the United States, the efforts were concentrated on reaching an agreement without them with only the Costa Rican partnership.

Pending Problems and Further Initiatives

The adoption of the unilateral legislation by Mexico, as a result of the failure in the negotiations, had more than negative results for the fisheries relationship between the two countries, especially because the United States denied Mexico sovereign rights over its tuna resources in the EEZ. When the legislation was first applied through the seizure of some United States tuna vessels, the United States, in application of its legislation, embargoed all tuna products from Mexico. The embargo has been in effect for almost 2 years now, resulting in an expansion of the tuna internal market in Mexico and the gradual fall of the price of this product. Ironically, this is one way in which the Mexican population has directly benefitted from the establishment of the EEZ.

Any further initiatives will have to contain new elements. However, it is highly unlikely that Mexico will ever let its sovereign rights over highly migratory species be vanquished in an international negotiation.

Reasons for Failure, Denunciation, and Unilateral Action

The reasons for all these unfortunate developments can be found in a conceptual divergence between the parties involved. The implementation of the new Law of the Sea in practice is already suffering the effects of having reached agreements at all costs, even through the use of vague or contradictory formulas. Article 64 of the Draft Convention is a good example. It is equally easy for both the United States and Mexico to substantiate their reasoning and their interpretation of its provisions. For the United States, no state can unilaterally and outside a regional system adopt measures for the exploitation and conservation of these resources.

For Mexico, when an attempt has been made at regional cooperation with no success, such unilateral action is amply justified.

Conclusion

Implementing the new Law of the Sea has been, in the Mexican experience, a process which has led it to very difficult confrontations with foreign fishing states. Despite its moderate positions in the international negotiations at the Law of the Sea Conference, when the agreements reached there are put in practice, Mexico has encountered the same radical intransigence that was apparently overcome through compromise formulas. In other words, the agreements reached at the Conference have not translated themselves in effective bilateral fishery relationships.

In order to overcome most of the problems deriving from differing interpretations and positions on the results of the negotiation at the Conference, perhaps the only recipe would be for a country like Mexico to launch an all-out effort to rationally and optimally exploit the living resources within its Exclusive Economic Zone, so that all surpluses will vanish and, with them, all vestiges of foreign interests and participation.

References

Anuario Estadistico del Departamento de Pesca. 1981. (In press.)

Carranza Frazer, 1978. La Investigacion Cientifica y Tecnologica en el Desarrollo Pes quero del Pais. *In* Vigesima Serio de Mesas Redondas: Panorama Pesquero Nacional, Analisis de Tres Lustros. Instituto Mexicano de Recursos Naturales Renovables, Mexico, p. 17.

Departamento de Pesca. 1977. National Program for Fisheries Development. Mexico City, 2 vols.

Diario Oficial, 18 December 1902.

Diario Oficial, 5 February 1917.

Diario Oficial, 31 August 1935.

Diario Oficial, 20 January 1967.

Diario Oficial, 26 December 1969.

Diario Oficial, 5 November, 1975.

Diario Oficial, 6 and 13 February 1976a.

Diario Oficial, 7 June 1976b.

Diario Oficial, 21 January 1980.

Garcia Robles, 1981. The Second United Nations Conference on the Law of the Sea. A Reply. 55 AJIL, pp. 660-675.

Joseph, J., and Greenough, W. 1979. International Management of Tuna, Porpoise and Billfish. University of Washington Press, Seattle and London. 1979.

Neptune. 1975. Independent News at the Law of the Sea Conference, Geneva, vol, 3, 14 April 1975, p.5.

Scripps Institute of Oceanography. 1976. Marine Resources Off the West Coast of Mexico. La Jolla, California, p. 1.

Székely, 1976. Study on the Contribution of Latin America to the Development of the Law of the Sea. (Oceana Publications, Dobbs-Ferry, New York. United Nations. 1900.

Székely, 1979. Mexico y el Derecho Internacional del Mar. Universidad Nacional Autonoma de Mexico, Mexico, p. 57.

Truman, H. 1945a. Presidential Proclamation N. 2667, of 28 September 1945, 59 *Stat.* 884.

Truman, H. 1945b. Presidential Proclamation N. 2668, Of 28 September 1945, 59 *Stat.* 885.

UNCLOS I, *Official Records,* Vol, III, p. 177.

Vargas Carreño, 1973. America Latina y el Derecho del Mar. Fondo de Cultura Economica, Mexico.

4

Jeffersonian Democracy and the Fisheries Revisited[1]

J. L. McHugh

Introduction

Thomas Jefferson believed that the new nation should be divided in such a way that every man should have a say in the common government (Dewey, 1940). However, as Schlesinger (1971) has pointed out, local government has become the government of the locally powerful, not the locally powerless, and to retain their rights, people have resorted, rightly or wrongly, to the national government. There has been tugging in one direction or the other as Republicans have succeeded Democrats, and vice versa. However, Jefferson expected citizens to be more reasonable than they are likely to be (Malone, 1981).

Fisheries in the United States have adhered more or less to the form of Jeffersonianism, although the trend has been toward more federal control. Coastal fisheries are still subject to state laws, and in some cases to county or even town

---

J. L. McHugh has been Head of the Office for the International Decade of Ocean Exploration, National Science Foundation; Acting Director, Office of Marine Resources, Department of the Interior; Chief, then Assistant Director, Division of Biological Research, Bureau of Commercial Fisheries, and finally Deputy Director of the Bureau; and Director of the Virginia Fisheries Laboratory. He has served in numerous international commissions and has represented the United States on the Inter-American Tropical Tuna Commission and the International Whaling Commission. In 1971 he was a Fellow of the Woodrow Wilson International Center for Scholars. From 1976 to 1979 he was a member of the Mid-Atlantic Fishery Management Council.

[1]Contribution 336 from the Marine Sciences Research Center. Parts of the work on which this chapter is based were sponsored by the New York Sea Grant Institute under a grant from the Office of Sea Grant, National Oceanic and Atmospheric Administration, U.S. Department of Commerce.

jurisdiction. Beyond 3 miles and out to 200 miles, however, they are now subject to federal law through the provisions of Public Law 94-265, the Fishery Conservation and Management Act of 1976 (FCMA; United States Congress, 1976). Even here some form of Jeffersonianism has been preserved, by placing major responsibility in the hands of regional Fishery Management Councils, which include in their membership the heads of the fishery management agencies of each of the coastal states. The Act also provides that each Council shall conduct public hearings with respect to development of fishery management plans and with respect to administration and implementation of the provisions of the legislation. Thus the Councils are unique in that they appear to preserve the basic tenet of the Jeffersonian doctrine, that each man can have a say in the government. The federal government, in the end, however, retains much of the power, because it has the option to approve or disapprove management plans. It would be instructive to know what Jefferson would have thought of this constraint. This arrangement has been in effect for only 5 years, and it may be too early yet to see trends clearly.

In the United States the Fisheries Conservation and Management Act (FCMA) contains a provision for takeover of state responsibilities within 3 miles under certain conditions. Section 306(b) provides that if any state has taken or omitted to take any action that would alter adversely the carrying out of a fishery management plan, the Secretary of Commerce shall notify the state of his intention to regulate that fishery. This appears to ensure that a fishery management plan cannot be rendered inapplicable by the inaction or action of a state which does not within a reasonable time, and with proper notification, adopt laws consistent with management of a species covered by a plan farther offshore. Note, however, the phrase "other than its internal waters," which in a state such as Maryland or Virginia, with extensive internal waters within Chesapeake Bay and its tributaries, could, if it so desired, nullify or make ineffective a plan promulgated by the Council.

In the state of Maine this has already happened. The fishery for young herring (*Clupea harengus*) produces fish which are canned as sardines. This fishery is not managed by the New England Council because it takes place within 3 miles for the most part. There is a substantial fishery for adults of the same species, which takes place for the most part outside 3 miles, and this split has created chaos in the fishery. Quotas have been rendered meaningless. The state has not been able to manage the fishery and, in fact, has no effective measures planned. It appears unlikely that the Council will try to take over within the zone of state jurisdiction, and the plan probably will be held in abeyance until the state wants to develop something that works. The situation in Maine, where the State Legislature can override decisions of the Director of Fisheries, is fairly typical of many states along the Atlantic coast.

"Jeffersonian Democracy and the Fisheries" (McHugh, 1972), to which this chapter is a sequel, considers fisheries that lie within state jurisdiction, either wholly, as oyster (*Crassostrea virginica*) and blue crab (*Callinectes sapidus*), or principally, as Atlantic menhaden (*Brevoortia tyrannus*). It is appropriate to bring the discussions of these three fisheries up to date, to see whether anything has occurred in the past 10 years to improve their management. This will be done relatively briefly.

The greater part of the chapter examines the work of the Councils, which have come into being since that time. I have chosen to consider only two Councils, and only one management plan of each: the sea clam plan of the Mid-Atlantic Council, which includes two species, surf clam (*Spisula solidissima*) and ocean quahog (*Arctica islandica*); and the groundfish plan of the New England Council, which includes three species, Atlantic cod (*Gadus morhua*), haddock (*Melanogrammus aeglefinus*), and yellowtail flounder (*Limanda ferruginea*). How well have the Councils been able to develop plans that are effective in managing these resources? What plans are developing to improve present performance and reach Council objectives?

In value these eight fisheries make up a considerable amount of the total for these two regions. In 1980 total commercial fishery landings for the area from Maine to Virginia inclusive was about 1.75 billion pounds or 794 thousand metric tons, worth nearly \$554 million to fishermen. The eight fisheries treated here accounted for 1.5 billion pounds (680.4 thousand metric tons) worth about \$250.6 million. This was about 86% of the weight and 45.2% of the value of all fisheries of the area. The discrepancy between weight and value was caused primarily by menhaden, which is dominant in the fisheries of the area but relatively low in value. Nevertheless, this demonstrates that these eight species make up a major part of the fisheries of the area, and also that much of the weight comes from state waters within 3 miles.

The Chesapeake Oyster Industry

The oyster industry in Chesapeake Bay is still the most important along the Atlantic coast. I will assume that the earlier chapter (McHugh, 1972) is readily available so that it will be unnecessary to repeat the detail. It is appropriate, however, to include a figure showing oyster landings in the Chesapeake since 1880 (Fig. 1), which shows at a glance the trends in gross production (Pileggi, 1980; U.S. Department of Commerce, 1980). Landings have fallen from 1880 to 1980, and total production now is less than one-fifth of what it once was.

Virginia has a fairly large extent of private grounds, but Maryland is mostly public. Production was greatest in Virginia for most years from 1930 to 1963 but began to rise in Maryland from a massive and costly replenishment effort, and from 1963 to 1973 increased from less than 8.0 million (3.6 thousand metric tons) to 20.4 million pounds (9.25 thousand metric tons). However, production as a whole has not risen for more than 15 years.

Despite an initial rise in unit price of oysters, which, adjusted by the Consumer Price Index rose from less than 25¢ per pound in 1929 to over 90¢ in 1962 in Virginia, prices then began to decline and are now between 50¢ and 60¢ per pound (Fig. 2). The adjusted total value of landings in Chesapeake Bay reached a

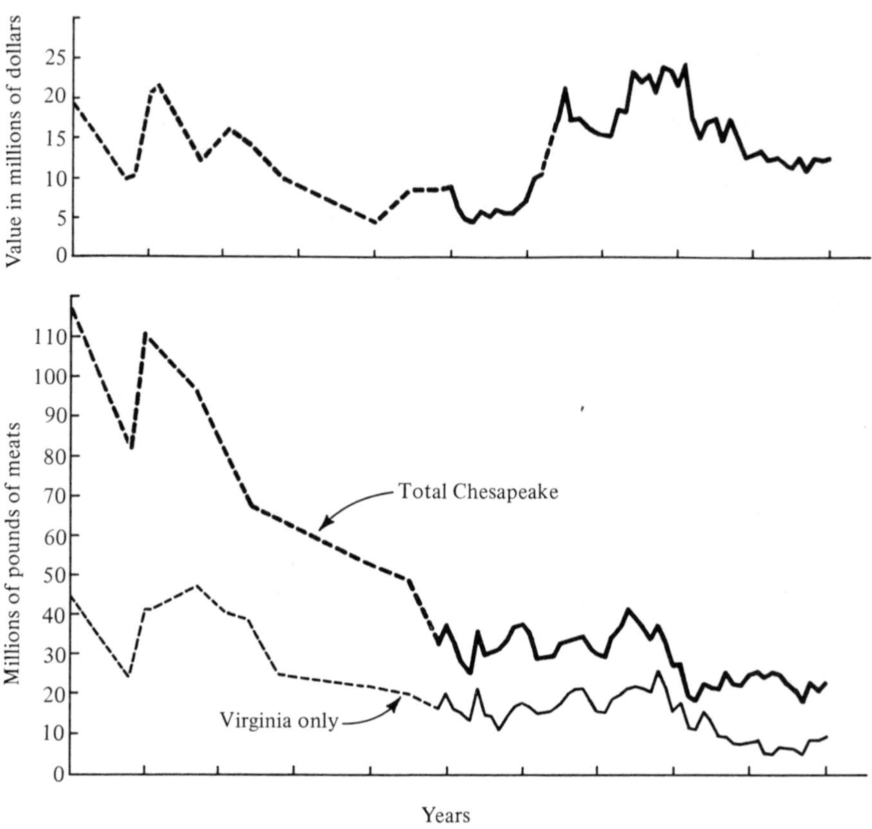

Figure 4.1. Landings of American oyster in Chesapeake Bay and total value of the catch adjusted by the Consumer Price Index, 1880-1980. Broken lines join points where one or more years are missing.

peak in 1961 and has fallen off to about half that amount by 1979 (Fig. 1). Prices have not held up, despite falling landings, which suggests that oysters are not as desirable any more, at least at the high prices that prevailed in the 1960s. Note that the price of oysters in New York, where the best half-shell oysters are grown, is considerably higher, but for the most part the fluctuations are generally consistent with Virginia (Fig. 2).

In 1971 the Virginia Institute of Marine Science began a detailed study of the major fisheries of the state, which is still under way. The study of the oyster industry was published in May 1978 (Haven *et al.*, 1978). This comprehensive and massive work forms the basis for most of the conclusions drawn here. A second study, dealing with the Baylor grounds, was published in April 1981 (Haven *et al.*, 1981). The following major recommendations are included:

1. Unproductive grounds within the Baylor survey (a survey made late in the last century by Lt. Baylor of the Coast and Geodetic Survey, which delineated the naturally producing grounds in the state) should be leased.

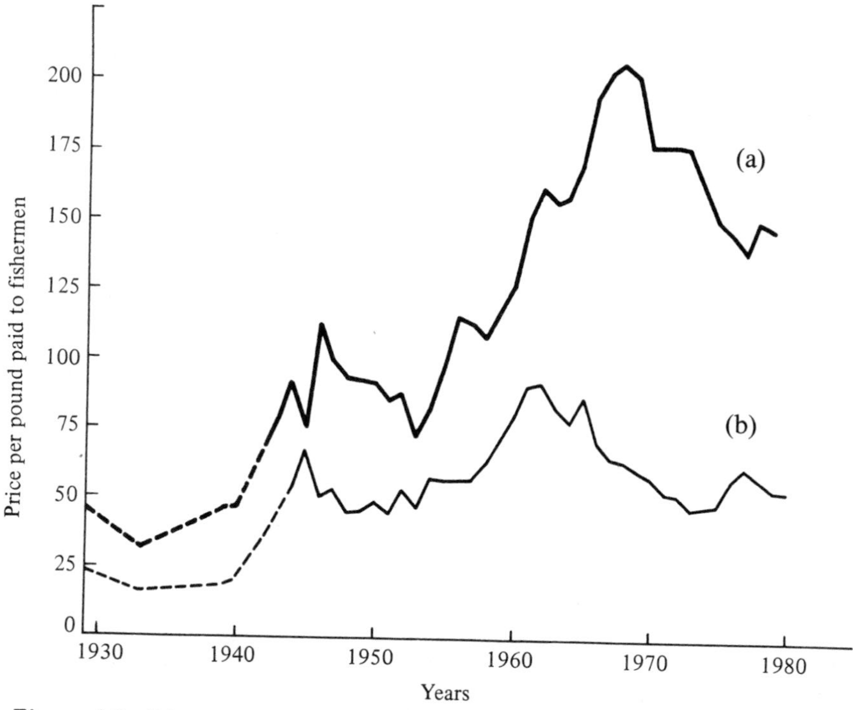

Figure 4.2. Price per pound paid to fishermen for American oysters in New York (a) and Virginia (b), adjusted by the Consumer Price Index.

2. Improve production of seed oysters (oysters below legal size which can be translated from public grounds where growth is slow to private grounds where growth is rapid). This would seem to be an obvious move, yet it has never been made.
3. Improve the repletion program on public grounds. Various ways are known to do this, but all have been ignored.
4. Evaluate research and improve utilization. The review showed that the yield per acre of ground has not increased for 45 years, which means that no improvement in efficiency has taken place.
5. Great benefits to public and private participants in the oyster industry can be derived from research. Much of the research has already been done, but apparently little, if any, has filtered through to industry.
6. Joint use of the Baylor bottoms by private and public sectors of industry.

These recommendations are backed by detailed examination of available records, and within each major heading are numerous subheadings containing recommendations. Scientific research has been done successfully on many of the problems, but industry has not been willing to apply the results. The problems include (1) archaic practices and attitudes within the industry; (2) economic and political conflicts between segments of industry, and between the fisheries and other users and uses of the environment; (3) lack of firm and consistent purpose and practice by industry and by the state toward achievement of realistic and improved management; and (4) continuation of legal restrictions and economic practices which mitigate against and prevent improvements in the industry. (It appears that these problems, and the lack of utilization of scientific findings, would be subject to attention by appropriate Sea Grant people.)

The Blue Crab Industry

The blue crab fishery in Chesapeake Bay has grown irregularly, but in the long run steadily, since landings were first recorded in 1880 (Fig. 3). I have not shown landings prior to 1929 because they are scattered. There were dips and increases, with some indication that production has now leveled off at about 67 million pounds (30.4 thousand metric tons). There are no clear indications from these data that hard crab production has been hurt, either by heavy exploitation or by environmental changes. It has been demonstrated (McHugh, 1969) that survival was better when relatively small numbers of spawners were present, so that about the same numbers of recruits were produced by any number of spawners within the limits observed (Fig. 4). On the other hand, soft crab production has been going down for nearly 80 years (Fig. 5), also with wide fluctuations, but there is no clear reason for the decline. This is a short-lived species with maximum age probably not more than 3 years (Van Engel, 1958).

Blue crab prices, adjusted by the Consumer Price Index, have been rising more or less steadily since the early days. There was a deviation from this trend in the mid-1940s, when a shortage of red meat at the end of the Second World War caused prices of almost all seafoods to rise sharply and fall almost as rapidly. The total value of the catch has remained high, even though catches are down from the peak in 1966. It appears that the blue crab resource is still in good condition, despite environmental stress. The industry may not be doing as well, for fishing effort appears to have increased without concomitant increase in abundance of crabs.

It appears that most of the measures utilized in the past to protect the blue crab resource, with possible exception of minimum size regulations, have had no effect. The arguments which have raged between different exploiters of the resource, or between Maryland and Virginia, appear not to have had any valid basis. Regulations have simply increased the cost of harvesting, without any gains. Biological research has not helped, because the results have not been utilized. Sea Grant perhaps has a role here also, which it has not yet fully exploited.

The Menhaden Industry

In weight landed along the Atlantic coast menhaden has been the most important species for at least 100 years. North of Chesapeake Bay, landings peaked in 1956 (Fig. 6) at over one billion pounds (467,200 metric tons). They remained fairly

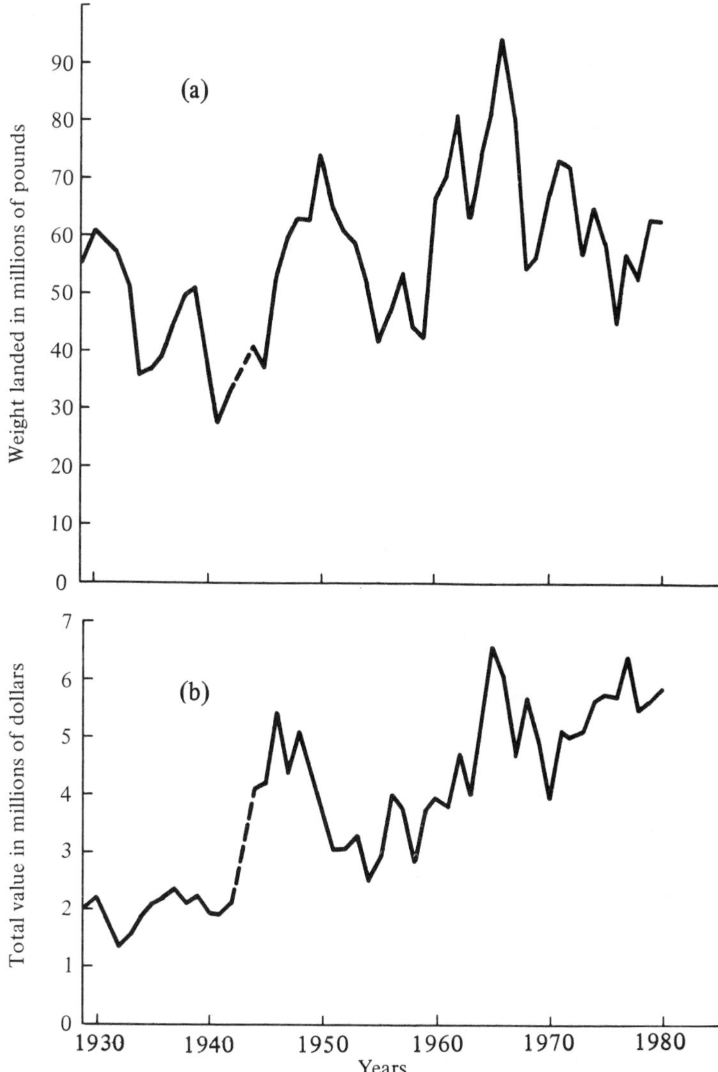

Figure 4.3. (a) Landings of hard blue crab in Chesapeake Bay 1929-1980. (b) Total value of catch adjusted by Consumer Price Index. (c) Price per pound adjusted by Consumer Price Index.

high from 1953 to 1962 then plummeted, and have exceeded 200 million pounds (90,700 metric tons) only once since that time, in 1973. There is evidence that water temperature may have some influence, menhaden migrating farther north in warm periods (Taylor et al., 1957) although the evidence is not conclusive. As long as the fishery remains intense in Virginia and North Carolina landings probably will remain fairly low farther north, because few fish survive long enough to migrate farther north, and spawning is not very intense north of Chesapeake Bay.

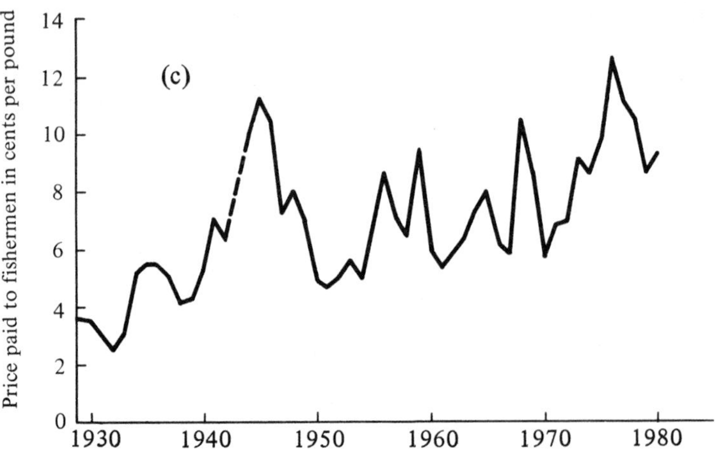

Figure 4.3. (*continued*)

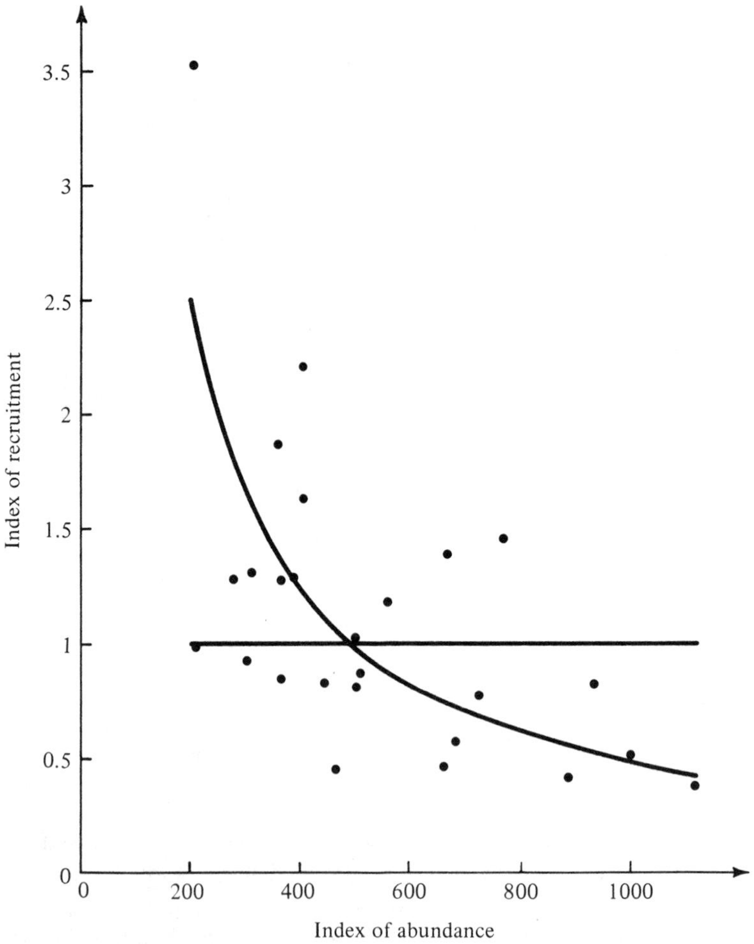

Index of recruitment

Index of abundance

Figure 4.4. Relation between recruitment and abundance of blue crab in Chesapeake Bay. The horizontal line at unity is the replacement line.

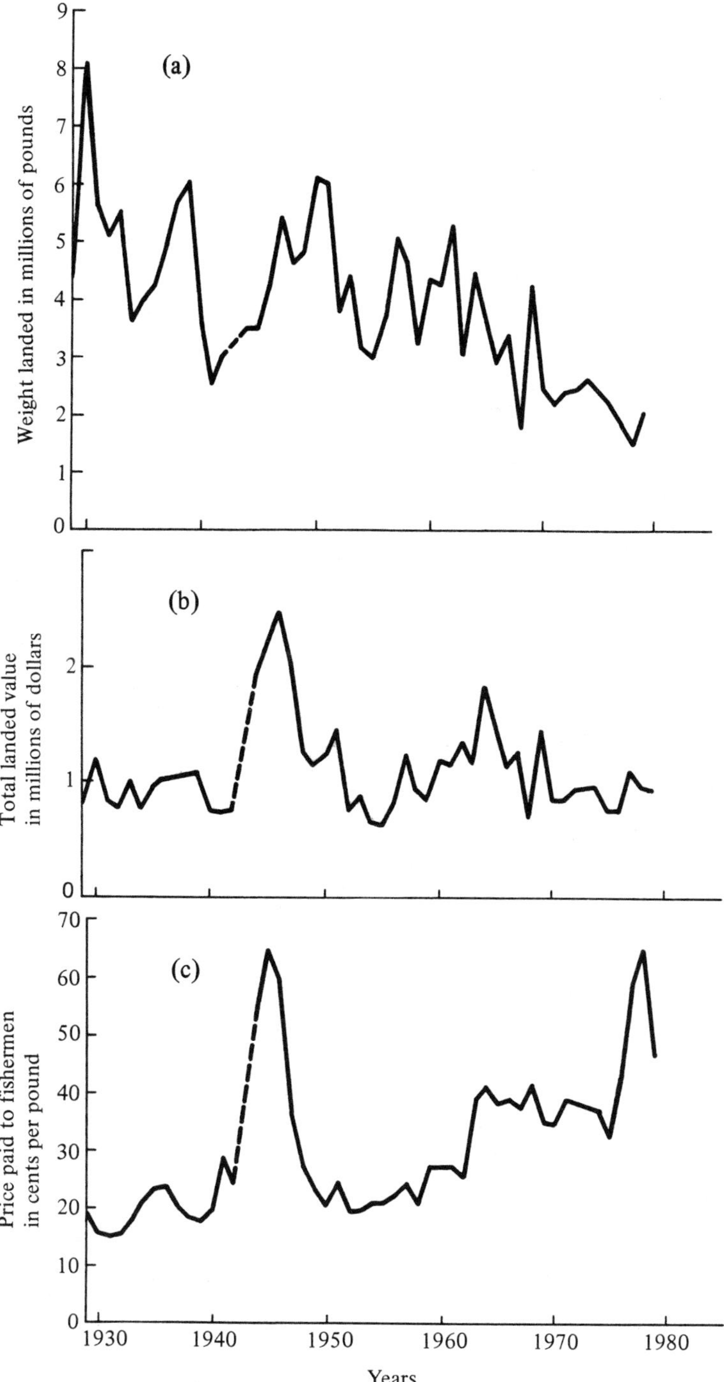

Figure 4.5. (a) Landings of soft blue crab in Chesapeake Bay, 1929-1979. (b) Total value of catch adjusted by Consumer Price Index. (c) Price paid to fishermen adjusted by Consumer Price Index.

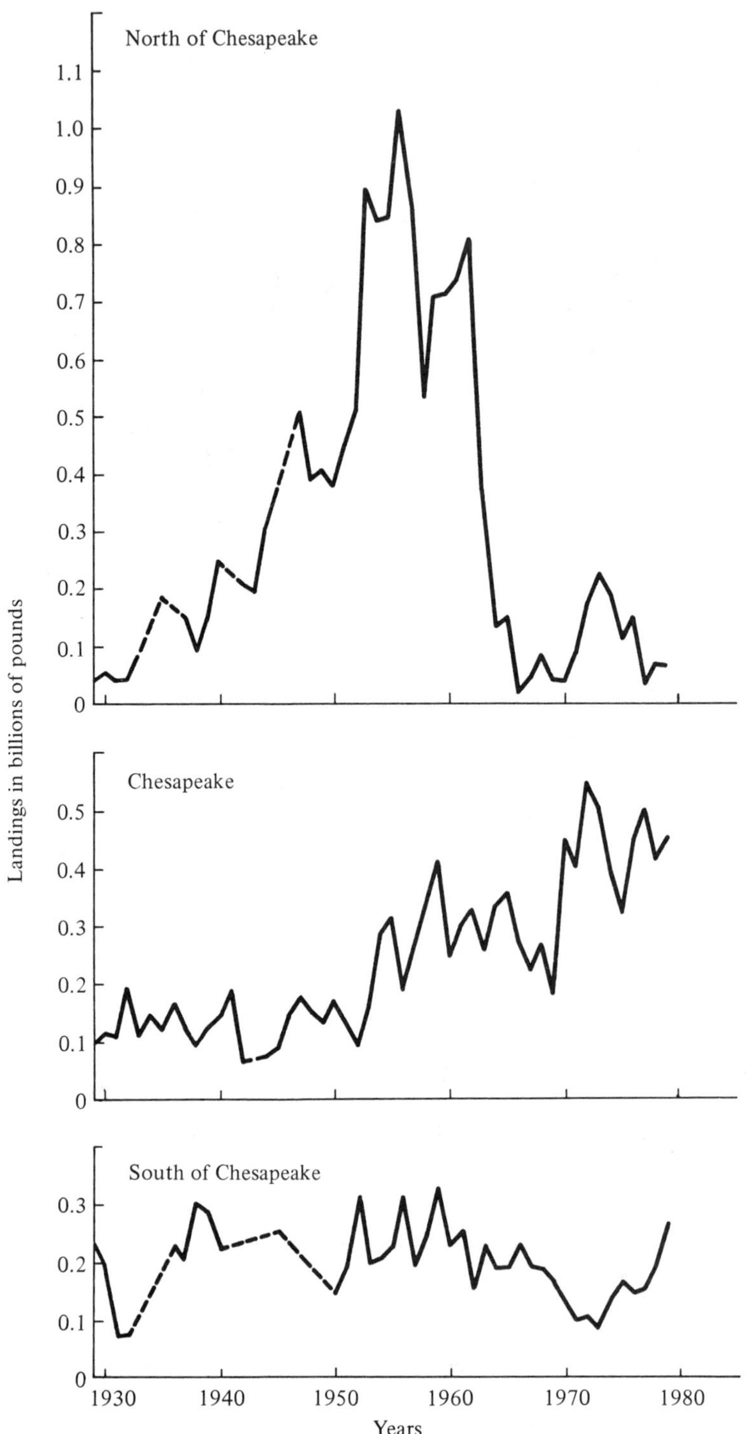

Figure 4.6. Landings of menhaden along the Atlantic Coast, 1929-1979, by major regions.

In Chesapeake Bay landings reached an all-time high (Fig. 6) of over 550 million pounds in 1972 (252,200 metric tons). The decade from 1970 to 1979 produced the highest average landings ever, and there is no sign yet that they are declining. The fishery in Chesapeake Bay and North Carolina now takes mostly immature fish which have not spawned. Landings here have been higher than ever before, but total landings along the entire Atlantic Coast have been less than half what they were at their peak in 1956.

Catches south of Chesapeake Bay (Fig. 6) peaked in 1959 at about 330 million pounds (149,700 metric tons), mostly taken in North Carolina. Subsequently they dropped sharply but have been increasing again since 1973. Whether they will exceed the 1959 level again is unknown, but it does not appear likely.

After about 1940, and until 1963, major catches of menhaden were made north of Chesapeake Bay. Subsequently, Chesapeake Bay has been the major producer. The fishery there is now so intense that it takes mostly fish which have not spawned. From Delaware north the fish were older and had spawned at least once, which would appear to be a good hedge against excessive reduction of the spawning stock. Dryfoos *et al.* (1973) concluded from tag returns that the exploitation rate on the average was about 50%. Schaff (1975) concluded that the maximum sustainable yield ranged from 400,000 to 620,000 metric tons and felt that management should aim at 370,000 metric tons (815.7 million pounds). He also said that the maximum sustainable yield would be achieved if fishing effort were distributed in the same way as in the peak fishing years 1953-1962, namely, with major effort north of Chesapeake Bay.

The menhaden industry is vertically integrated to a much greater degree than most fisheries. Prices derived from official statistics may not readily identify the actual price paid to fishermen.

A New Approach

In 1976 the Congress of the United States passed the FCMA. Something like this had been in the minds of some people for a considerable time. Congress believed that the international fishery commissions were not working well. When the Law of the Sea Conference appeared to be failing consistently to deal with the matter expeditiously, it appeared to be necessary for the United States to take a conservative position, before the resources virtually disappeared. This was perhaps too simple a view to take of a very complex subject. There was doubt in the minds of some that the international commissions were doing such a bad job, but that was not the prevailing view and the time appeared ripe to try something new. The mechanism by which FCMA was put into effect was to create eight regional Fishery Management Councils, with authority to prepare management plans for each major fishery within their jurisdiction, and to license foreign fishermen to fish within 200 miles of the coast only if there was clearly a surplus production which American fishermen could not take. Most people agreed that this move was desirable. Only a few were dubious.

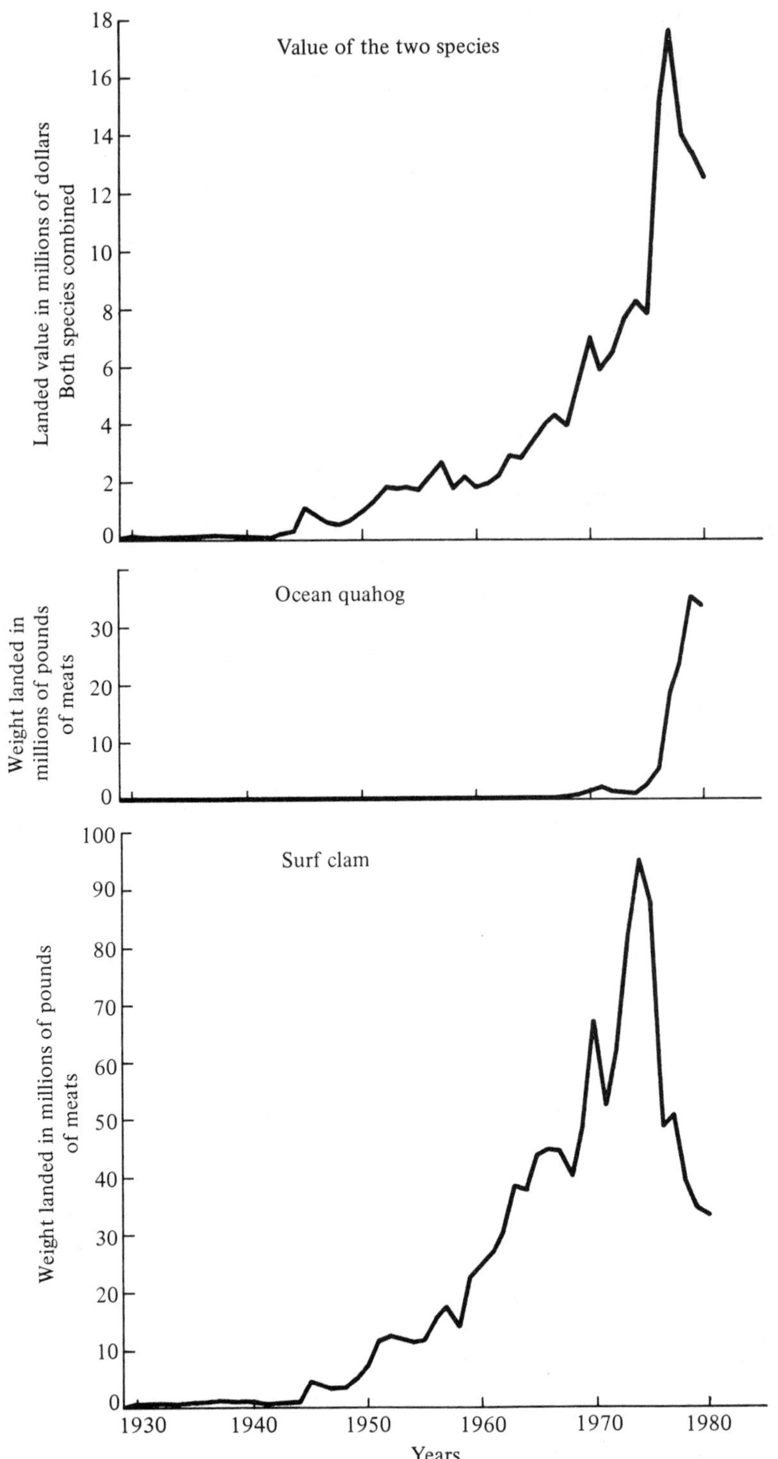

Figure 4.7. Landings of surf clam and ocean quahog along the Atlantic Coast, 1929-1980, and total landed value of the two species adjusted by Consumer Price Index.

The FCMA applied only to the fishery conservation zone (FCZ), from 3 to 200 miles. Anadromous species were to be managed beyond 200 miles wherever they might roam. Highly migratory species, chiefly tunas and whales, were exempted on the ground that they must remain under international control. Fisheries within 3 miles and in inland waters remained under the jurisdiction of the individual states, although it was obvious that the states must cooperate with decisions of the Councils if important species migrated substantially between the two jurisdictions.

Fisheries in the FCZ also may be difficult to monitor and enforce, especially if separate populations exist, so that quotas must be set for different areas in the FCZ. Fishermen may evade regulations by fishing in one area and reporting the catch from another, or by fishing within 3 miles and reporting to catch from the FCZ, or vice versa. Courts may also make matters difficult by not cooperating by setting adequate fines and other deterrents, or in not acting promptly.

The Sea Clam Management Plan of the Mid-Atlantic Council

Sea clams belong to two species, surf clam and ocean quahog. The surf clam has been overfished for some years, in the sense that it was becoming harder each year to make a catch, took more searching, and required constant hunting for new grounds, mostly toward the south; after 1974, despite increased effort, the catch declined. When the Mid-Atlantic Fishery Management Council came into existence this was the first fishery to occupy its attention.

The surf clam sub-board of the State-Federal Fisheries Management Board had a management plan almost complete when the Council was formed (Rinaldo, 1977) and this soon became the Council's plan. It placed a quota on surf clam landings of 30 million pounds (13.6 thousand metric tons) of meats and in the interest of processors, to spread landings over a full year, broke the quota up into four quarters of the year. It also placed a moratorium on entry of new vessels into the fishery, to protect vessels already in the fishery against diluting the reduced catch further. Almost immediately the vessels in the fishery began to improve their catching efficiency by obtaining dredges with wider blades, improving their pumping efficiency, and in some cases adding another dredge, to attempt to get as large a share as possible of the quarterly quota. To counter this, it was necessary to reduce the fishing week until a boat could operate only one day per week, and finally for only 12 hours in that day. This was hardly an ideal arrangement, but it was necessary if all segments of the industry were to be generally satisfied. Meanwhile, because clams were becoming scarcer, the price increased substantially, and this allowed the fleet to continue to harvest clams at a profit (Fig. 7) despite constraints. Industry went along with these arrangements, although the regulations were often broken. The situation was further exacerbated by a major kill of surf clams in a wide area off the coast of New Jersey in the summer of 1976 by unusually low oxygen conditions. If a moratorium on vessels had not been imposed, the industry probably would have been in much worse condition.

More recently, as the new year-class that set immediately after the low oxygen kill began to reach harvestable size, the quota was increased to 40 million pounds (18 thousand metric tons). Despite this raise, the quota for the entire year was almost caught by August 1981, and it was agreed to raise the quota once again to 45 million pounds (20.4 thousand metric tons).

The present plan, now approved by the U.S. Department of Commerce, has certain other provisions. Annual quotas will be set through consultation annually, depending upon the condition of the stocks. For surf clams the values are between 30 and 50 million pounds of meats for the Mid-Atlantic area and 425,000 to 1.7 million pounds of meats for the New England area. For ocean quahogs the values are between 40 and 60 million pounds of meats. The plan provides for an indefinite extension, with equal quarterly quotas for surf clams in the Mid-Atlantic area, and a 5.5-inch minimum size limit. Areas closed because they contain concentrations of undersized surf clams, when reopened, will have allowable catches separate from the overall quota. There are also provisions to ensure that the harvest will extend through a predetermined time period. Together, these provisions will make the plan easier and quicker to change, as necessary.

Early details of the plan and a review of the biology of the two species were given by McHugh (1978). Important among the conclusions was that this was a "textbook" example of what happens when a fishery develops uncontrolled and overbuilds. It was never complicated by foreign participation, and there was no recreational fishery, other than a specialized fishery for bait. Thus, the fishery should have been relatively easy to manage if it had been attempted soon enough. That optimum time was about 15 years earlier, when some of the symptoms were already evident (United States Congress, 1963; McHugh, 1978). The same mistake should not be allowed to happen with the ocean quahog resource.

Nevertheless, it appears that these words were not listened to. The ocean quahog industry began to develop, as problems related to their utilization were solved. The quota for ocean quahog, which had originally been set at 30 million pounds (13.6 thousand metric tons) of meats, was raised to 40 million (about 18 thousand metric tons), with no biological data to support this increase. Ocean quahog is clearly very long-lived and slow-growing, some individuals surviving to 100 years or more. Thus, it is wise to be concerned about its ability to withstand a heavy harvest. Even 30 million pounds might be too large, although it is too early yet to be sure. By 1980 the catch was already nearly 34 million pounds (15 thousand metric tons), with relatively few vessels harvesting it. There are presently no constraints on the number of vessels or on fishing days, and the increase to 40 million pounds and the most recent decision that it might range from 40 to 60 million pounds (27.2 thousand metric tons) are based on inadequate knowledge of the biology of the animal. Ocean quahog brings a lower price per bushel than surf clam, but its greater abundance and availability may outweigh that difference. It would be wise to place a limit on the number of vessels, and to watch very carefully the development of this fishery so that further constraints can be imposed if necessary before their impact becomes too severe. The Mid-Atlantic Council may be paying too much attention to short-term economics and not enough to biology in developing the ocean quahog part of the plan.

The price of both species has been trending upward, and the total value of landings of the two species, adjusted by the Consumer Price Index, has been considerably higher since the Council began to manage the resource. The moratorium on entry of new vessels into the surf clam fishery, and constraints on the number of vessels taking ocean quahogs imposed by the few buyers of this resource, appear to have been advantageous to fishermen. There is no moratorium at present on ocean quahog vessels except that applied by industry, and no guarantee that it will last. The Council should address this question now, before it becomes worse.

There is a further potential problem in the surf clam fishery, when areas presently closed are ready to be opened. This is imminent, and there may be more surf clams available then than ever before. The Council should be planning more carefully what it intends to do when these clams are large enough, to prevent excessive catches and lowered prices. This will not be easy, because there is always a threat that a kill will occur again, and wipe out everything. On the other hand, an increase in availability of surf clams will ease the pressure on the ocean quahog resource for a while, allowing more controlled development of this fishery as new information accumulates. It will take some decisions equal to those made by Solomon to solve these problems, with about as little information to go on.

The Groundfish Plan of the New England Council

On 14 March 1977 the Secretary of Commerce approved an emergency groundfish management plan, one of the first under FCMA. The species were Atlantic cod, haddock, and yellowtail flounder, all of which were overharvested at the time. Foreign fishing for these species was prohibited, except for small incidental catches taken when fishing for permitted species. The plan set restrictive quotas for the 1977 fishing year and set two long-term objectives: (1) to provide for a sustained optimum yield of biomass based on increased and stable stock levels; and (2) to provide long-term economic stability in the fishing community harvesting groundfish species. The methodology used was essentially that developed under the International Commission for the Northwest Atlantic Fisheries (ICNAF). For all three species 1975-1976 population levels were well below maximum sustainable yield (MSY) levels, and the plan set quotas roughly corresponding to the previous year's domestic catch, although the quota for yellowtail flounder was substantially higher than the predicted MSY level. There was an urgent need for a plan, because ICNAF regulations had lapsed at the beginning of 1977, although the preliminary fishery management plan imposed by the National Marine Fisheries Service (NMFS) and carryover agreements from ICNAF held catches down. Passage of FCMA the previous year had encouraged new boats to enter the fishery. Evidence continued to mount on the depleted nature of the groundfish stocks. The plan was far from perfect, and changing conditions and fishermen's reactions to the regulations, e.g., catching fish in one part of the area and reporting it from another, led to almost daily tinkering by the National Oceanic and Atmospheric Administration (NOAA) and the Council in an effort to respond adequately to the situation.

Only 11 weeks after emergency regulations went into effect, domestic fishermen exceeded the entire 1977 quota for cod in the Gulf of Maine and the fishery had to be closed. The closure occurred before the inshore fishermen of Maine began their inshore summer fishery for cod. Soon afterward the domestic catch of cod on Georges Bank reached the estimated optimum yield and the entire New England offshore fleet faced closure for the remainder of 1977. There followed an intense political uproar that led to several emergency extensions of the 1977 quarterly quotas. Fishermen from Cape Cod and Gloucester were at the same time reporting phenomenal amounts of cod in the Stellwagen Bank area of Massachusetts Bay. By November 1977 NMFS determined that an "economic emergency" existed, and the optimum yield was increased to permit fishing on a limited basis. The biological basis for this increase was slim, and preliminary stock assessments showed only a slight improvement in recruitment for 1977. The original plan contained no long-term projections on the impact of increased quotas or fluctuations in recruitment on stock size; thus the Council and NMFS had little scientific guidance, and the increase was justified solely on economic grounds.

By the middle of July 1977 the haddock and yellowtail flounder fisheries in this area were also closed, which meant that vessels in all three fisheries were limited to 2 metric tons or 10% of the total catch for each species per trip. This was called unfair and unrealistic by fishermen because it favored those making short trips. It also created hard feelings between fishermen and NMFS assessment personnel which has not eased much 4 years later.

Biologists had recommended that no haddock at all be taken in 1977 to aid the stocks in rebuilding. The 6200-ton quota was set because an unavoidable by-catch of haddock was inevitable as they pursued other fisheries. Fishermen said that haddock were so thick on Georges Bank "that you could walk on their backs." The biologists said they were aware of the tremendous year-class of haddock born in 1975 which fishermen were seeing as scrod in 1977, but believed that these fish should be allowed to accumulate. Fishermen claimed that this knowledge was not available to them at the time they were reporting large numbers of haddock. They were at odds with officials on numbers, quotas, and the early guess that quotas would not prove to be limiting in 1977.

This led to almost total confusion. Offshore vessels began to invade inshore waters where they had rarely fished, making short trips to circumvent quotas. "Piggybacking" became a common practice, yellowtail flounder boats catching a trip limit east of 69°W longitude and a second trip west of this line, or claiming they had. The 3-mile inshore zone became a favorite fishing ground, or was said to be, allowing large catches of the three species which would be illegal outside 3 miles. Fish smuggling became common in the region's major ports. Smaller landing places, not covered adequately by NMFS enforcement people, became dropoff points for illegal catches. Getting around the rules became a game, by and large a successful one from the fisherman's point of view.

This pattern continued in 1978. The Secretary proposed an increase in the haddock quota from 6200 metric tons in 1977 to 8000 tons in 1978, an increase in the cod quota from 25,000 metric tons in 1977 to 28,000 in 1978 (commercial catch only), and a reduction in the yellowtail flounder quota from 14,000 metric tons in 1977 to 8100 tons in 1978. Throughout 1978 commercial fishermen consistently caught most of their quotas in a matter of weeks and operated under constant threat of closures. In response NMFS revised quotas in July 1978, more than doubling the quota for haddock to 20,000 metric tons and raising the cod quota by 2000 metric tons, citing economic and social factors which must be taken into account. Again, before the year was finished, NMFS changed the regulations again, creating a new "fishing year" beginning with the last quarter of 1978, in effect allowing fishermen to catch fish in late 1978 that had been allocated to the first three months of 1979.

Early in 1979 the Council released a further amendment to the plan, which increased the optimum yield for all three species by amounts ranging from 23% for yellowtail flounder to 62% for haddock. These were viewed by some critics as based on incomplete and misleading data, not supporting an increase in quotas; that 1979 quotas did not take into account natural variations in abundance of groundfish stocks; and that the size of the proposed increases appeared to be dictated by increased fishing effort. In the view of one important critic of the plan, the latest amendment was proposed in response to short-term economic pressures, and the long-term conservation goals of FCMA were being ignored.

From the fisherman's point of view, as more vessels entered the fishery, quotas per boat, in effect, were lowered. They threw away protected species to avoid charges of violation. Recognizing the pointlessness of throwing away valuable food fishes that were dead, the Council recommended and the Secretary approved a ban on discards. Fishermen then found themselves with protected species in illegally high quantities which they could neither keep nor throw away. This further eroded confidence in the system. The complex rules mystified fishermen and confused law enforcement officials, so that even the simplest provisions of the regulations could not be enforced. After much thought and discussion the Council decided that it should take interim action to relieve stress in the management system until a more comprehensive management policy could develop. It therefore chose to adopt an interim management approach, focused on three limited, but according to the Council, attainable goals: (1) it would provide basic conservation measures to protect the stocks in the interim period; (2) it would provide time to get the data to develop a more comprehensive plan; and (3) it would provide time for the Council to make the necessary policy decisions.

At present, landings of cod, haddock, and yellowtail flounder are all down. Cod reached its lowest point early, in 1953; haddock reached a low in 1974, and yellowtail flounder in 1978 (Fig. 8). Cod landings since have more than doubled, the greatest rise coming after 1976. Haddock also has risen in the last few years, and yellowtail flounder is also up slightly, although not yet significantly. Prices of all three have risen, even when adjusted by the Consumer Price Index, so that the total value of the catch to the fishermen is at a maximum, taking inflation into account (Fig. 9) except for a short period during the Second World War. The number of vessels in the fishery, however, has increased substantially, and average earnings per vessel almost certainly are down. Moreover, in the last few years prices have fallen somewhat from maxima reached in the late 1970s. Failure

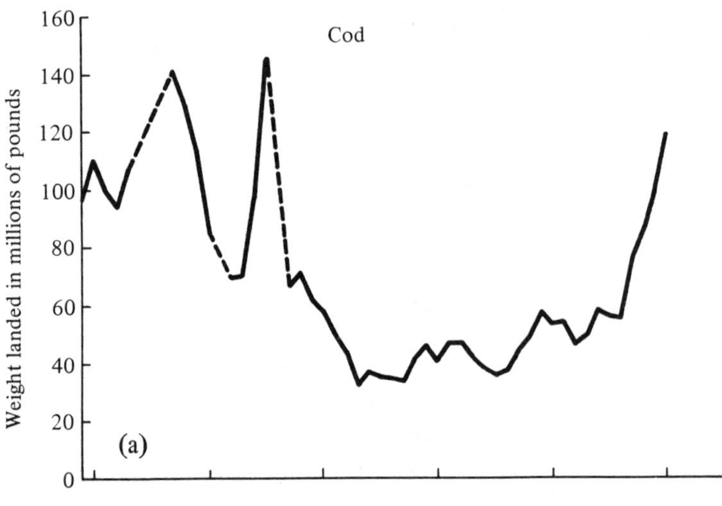

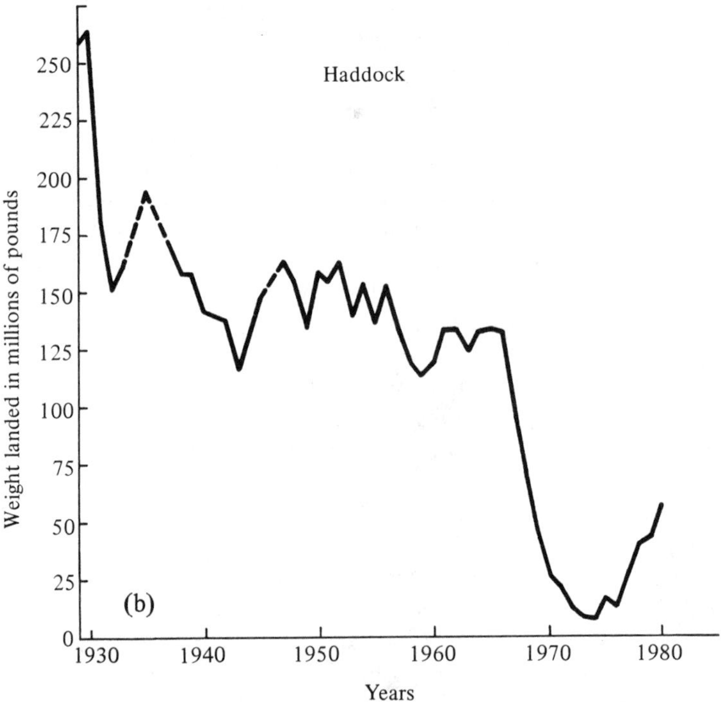

Figure 4.8. Landings of (a) cod, (b) haddock, and (c) yellowtail flounder along the Atlantic Coast 1929-1980 (yellowtail flounder not separated from other flounders prior to 1938).

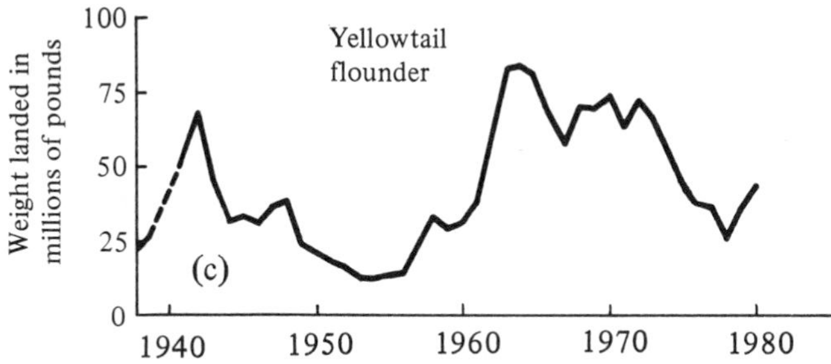

Figure 4.8. *(continued)*

of the New England Council to place a moratorium on vessels in this fishery may have nullified any gains that might have been made and has put the industry in a worse situation than might otherwise have been. Lower prices for these species have to some extent been caused by actions of the industry itself, in attempting to avoid regulations. However, the relative gains or losses from alternate management strategies are not known.

On 28 April 1981 the Groundfish Oversight Committee of the Council issued a report which outlined the provisions of the interim plan. In effect it designated an area, principally on Georges Bank, in which the minimum mesh size will be $5\frac{1}{8}$-inches (13 cm) during the first year and will be increased to $5\frac{1}{2}$-inches (14 cm) 1 year later. To allow vessels to fish squid, silver hake, redfish, and other species in this area, a smaller mesh will be allowed, and vessels participating in the optional settlement program will be allowed a limited percentage of their total catch, over a reasonably flexible settlement period, to consist of regulated species. Any vessel fishing small-mesh gear in the waters of the large-mesh area must be registered under the optional settlement program or be in violation of the plan.

Furthermore, under the plan there will be established minimum fish sizes for regulated species as follows: 17 inches (43 cm) for cod and haddock, and 11 inches (28 cm) for yellowtail flounder. There will not be a specified tolerance for undersized fish, but the Council will expect enforcement agents to exercise reasonable discretion in implementing the minimum size.

Haddock spawning closing areas presently in force will be continued under the interim plan, to enhance the spawning activity of the three species. No additional closures are contemplated at present.

The Committee decided that it is preferable not to include trip limits in the interim plan. Nevertheless, the Committee was aware of concern in some sectors of the fishing industry over the immediate impact of substantially increased landings of cod, haddock, and yellowtail flounder. To allow for this the Committee will propose within the environmental impact statement and at public hearings an alternative to the proposal for no trip limits. This would place an initial 60,000-lb (27.2 metric ton) trip limit of any combination of cod, haddock, and yellowtail flounder, except that yellowtail flounder will not exceed 33% of the limit. The trip limit would be increased by 5000 pounds (2.3 metric tons) every 3 months and would be removed completely when the 5½-inch

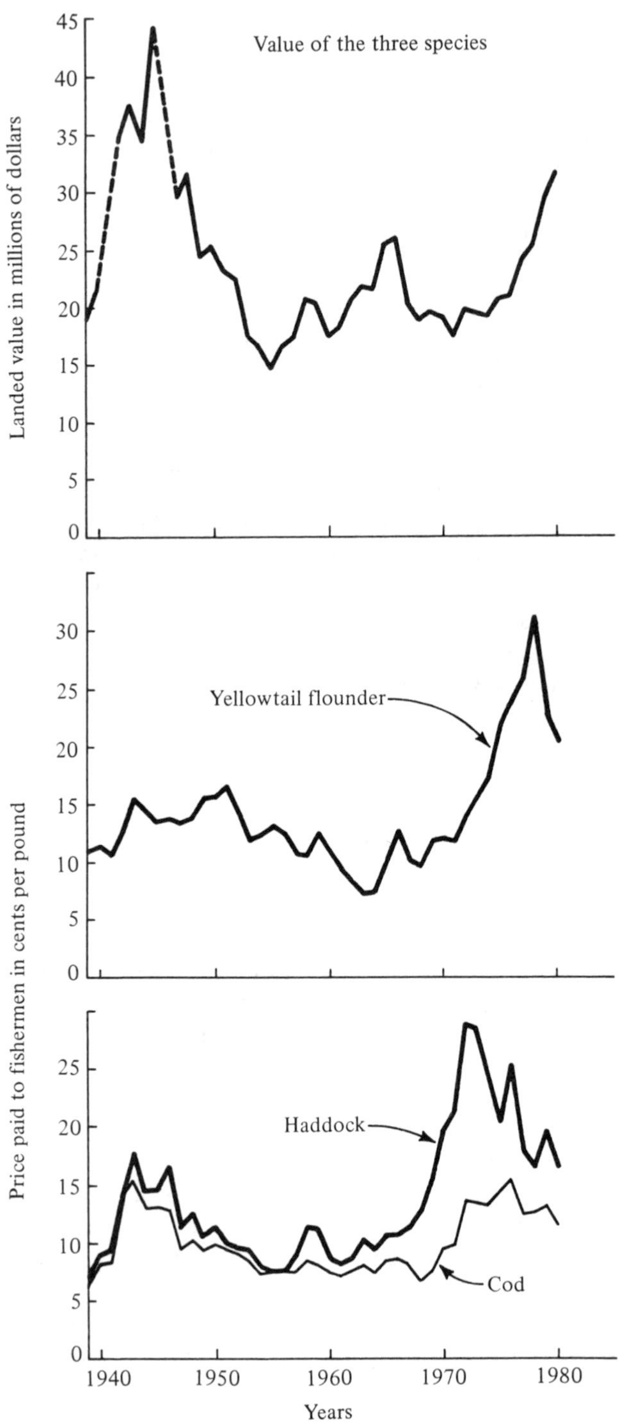

Figure 4.9. Price paid to fishermen for Atlantic cod, haddock, and yellowtail flounder, 1939-1980, and total landed value of the three species adjusted by Consumer Price Index.

(14-cm) mesh goes into effect. This would provide a transition period for industry to adjust to changing market situations should such occur.

The Committee recognized that there might be a need for contingency or failsafe measures in the event of an emergency during the period of the interim plan. It felt that this issue was most complex and its resolution would take more time. Meanwhile, it felt that this interim plan should be put into effect until a more permanent set of arrangements could be developed.

The interim plan was referred to the Scientific and Statistical (S & S) Committee of the Mid-Atlantic Council by that Council, for an assessment of its effect on fishermen in the Mid-Atlantic area, who also fish for these species, especially cod and yellowtail flounder. The fishery for cod is seasonal, especially for commercial fishermen, and to a somewhat lesser extent so is that for yellowtail flounder. There is also considerable interest in cod by recreational fishermen in the area. The S & S Committee, at its meeting of 6 May 1981, was of the opinion that the measures proposed were not a satisfactory means of resource conservation. They believed that the recommended mesh sizes would not necessarily prevent the capture of significant numbers of fish less than the minimum sizes. They believed also that the absence of an over-all catch limit would likely result in an uncontrolled harvest of cod, haddock, and yellowtail flounder and lead to stock reduction. The minimum fish size may be the only measure having a possible effect on fishermen in the Mid-Atlantic area, but if fishing in New England leads to a reduction in stock size, this will have an adverse effect on Mid-Atlantic fishermen.

The Secretary of Commerce has at last acted, and on 31 March 1982 emergency regulations went into effect. Optimum yield will be the amount of fish actually harvested by United States fishermen in accordance with measures adopted. These include minimum mesh sizes for trawls and gill nets in a part of the Gulf of Maine and Georges Bank where about 90% of the cod, haddock, and yellowtail flounder are normally taken. Minimum fish lengths are prescribed. There is an optional settlement program under which a vessel may fish in the large-mesh area for species requiring the use of smaller mesh nets, with certain restrictions. Spawning area closures in spring are about the same as under the previous management system. Collection of catch data is based on a voluntary system. The emergency regulations will remain in effect for 90 days, during which the public will be able to comment. The Secretary also will use the emergency period to monitor the fishery and see how well the plan is working. The goal is to reduce catching of small fish, enhance spawning, and collect accurate catch data. It was stated that the fishing regulations were able to be relaxed because the fish stocks had increased in abundance. When finally adopted the plan will be effective for 3 years. Meanwhile the Council will define its long-term management goals and determine whether certain levels of stocks should be maintained and the measures necessary to reach those levels. It remains to be seen whether these interim measures will have the desired effect, and whether the Council will be able to come up with acceptable management goals and can put them into effective action. The New England Council has a difficult few years ahead of it.

Conclusions

The picture given here by the review of the eight species (oyster, blue crab, menhaden, surf clam, ocean quahog, cod, haddock, and yellowtail flounder) does not present an encouraging forecast of our ability to manage. Oyster production continues to decline, seriously hampered by mistaken attitudes among the people concerned, and by differences between segments of industry. Despite a careful study of what is wrong with the industry in Virginia, and a series of recommendations that could turn the industry around, there is little evidence that anyone is about to act.

The blue crab resource appears to be in better condition. There is no evidence as yet that the resource has been overfished. In fact, over the range of stock sizes experienced in the past, recruitment has been remarkably constant. There are deep differences between segments of industry, with no evidence that they are valid. Increasing loads of pollution entering the bay, and the steady growth of the human population, suggest that at some point in the future the stocks will be affected. There are no clear plans to act, when and if necessary.

The menhaden resource has been equally devoid of management. Production along the Atlantic Coast as a whole is down to less than 50% of the all-time maximum, and although the last 10 years in Chesapeake Bay have produced an all-time high in average landings for that area, these have been principally immature fish. The long-term prospects are not encouraging. Distribution of fishing effort along the coast would be optimal if it were farther north, but a shift of effort is unlikely. The Zapata Corporation, the major producer now in Chesapeake Bay, is moving out of the menhaden business, and one wonders why.

It appears that the three resources located primarily in the Chesapeake Bay area could profit by increased attention from the Sea Grant Advisory Service. The primary need appears to be a will to use research results already at hand and to solve some difficult social and political problems.

When the FCMA came into being late in 1976, it raised high hopes in many segments of industry. After 5 years of experience it appears that these hopes were too optimistic. One industry much in need of help was the surf clam industry, which was already overfished and the fleet overbuilt. The Mid-Atlantic Council wisely froze the number of licenses in the surf clam fishery and also paid some attention to the ocean quahog, which was a potential substitute, although not important at the time. These moves more or less stabilized the industry in the short run, and although many problems arose as the plan developed, most of them were solved in one way or another. The Council now faces some difficult problems as the time approaches to open up closed areas off the New Jersey coast, problems that may affect prices if too many clams become available at once. Equally difficult threats of another massive kill from low oxygen arise as each year comes along. There is also a need to consider whether it would not be wise to place a moratorium on the ocean quahog fleet and to limit the quota more strictly while there is still time. A species as long lived as ocean quahog appears to be may have a limited optimum sustainable yield. The mistake made with surf clam should not be allowed to be repeated.

The New England Council has an even greater problem on its hands, and there is no sign that it is close to solution, despite 5 years of trying. The present management plan for groundfish is not working and practically everyone thinks it should be scrapped. An interim plan has been proposed but there is considerable doubt that it is adequate. The future looks bleak, and there is nothing on the horizon that offers anything better.

There are proposals to do away with the present moratorium on surf clams. Some members of the Mid-Atlantic Council believe that it was a mistake in the first place, that the Councils should not concern themselves with the economics of the fisheries. In my view it would be a great error to remove the moratorium now, when the supply of surf clams is about to increase greatly, but probably temporarily. Rather, there should be serious consideration to the advisability of extending the moratorium to the ocean quahog fleet also, while there is still time.

The groundfish management plan is at present in chaos. The New England Council has decided that the present plan cannot work and is proposing an interim plan, which has just recently been adopted in most of its details by the Secretary of Commerce. This plan has a number of deficiencies and is useful as an interim plan only because the stocks are presently in better condition than before. There is no limit on the number of vessels that can enter these fisheries, and the number has almost doubled since management was first attempted. Thus, any opportunity to gain by increased recruitment has been dissipated by spreading the catch among more vessels. There is talk about including more species within the plan, which would appear to make a move toward limited entry even more unlikely. At any rate, it is probably too late now to do much good. The interim plan will not be adequate in the long run, and the New England Council appears to have a limited number of options.

References

Dewey, J. 1940. The living thoughts of Thomas Jefferson. Longmans, Green and Co., New York.

Dryfoos, R. L., Cheek, R. P., and Kroger, R. L. 1973. Preliminary analyses of Atlantic menhaden, *Brevoortia tyrannus,* migrations, population structure, survival and exploitation rates, and availability as indicated from tag returns. Natl. Marine Fish. Serv., U.S. Dept. Commerce, Fish. Bull. 71(3):719-734.

Haven, D. S., Hargis, W. J. Jr., and Kendall, P. C. 1978. The oyster industry of Virginia: Its status, problems and promise. Va. Inst. Mar. Sci. Spec. Pap. Marine Sci. 4.

Haven, D. S., Whitcomb, J. P., and Kendall, P. C. 1981. The present and potential productivity of the Baylor grounds in Virginia. Va. Inst. Mar. Sci. Spec. Rept. Appl. Sci., and Ocean Engineering 243, vols. 1 and 2.

Malone, D. 1981. The Sage of Monticello, vol. VI. Little, Brown and Co., New York.

McHugh, J. L. 1969. Fisheries of Chesapeake Bay. Proc. Governor's Conf. on Chesapeake Bay, 12-13 September, vol. II, pp. 135-160.

McHugh, J. L. 1972. Jeffersonian democracy and the fisheries. *In*: B. J. Rothschild (ed.), World Fisheries Policy–Multidisciplinary Views Univ. Washington Press, Seattle, pp. 134-155.

McHugh, J. L. 1978. Atlantic sea clam fishery: A case history. *In*: K. M. Jurgensen and A. P. Covington (eds.), Extended Fishery Jurisdiction: Problems and Progress, 1977. Proc. N.C. Governor's Conf. on Fish. Mgmt. under Extended Jurisdiction. U.S. Dept. Comm., NOAA, Off. of Sea Grant, and Dept. Admin., State of N.C. UNC SG-77-19, pp. 69-89.

Pileggi, J. 1980. Fishery statistics of the United States 1976. Natl. Mar. Fish. Serv., U.S. Dept. Commerce, Stat. Dig. 70 (and previous numbers in this series).

Rinaldo, R. G. 1977. Atlantic clam fishery management plan. Environmental impact statement. Prepared by Ronald G. Rinaldo and support staff from National Marine Fisheries Service, Northeast Regional Office, and State-Federal Clam Management Sub-Board and Technical Committee, Mid-Atlantic and New England Regional Fishery Management Councils. Dover, Delaware and Peabody, Massachusetts.

Schaff, W. E. 1975. Status of the Gulf and Atlantic menhaden fisheries and implications for resource management. Mar. Fish. Rev. 37(9):1-9.

Schlesinger, A. Jr. 1971. "Is it Jeffersonian?" The New York Times, Saturday, 30 January 1971.

Taylor, C. C., Bigelow, H. B., and Graham, H. W. 1957. Climatic trends and the distribution of marine animals in New England. U.S. Dept. Interior, Fish and Wildlife Serv., Fish. Bull. 57(115).

United States Congress. 1963. Molluscan shellfish. Hearings before the Subcommittee on Fisheries and Wildlife Conservation of the Committee on Merchant Marine and Fisheries, House of Representatives, 88th Congress, 1st Sess., Washington, D.C.

United States Congress. 1976. The Fishery Conservation and Management Act of 1976. Public Law 94-265, 94th Congress, 2nd Sess., H.R. 200, 90 Stat. 331-361. U.S. Government Printing Office, Washington, D.C.

United States Department of Commerce. 1980. State landings, annual summaries 1979, Current Fish. Stat. various numbers (and previous issues in this series to 1977).

Van Engel, W. A. 1958. The blue crab and its fishery in Chesapeake Bay. Part 1–Reproduction, early development, growth, and migration. Comm. Fish. Rev. 20(6):6-17.

5

Practices and Prospects for Fisheries Development and Management: The Case of Northwest African Fisheries

Jean-Paul Troadec

Introduction

Northwest African countries, from Morocco to Guinea, represent probably the group of countries for which the extension of national jurisdiction has had the most significance. While it is true that countries bordering the northern Atlantic and the northern Pacific also acquired control over fishery resources of comparable importance (in absolute terms), the present state of their economies and the other natural resources available for their development do not render fisheries as critical a factor of economic progress as they are for the Northwest African states.

Reported catches, excluding tuna, reached 2.7 million metric tons during the period 1975-1977 (FAO, 1981a,c) for a gross economic value at landing close to $ US 1 billion. Three-quarters of this amount, in economic as well as in physical terms, accrued to long-distance fishing countries—principally Spain, USSR, Japan, and Korea (Christy, 1979).

J.-P. Troadec is Director of the Scientific and Technical Institute of Marine Fisheries (France), he previously served in the Department of Fisheries, Food and Agricultural Organization of the United Nations, first as a resource evaluation specialist and then as Chief of the Fisheries Development Planning Service. He started his career in West Africa (Congo, Ivory Coast) as a stock assessment expert. From pure resource evaluation problems in tropical fisheries and in developing countries, he progressively became concerned by the wider problems of fisheries development and management in Third World Countries and the need to revise the strategies and approaches developed in the context of large-scale international fisheries.

This immediate change in ownership may appear spectacular; however, in terms of real benefits to the Northwest African countries, it has not been very significant. Owing to the high pressure of fishing exerted upon the major stocks, net benefits presently derived from their exploitation are, at best, minimal. Although coastal countries do derive certain benefits either in kind (equipment, infrastructure, loans at preferential interest, technical assistance, etc.), in cash (Mauritania, for example, received $ US 29 million from fishing licence fees and fines in 1978 (Doucet *et al.*, 1981), or of a political nature (outputs of overall bilateral cooperative arrangements) from the operations of foreign fleets, such gains are modest compared to the net hypothetical benefits–likely to reach several hundred million dollars if exploitation could be fully rationalized (Christy, 1979; Doucet *et al.*, 1981; Griffin *et al.*, 1979). On the other hand, the fish production of these coastal countries, confronted by the handicaps of economic underdevelopment, has been fluctuating between 600 and 700 thousand tons since 1972 without signs of any evident upward trend (FAO, 1981a,c).

The past ocean regime of open competition tended to prescribe a uniform set of exploitation objectives and fishing patterns among the various countries engaged in international fisheries. By releasing the competition for the exploitation of stocks which in principle were accessible to all but, in practice, more to those in possession of technological and economic capabilities, the coastal countries covered in the present study can now envisage developing their participation according to their respective interests, strengths, and weaknesses. For that purpose thay have to design and implement more appropriate fishery development strategies, taking into account not only the implication of the new ocean regime but also the experience that can be drawn from three decades of fishery development and technical assistance. The limitations of those programs, notably in their lack of relevance to the economic situation of the countries concerned, now appear more clearly.

Past efforts in promoting the development of world fisheries, including the provision of technical assistance to the fishery sector, were explicitly or implicitly based on two main strategies:

The worldwide expansion of the activities of long-distance fleets operating from countries in the northern hemisphere which started, some decades ago, to industrialize their fisheries (Troadec, 1976);

The transposition of this expansion model in the promotion of fishery development to Third World countries.

Insofar as it can stimulate nationally determined development, the new ocean regime can facilitate a long-due evolution in fisheries development strategies appropriate to Third World countries and, more specifically, to the states covered in this chapter.

Coastal countries also have, under the new ocean regime, the authority to control the amount of fishing of both domestic fleets and foreign partners; they can now use and manage the stocks in such a way as to prevent the dissipation of economic and social benefits and the biological overfishing which characterized the past regime of open access and free competition.

The task faced by the Northwest African countries is complex in two respects: they simultaneously have to (1) expand their participation in fisheries from which they have been partially excluded by the combined effect of reduced stock sizes due to heavy foreign fishing and their present limited fishing efficiency, and (2) progressively implement adequate management measures to increase the benefits they can expect to derive from those fisheries. Everywhere in the world expertise and experience in each of these complementary fields are notably insufficient.

The review in this chapter does not pretend to provide answers to these difficult issues as the quantitative analyses required for such are seriously lacking. The following section simply intends to examine, using the Northwest African fisheries as an example, past performances and new opportunities offered by the new regime for a more coherent development and effective management of Third World fisheries. Reference will also be made occasionally to the fisheries in the Gulf of Guinea where the resources are markedly less abundant. The comparison of situations and trends between these two geographical sectors enables interesting hypotheses on the dynamics of fishery development to be put forward.

Development of the Region's Fisheries

Brief Historical Review

Although the commercial exploitation of the Saharian Banks started at the end of the last century and the pilchard fishery in Morocco was initiated during the period between the two world wars, it was only in the late 1950s that commercial fisheries began to expand in this area as well as further south in the Gulf of Guinea (Gulland *et al.*, 1973).

Fleets of medium-sized trawlers and purse seiners started to operate south of Senegal from the major ports of the Gulf of Guinea. However, their development remained slow compared to the soaring expansion of the foreign large-scale fisheries which started in the Saharian area in the late 1960s. From 400,000 tons in 1964, catches made by long-distance fleets reached a peak of 2.3 million tons 20 years later; from 1964 to 1970 their annual growth rate never fell below 30% (FAO, 1981a,c).

This development corresponded to the activities of long-distance fleet operations by industrialized countries spreading over subtropical and southern temperate areas (North and Southwest Africa, Southeast America, tropical-ocean (tuna). The factors contributing to this expansion can be summarized as follows: the possibility to increase the resource base by developing new fisheries in increasingly remote areas; the existence of large home and export markets; cheap energy; a relatively expensive manpower base which was, however, apt to assimilate the technological innovations required by more autonomous operations (involving fish detection and catching devices of growing sophistication); experienced managerial capabilities; and efficient research and administrative institutions. Under those circumstances, the race for open-access, remote resources depended primarily on investments and the acquisition of technological innovations required to expand fishing operations in increasingly distant waters.

It was also the period (mid-1950s to mid-1970s) of such large international oceanographic programs as the International Cooperative Investigations of the Tropical Atlantic (ICITA) and the Cooperative Investigations in the Northern part of the Eastern Central Atlantic (CINECA) and such fishery resource surveys as the Guinean Trawling Survey (GTS) and the FAO/UNDP Sardinella Projects.[1] Although some of these surveys contributed to the identification of new resources and the development of new fisheries (for example, the French tropical tuna fishery, the sardinella fishery in the Congo, various shrimp fisheries in the Gulf of Guinea), others merely confirmed the existence of stocks already successfully developed by the industry (GTS and demersal stocks, Sardinella project, and coastal pelagic resources in the Gulf of guinea, for example). The convergence which prevailed then between fishery research investigations (essentially under bi- and multilateral programs) and long-distance fisheries (involving almost exclusively world fishing powers) equated the principle of open access to the resources to that of freedom for oceanographic research.

Fleets of medium-sized trawlers and purse seiners developed slowly, south of Senegal, as domestic markets permitted. Their expansion was primarily the result of the initiative taken by small, locally based European fleet owners, largely outside national and multilateral technical assistance programs. A few countries bordering the Gulf of Guinea—the Cameroons, Ivory Coast, but mainly Ghana—also acquired long-distance vessels to operate further north and south for the fish they could not find in sufficient quantities off their shores. Little attention was then paid to small-scale fisheries: governmental aid per ton-caught was markedly below that received by the commercial sector (FAO, 1979); similarly, research investigations were largely restricted to the assessment of the resource base and of the prospects offered for expanding large-scale fishing operations. The specific opportunities of small-scale fisheries and the possible conflicts between artisanal and commercial operations were largely neglected. It seemed to be accepted then that artisanal fisheries would be replaced eventually by larger scale operations for which the former could possibly provide skilled manpower.

In this region, as in other parts of the world, the philosophy explicitly or implicitly prevailing in the 1960s and 1970s was essentially based on the extension of long-distance fishing operations to the world's richest fishing grounds. This strategy was continued under the new regime by joint ventures. Similarly, the promotion of fisheries in developing countries was greatly influenced by the model through which large-scale fisheries expanded in industrialized nations: fishery development and technical assistance programs concentrated their actions on the production sector by adding new boats and increasing their fishing power by the adoption of technological innovations.

[1]FAO/UNDP Regional and National (Congo, Ghana, Ivory Coast, Senegal, and Sierra Leone) Projects for the Survey and Development of Pelagic Fishery Resources (*Sardinella*) in West Africa, to mention only international programs.

The Crisis

In the early 1970s, however, several developments cast some doubts on the relevance of the strategy:

Estimates of the world fishery potential yield published at the beginning of that decade (e.g., Gulland, 1971; Moiseev, 1971) suggested that it was unlikely that the effective sustainable yield from conventional resources would exceed twice the current catches; such a ceiling would be reached within 10 years if current growth rates were to be maintained; it was, however, expected that unconventional resources (krill, mesopelagic fishes, oceanic squids, etc.) would permit the expansion.

The year 1972 not only marked the collapse of the Peruvian anchovy fishery—in terms of weight, the largest fishery in the world—but also the sudden decline in the incremental rate of world fish production: from an average of 6% annually observed during the past two decades, it fell to below 2%, thus confirming previous pessimistic forecasts.

From 1973, the prospects of the expansionist vision became even more gloomy with the sudden and sustained increase in the cost of fossil fuel.

Finally, in 1974, at the First Conference on the Law of the Sea (Caracas, Venezuela), the principle of free access to the majority of world fisheries resources was, for the first time officially challenged. Although they did not stand to acquire the largest share of world resources, as those were primarily concentrated in high and medium latitudes, this initiative was taken primarily by developing countries.

However, all during the 1970s, the same expansionist vision and the same development strategy continued to prevail. For the group of countries under consideration, the progress made by their fisheries remained, at best, modest, despite the exceptional opportunities offered by the new regime.

Limitations of the Conventional Development Model

Balance of Production and Processing Inputs. In a few regions (e.g., the advanced Latin American countries) where conditions were grossly similar to those prevailing in the northern hemisphere during the past century, notably with respect to the nature and abundance of the resources and state of economic development, or in certain countries (Korea, Thailand, etc.)[2] where fishery development could be triggered by an initial technological input, recourse to this model was justified. Unfortunately, in the majority of other developing countries, it proved to be ill matched to the local environment. Even prior to the increase in the price of oil, the development schemes modeled after the expansion of large-scale operations in advanced countries could have been expected to prove irrelevant and of

[2] In these countries, however, as well as in world fishing powers, this vision did not facilitate an official recognition of the implications of the limited nature of fishery resources and the opportunities attached to the early introduction, at national and regional levels, of proper management schemes.

limited efficiency in countries where the factors and conditions for fishery development were characterized by abundant and cheap labor; often scarce fishery resources made up of smaller fish, which does not facilitate the mechanization of processing; limited or little-accessible domestic and export markets; lower over-all level of economic and technological development impeding the assimilation of foreign technology; lack of managerial skills as well as administrative and research capabilities; limited investment capacities; etc. In Morocco, for example, larger trawlers recently introduced in the coastal fishery do not achieve catch rates higher than those of medium-sized vessels. This suggests that the constraints are more likely to be related to factors such as crews' ability to master sophisticated equipment or market gluts, rather than to insufficient fishing capacities (Gréboval, in preparation). As compared to the emphasis given to investments and the adoption of technological innovations in the production sector, considerably less attention was devoted to other aspects of the fishery continuum as well as to the objective appreciation of local conditions and their implications in determining appropriate interventions as a result of establishing priorities for development.

Although quantitative analyses are lacking, the adoption of inadequate balances of basic inputs (manpower, capital, energy) could only have added to the production and processing costs already taxed by the inefficiencies of underdevelopment. The resulting over-all cost can reach very high levels: thus, in Mauritania, local processing (deep freezing) was in 1979 three times more expensive than in other ports of the region (Doucet *et al.*, 1981). The resulting excess cost becomes particularly critical in international fisheries when most competitive foreign fleets reduce stock sizes to levels that can hamper the participation of the less efficient coastal fleets and when the production has to be disposed on world markets to which extra catching and processing costs cannot be passed on. For high-value species, such as cephalopods, fishing by coastal countries can still be profitable, though at a potential loss when viewed against the gains that may be achieved by having recourse to the most efficient foreign fleets and the extractions of fees close to levels of economic rent. However, exploitation of stocks, for which the fishing cost by even the most efficient fleets is close to their market value, will not be feasible for coastal countries, at least without a substantial reduction in the average amount of fishing. This is probably the case in the region for the exploitation of offshore, abundant concentrations of small pelagic stocks (Doucet *et al.*, 1981).

Markets. The comparison of rates of increase in the production achieved by Northwest African countries on the one hand, and by countries bordering the Gulf of Guinea on the other, provides some indication of the constraints imposed by markets on fishery development. The increase was comparable between the two areas during the second half of the 1960s but, according to the statistics available, slightly higher in the Gulf of Guinea during the 1970s (FAO, 1981a,c). Inasmuch as the effect of other relevant factors is comparable, the advantage of more remunerative prices on account of a higher domestic demand in the Gulf of

Guinea countries seems to have amply offset the lower abundance of fishery resources in that sector.

The situation is basically opposite in the nothern subtropical sector where domestic markets, in relation to the combined effects of locally sparse populations and more abundant meat supply, can only absorb a small fraction of the potential yield of the abundant marine resources: the bulk of catches has to be disposed of in world markets at world prices. This does not imply that their domestic markets are fully developed as yet. Although some countries, such as Morocco or Senegal, have undertaken efforts to improve fish distribution to their hinterlands, in many countries technical assistance programs on fishery development still concentrate their interventions on the primary sector and neglect the importance of markets, both domestic and export. Thus, in Mauritania, emphasis in aid programs was given to the improvement of fishing techniques in the small-scale fisheries. Such a strategy did not reflect the fact that the local fisheries are not connected neither to domestic markets, owing to the lack of feeder roads and harbor infrastructure, nor to the export markets, as the processing and export industry is still oligopolistic in nature (Doucet *et al.,* 1981).

Actually, the development of Northwest African fisheries seems to depend on readjustments in international trade patterns. In the past, long-distance fisheries were able to expand due to the existence of developed markets in their home countries; e.g., southern European and Far East countries looked for high-value bottom species, including cephalopods and shrimp, while eastern European countries were primarily interested in coastal pelagic species. Even if these countries lost access to fishing grounds, they have kept control of the markets. For example, the cephalopod market remains dominated by Japan, which consumes half of the world production. This forces the Moroccan cephalopod fishery to land its catches and operate from Las Palmas (Canary Islands). Nevertheless, half of the world production and three-quarters of total imports of octopuses come from the Saharian area (Gréboval, in preparation). An increase in the participation of coastal countries in the development of fisheries off their shores, and in the benefits that can be derived from them, will thus depend on the arrangements they will be able to negotiate for the harvesting as well as for the marketing of their production, including that of their artisanal fisheries. Agreements of this nature, though still of limited scope, have already been signed by countries such as Senegal for the sale and export of high-value species (shrimp, lobsters, cephalopods) harvested by their small-scale fishermen.

Moreover, new markets have to be opened and developed. The opportunity offered by intertropical Africa, where populations are dense, fish and meat resources are limited, and fish consumption is popular, is particularly relevant. The development of long-distance fisheries in the subtropical (north and south) sectors did not initially reduce this growing gap in the supply to the equatorial sector since the catch was taken back to supply the home markets of the fishing nations. Traditionally, this supply gap has been filled by imports of artisanally cured fish from the Sahelian area (inland fisheries), Angola, and Mauritania and by cured fish from the North Atlantic (stockfish from Norway). It was only in the 1970s that vessels, initially Japanese, later from eastern European countries, began to land in the Gulf of Guinea blocks of deep-frozen, small pelagics and sparids caught in the Saharian sector. Still, the Gulf of Guinea depends on imports from other regions, notably from the northeast Atlantic and the southeast Pacific.

Fishing and import agreements signed between countries (e.g., Ghana and Gambia, Ivory Coast and Senegal) are important steps toward a desired development of intraregional trade. The project which FAO would like to initiate in West Africa to disseminate information on products and markets should contribute to such development. However, a reduction in the overall fishing costs will probably contribute more to the development of the African fisheries than the mere dissemination of information. The progressive rise in the absolute price of fish, which goes with the relative decline of world production (FAO, 1981c), should also facilitate, in the long run, the development of fisheries whose profitability is still marginal.

Research. In the 1950s, the major thrust of research investigations was on the environment and the inventory of fishery resources. This was followed, in the 1960s and the 1970s, by the assessment of their potential yield. During the 1970s, attention was given to management issues, notably the questions of mesh size limitation and effort regulation in major trawl fisheries. Although research capabilities of coastal countries are generally still modest, efforts deployed under bi- and multilateral assistance schemes and integrated through the FAO Committee for the Eastern Central Atlantic Fisheries (CECAF) have led to a satisfactory knowledge of the resources and their potential yields (Belvèze and Bravo de Laguna, 1980; Troadec and Garcia, 1979) *vis-à-vis* other tropical areas.

What is most striking, given a time perspective, is not so much the need often stressed to adapt resource evaluation methodologies developed in higher latitudes to the peculiarities of tropical stocks, but the direction and scope given to fisheries investigations. Research programs were, to a large extent, similar to those executed in support of large-scale fisheries in advanced countries. There, fisheries development concerned itself mainly with the identification of new exploitable stocks and the determination of their optimum fishing conditions; investment planning remained largely, in market economy countries, the concern of the private sector. According to this model, the Northwest African countries were invited to join in the research effort for resource evaluation and monitoring, including data collection. This work, of immediate relevance for foreign fleets, presented less immediate urgency for the coastal states than the investigations on the conditions, options, and priorities of their fishery development as long as their participation in the fisheries remained modest and stagnant. It was only recently that some local laboratories in the region took the initiative to objectively analyze fisheries development issues, notably through the overall assessment of the biological, economic, and social aspects of the fisheries continuum, from the resource base to marketing (Fréon and Weber, in preparation; Weber and Fréon, in preparation). The weaknesses of past development strategies, as evidenced by the arbitrary selection of priorities for development and assistance programs and the disappointing progress made by fishery development, would justify that such investigations be considered an area of emphasis in local programs of fishery research and technical assistance.

The Case of Artisanal Fisheries

In many developing countries, as was earlier the case in industrialized nations, the possibilities and vocation of traditional, local fisheries have often been underestimated in national and international development programs. These fisheries, in addition to a handicap in the race for remote resources, suffer from an economic and, sometimes, cultural isolation which affect their ability to vindicate their rights. Moreover, their promotion requires more varied and long-lasting interventions.

However, toward the late 1970s, official strategies started to evolve. In spite of the priority given to large-scale fisheries, traditional fisheries were still responsible for almost half, and probably more in value, of the world catches used for direct human consumption and directly supported many million fishermen families (Indo-Pacific Fishery Commission, 1980). In Ghana and Senegal their strength was particularly evident: in the 1970s the Ghanaio-Ivorian stock of sardinella collapsed twice in relation to the intense fishing pressure exerted by the canoe fishery (Fishery Research Unit, Tema/ORSTOM, 1976; Troadec *et al.*, 1980); in Senegal economic analyses indicated that the profitability of canoe fisheries compared well to that of commercial boats (Jarrold and Everett, 1981).

Several factors called for a revision of priorities in favor of small-scale fisheries: the new prospects offered to them with the sedentarization of fisheries made possible by the extension of national jurisdiction; the difficulties facing the redistribution of national wealth to underprivileged population classes and the need to directly promote rural development for a more even geographical distribution of economic activity and for counteracting urban migration; their better adapted balance of production costs (labor, energy, capital—especially the imported capital costs); etc.

The number of projects aimed at uplifting small-scale fishery communities has increased. However, most aimed at reaching this economic and social goal through the improvement of fishing methods, notably the motorization of canoes, and the adoption of more effective gear and boats. Some projects undeniably had positive results: the introduction of purse seining in the Senegalese canoe fishery (Grasset, 1972) led to the establishment of a new fishery presently employing 6000 fishermen and where fishing units have annual yields close to those achieved by local purse seiners (Fréon and Weber, in preparation; Weber and Fréon, in preparation). Unfortunately, most of these projects neglected the external constraints affecting unmanaged inshore fisheries, viz:

The limited yield of fishery resources, and their frequently high level of exploitation in inshore waters.

The encroachment of the resource base available to traditional fishermen by large-scale fleets, notably trawlers, favored in the past national development programs.

The tendency toward overcrowding, notably in countries affected by unemployment, as long as access to local fisheries are not regulated.

Under such conditions, it is obviously illusory to equate gains in efficiency with gains in productivity. Unfortunately, this was the strategy behind many projects which resulted, for the fishermen, in increased production costs on account of the acquisition of the technological innovations not being counterbalanced by comparable gains in catches. In fact, several projects resulted either in a further impoverishment of fishermen or in a reduction in employment (Panayotou, in preparation; Smith, 1979).

In West Africa, overcrowding and conflicts between small- and large-scale fisheries have not reached the insurrectional levels recently observed in Southeast Asia which, for example, led Indonesia to totally ban trawling in Java, Bali, and Sumatra in 1980 (Unar and Naamin, 1981). Nevertheless, the few socioeconomic analyses available illustrate the deceptive and unexpected results of development plans: in Senegal, for example, the amount of subsidies given to the canoe fishery supplying a fishmeal plant is equal to the price paid to the fishermen for their supply (Fréon and Weber, in preparation; Weber and Fréon, in preparation). In the Ivory Coast, the lack of relevant investigations has probably precluded the timely forecasting of the probable collapse of the trawl penaeid shrimp fishery as a result of the development of small-scale fishing which directly cut down its recruitment.

A New Strategy for Fishery Development

The result of two decades of fishery development in Northwest Africa may appear deceptive, particularly when one considers the resources available. The major reasons for this situation seem to lie in:

1. A development strategy which was not, under the past regime, specifically geared to the direct benefit of coastal countries.
2. The uncritical transposition of a development model ill matched to the local environment, apparently because insufficient attention had been given to the appreciation of local conditions and the objective ranking of constraints, the emphasis during project identification being put on the straightforward transfer of foreign technical solutions; the lack of relevance of interventions affected both scientific and technological activities.
3. The priority given to the direct promotion of production through the increase in fishing capacity and efficiency, as opposed to strategies based on the stimulation of comprehensive development throught the improvement of the environment for fisheries (infrastructure, market development, administrative set up, training, etc.).
4. An insufficient recognition of the causes and mechanisms of development, especially those resulting from the peculiarities of the dynamics of a fishery

taken as a whole, viz., the limited nature of the resources and the open-acess nature of fisheries.

Probably some of these limitations could not have been avoided. The adoption of the model of expansion of large-scale fisheries may have been ineluctable as long as the regime of open access, favoring uniform exploitation schemes and the dominance of the most efficient, prevailed. Others could be considered as the price of acquiring direct experience which was lacking. In any case, an improvement in the performance of development programs will depend on the revision of the concepts and strategies underlying their formulation. Research on fishery development needs to be intensified for that purpose, i.e., determination of appropriate balances of production and processing inputs, assessment of over-all fisheries from catching to consumption for an objective identification of constraints and priorities, studies of international trade patterns and formulation of marketing strategies more favorable to developing countries, etc.

As compared to the other sectors of the economy, education, health, etc., the importance of nationally determined development has not been recognized in fisheries owing to the leveling nature of the past international ocean regime. National administrations now have the opportunity to formulate long-term development policies and strategies better tailored to local conditions.

The first step will be to clarify the objectives assigned to their various fisheries, including those in which foreign participation is to remain temporary. Most common objectives (maximum economic yield, maximum employment, optimum individual income, optimum food supply, optimum foreign currency earnings, etc.) are incompatible, each of them being achieved for a different level of fishing effort. They therefore need to be ranked. This can be done by referring to the goals in national development plans. The specific peculiarities of individual fisheries, notably their respective aptitudes to satisfy particular economic or social goals, also need to be taken into account. Thus, the Saharian cephalopod fishery, which if properly managed can produce net benefits estimated in 1977 at $ US 250 million (Christy, 1979; Gréboval, in preparation; Griffin *et al.*, 1979), can be considered primarily as a source of foreign currency, whereas the inshore artisanal fisheries would be used to stimulate employment and rural development.

Major development options (e.g., small- vs. large-scale fisheries, domestic vs. foreign fishing, various arrangements for foreign involvement) could then be compared to determine their respective merits to reach the selected goal mix. For example, defining the optimum level and scheme for foreign participation and its future evolution is likely to raise complex choices. Northwest African countries will not be able to derive the maximum benefits from some of their fisheries until they possess the variety of required means and capabilities. Confronted with that situation, most have recourse to joint ventures. At the cost of some immediate loss on revenue and liberty since their partners' interests necessarily diverge from theirs, coastal countries adopt joint-venture schemes in the expectation that, if properly conducted, these will accelerate the transfer of knowhow. This kind of arrangement is not necessarily the scheme which could best serve their interests. Alternatively, coastal states can sell, *in situ* and against the payment of fees, predetermined yield surpluses to foreign fleets most capable of harvesting them efficiently and, thus, of paying the highest fees. Governments could then use such revenues as they deem fit, including the acquisition of infrastructures and equipment, for organizing the training programs and other activities they require to promote domestic participation in offshore fisheries.

The choice of one scheme against another will be facilitated by an analysis of their short- and long-term economic and social effects. However, the final decision will also depend on the firm beliefs of the decision makers regarding the progress they expect their national fisheries to achieve and the immediate sacrifices that their sustained expansion would justify (Doucet *et al.,* 1981).

Management of Fisheries

Review of Past Actions

Measures Taken. In 1971 CECAF adopted its first management recommendation, inviting countries fishing in the northern subtropical sector to use mesh size larger than 70 mm when fishing for hake and sparids, and 60 mm in the cephalopod fishery (FAO, 1971). In 1979 the use of a single mesh size of 60 mm was recommended for all trawl fisheries, including the shrimp fisheries, throughout the whole area supervised by the Committee, i.e., including the Gulf of Guinea. The adoption of a unique mesh size regulation was intended to facilitate enforcement (FAO, 1980a). Though this legislation is not yet fully enforced, several coastal countries, such as Mauritania, prescribe it to their foreign partners and are actively engaged in its progressive implementation in their own fleets.

It was also in 1971 that the merit of effort limitation was discussed for the first time. However, no decision was taken at that session or at successive ones. Initially the measure did not appeal to foreign fleets, which could rely on the advantage of their higher efficiency, or to the coastal countries, which were still participating only marginally in the fisheries. Then the *de facto* transfer of jurisdiction reduced the chances of management measures of this kind being taken within the framework of the Committee. However, for their negotiations on the maximum amount of fishing to be allowed to their foreign partners, coastal countries continued to rely on stock assessment studies prepared by the Committee. For example, Morocco regularly considered conclusions of the CECAF Working Party on Resources Evaluation to negotiate the capacity of foreign fleets allowed to operate in the waters under its jurisdiction. Similarly, Mauritania, in formulating its new fishery policy in 1979/1980, determined the maximum processing capacities that it intended to acquire through joint-venture agreements on the basis of potential yield estimates issued by the Committee (Doucet *et al.,* 1981; Gréboval, in preparation).

Whatever economic or social criteria are considered, the amount of fishing remains inadequately controlled in many fisheries. For example, in the Moroccan inshore trawl fishery, the doubling of fishing power in the mid-1970s did not lead to any significant increase in sustainable yield. This situation has various causes. National administrations do not always enjoy full autonomy to set the limits of foreign fishing. When negotiating bilateral cooperation agreements, fisheries may bear the costs of the coastal states need to obtain advantages in other economic and political fields. Although little information has been published on that matter, it is likely that events affecting the political status of certain land areas in the Sahara have also affected the fishing agreements between coastal and noncoastal countries. Enforcement capabilities are frequently inadequate. Last, in the face of their fleets' present tonnage relative to past and present amounts of fishing by foreign vessels, national administrations may find it difficult to limit domestic fishing activities in areas where the fishing effort appears excessive.

Effects of the Past Regime on Fisheries Management Practices. In the past, neither in the CECAF region nor elsewhere in the world have fishing countries been able to reach even implicit agreements on the long-term sharing of the resources they were internationally exploiting. National administrations were basically confronted with a conflict situation identical to the one classically described for the harvesting sector. As national fleets constituted the essential instrument with which they could defend the shares they believe should or could accrue to their respective industries, they could not envisage controlling their fishing capacities.

In other parts of the world where the intensification of competition made the conservation of the resource a matter of urgency, countries came to agree on controlling their aggregate catches. However, no regulation of fishing capacities accompanied such decisions. At best, they agreed, to relax the immediate effects of competition, to split the overall allowed catch into annually revisable national quotas. Anxious to maintain for their industries the highest or least reduced quotas, they were understandably reluctant to fix the total allowable catch below maximum sustainable yield (MSY). Pressed between limited resources and the entry of newcomers, they pursued sharing arrangements aimed at maintaining past levels and patterns of participation. In the CECAF area, the disparity between historical performances and capacities of coastal and noncoastal countries actually precluded even limited agreements of this nature to be reached.

It is, therefore, no surprise that under conditions of open access and free competition, the MSY and the national catch quotas had been the commonly accepted management objective and method. The fact that the criticisms expressed by economists (e.g., Christy and Scott, 1965) and by biologists (e.g., Larkin, 1977; Sissenwine, 1978) on the relevance on MSY as a management objective received little official recognition was less due to an imperfect understanding of its limitations than to the conditions imposed on management practices by the regime of the sea.

The general adoption of the catch quota system appreciably increased the research needs: stock fluctuations had to be accurately monitored for setting the annual quotas and for preventing the risk of biological overfishing enhanced by the existence of excessive fishing capacities. Consequently, an appreciable part of this research effort was devoted to the routine activities of stock and fleet monitoring and the mechanical revision, year after year, of stock assessments. In comparison, topics of more basic interest such as the environment-induced variability of fish stocks, the stock-recruitment relationship and their combined effects on multispecies resources composition and size, the exploitation strategies for highly fluctuating multispecies stocks, etc., received less attention. As no restriction was imposed on the geographic location of fishing operations, investigations of stock distribution and migration patterns were in a similar situation. Owing to the *de facto* impossibility of controlling fishing costs, little emphasis was placed on fishery economics and the development of economic methods for fishery management. Effects on management of market fluctuations were no better investigated, possibly because of the current difficulties in controlling fishing effort.

In international fisheries, all participating countries were urged to equally share the basic tasks of data collection and stock assessment and monitoring, independently of the level of their participation and the progress acheived in the development of their fisheries. The research task of Northwest African countries is not likely to be reduced under the new regime. In addition to undertaking investigations on fisheries development prospects and conditions, they will now need information on the biological, economic, and social aspects of fisheries to determine the optimum ceiling and schemes of participation of foreign fleets as well as to formulate strategies for the effective development of their own fleets. Already fishing capacities would need to be regulated in certain domestic fisheries such as, in Senegal, the purse seine canoe fishery (Fréon and Weber, in preparation; Weber and Fréon, in preparation) or, in Morocco, on the inshore trawl fishery (Gréboval, in preparation). In summary, the scope of fisheries investigations falling upon coastal countries is likely to widen appreciably with the extension of national authority.

New Prospects

Under the new regime, coastal countries have the opportunity to give equal attention to the costs of fishing the stocks totally or largely under their jurisdiction as that previously given to the physical outputs. This should influence both the management objectives and approaches. Only national policies which give similar weight to inputs and outputs would be relevant in the future, as they are the only ones which maximize net benefits—ultimately the only relevant goals. Analysis of the biologic, economic, and social aspects of management indicates that, even if their respective positions differ somewhat in terms of fishing effort, most economic objectives (maximum economic yield, maximum net gain of foreign currencies, rational food supply, etc.) are situated on the left limb of the yield curve. Even for fisheries which are primarily considered as a source of employment, the level of fishing ultimately leading to the highest overall employment should not be as far remote from the top of the curve as is often assumed when one considers simultaneously the multiplier effects on ancillary activities (processing and distribution) and the potential jobs to be generated by the economic rent (Panayotou, in preparation; Troadec, in preparation).

Consequently, the achievement of most possible economic, social, nutritional, and even recreational objectives will very much depend on the capability to regulate the amount of fishing. The ability to control fishing should permit substantial reductions in research costs, as the complexities and risks associated with recruitment overfishing or amplifications in stocks and catch fluctuations generally become less critical for exploitation rates at which most economic and social goals are achieved (Troadec, in press). Northwest African countries have, thus, a possibility to work out a suitable balance of tradeoffs between the research costs and the authority required to maintain fishing effort.

Ideally, comprehensive management schemes should aim at simultaneously:

1. Maintaining the resource at levels corresponding to the sought objectives.
2. Minimizing the harvesting costs for the adopted level of exploitation.
3. As far as possible, promoting fishermen's concern and support for the effective implementation of the management scheme.

The importance of this last point has not always been properly appreciated. In the context of international, large-scale fisheries, management has been perceived from its coercive angle and applied to large fishing units. It is true that management does imply resource reallocations which, if not adequately compensated for in one form or another, cannot satisfy those whose activity is to be reduced. However, fishermen to be accommodated under new management schemes should derive some benefits from them if their support is to be obtained. This observation is important for domestic fisheries and especially for small-scale fisheries in which the numerous small, scattered units cannot obviously be controlled individually.

Under the new regime, a wider set of tools will be available to fishery administrators for performing fisheries management. The following methods will likely replace progressively the national quota system commonly used in international fisheries:

The allocation to individual fishermen, fleet owners, or processing plants, depending on the features of the fishery, of quantitative rights (individual catch quotas).

The licensing of a limited number of fishing boats, gear, or fishermen, whose individual fishing power will simultaneously be controlled by limitations imposed on certain physical characteristics of the fishing unit.

The allocation of the stock itself (biomass) to specific fishermen or fishermen groups.

The taxation of catches, levying of licence fees or, when possible, the leasing of the resource itself, in order to capture the economic rent and to prevent its reinvestment in excessive fishing capacities.

None of the above is likely to be fully satisfactory. Used alone or in combination, the suitability of these methods will depend on the characteristics of each fishery and the management objectives sought. Although experience is still largely lacking on the conditions and implications of their application, they open a new range of opportunities to fisheries management.

Small-Scale Fisheries

The relaxation of external pressures under which artisanal fisheries are operating, notably the progressive adjustment of fishermen populations to the productivity of the resources and the prevention of encroachment and reduction of their resource base by larger scale fisheries, raises very serious technical and political issues. This has led certain governments to adopt a *laissez faire* attitude and to consider artisanal fisheries as an activity of last resort designed to accommodate as much of the jobless and capital-deprived labor forces as possible. The socioeconomic status of small-scale fisheries would then largely depend on, and fluctuate alongside, the state of national economies.

It is less than certain that such an approach would enable the specific opportunities of small-scale fisheries to be taken advantage of, notably those resulting from their lower demand in energy and capital, notably in foreign component inputs. This is particularly important for Northwest African countries, considering their substantial potential for fisheries development and the role that local, traditional fisheries can play in that expansion. The transition, in industrialized countries, of such fisheries from a subsistence to a market economy shows that they are able to evolve and be the major tool for such a development (Cadoret *et al*, 1978). Conversely, fisheries in Southeast Asia, for example, show that the risk is real of having to renounce real opportunities for appreciable periods of time if proper attention is not given in due time to their management needs. The fact that many traditional communities, in both developing and developed countries, have succeeded in regulating the access to and exploitation of littoral fisheries provides an indication that, in small-scale fisheries as well, gains derived from management can well exceed the costs of their enforcement. Otherwise, it is hard to see why fishermen communities would have devoted considerable efforts to the improvement and retention of often highly complex schemes. It seems desirable, therefore, before overcrowding renders the application of long-term management schemes politically very difficult, to survey traditional fisheries management schemes and to analyze the contribution they could make to effective management of inshore fisheries. The example of Southeast Asian fisheries shows that, without appropriate regulation, overcrowding of littoral fisheries may well, as a result of over- and misfishing (e.g., dynamiting, poisoning, destruction of coral reefs, etc.), endanger the resource productivity and appreciably reduce the socioeconomic potentials of the fisheries.

Furthermore, the unequal terms of present and potential conflicts between small- and large-scale fleets make the *laissez faire* attitude unacceptable for domestic fisheries. The example of the Ivory Coast shrimp fishery has shown that the artisanal segment can cause considerable economic losses to trawl fisheries, possibly to the point of economic collapse (Griffin and Grant, 1981). In West Africa, several small-scale fisheries concentrate their harvest activities on the juvenile phase of various other stocks (e.g., sardinella) which are also exploited, when they are older, by large-scale fleets. The former thus enjoy a strategic position in the competition inasmuch as the latter cannot operate in littoral areas. Conversely, traditional fisheries are often handicaped in that competition by a lower modility of labor and assets, as well as with respect to the disposal of catches and access to markets (Panayotou, in preparation). In international fisheries, the regime of free competition did not allow for consideration to be given to the location, throughout the life cycle of exploited species, of the operations of the various fleets. The case of tropical tuna fisheries, for example, where the development of a surface fishery (pole-and-line and purse seine) had been detrimental to the previously established longline fisheries, is well documented. Under the new regime, the intervention of central administrations for allocating the resources and maintaing adequate balances in multisegment fisheries is possible and highly desirable.

To formulate appropriate management schemes, more information and investigations will be required on the following topics:

Distribution and migration pattern of stocks, notably across the shelf, to determine, when necessary, the best location of demarcations between inshore and offshore fisheries and to analyze the conditions and implications of their competition; these issues are anticipated to be of considerable importance, even greater than that of internationally shared stocks, given the enhanced capabilities now available to national administrations for managing domestic fisheries.

The multiplier effects on employment in ancillary activities, in relation to the rate of fishing, in order to estimate the likely position, in terms of overall fishing effort, of maximum employment in individual domestic fisheries.

Fishermen's employment and income alternatives to evaluate their mobility and the problem of regulating the size of fishermen populations, etc.

As stressed earlier, surveys of traditional management schemes—either the property rights over the use of the resources when these are of sufficiently low mobility or through controlled fishing rights given to specific individuals, families, villages, or larger communities over well-demarcated fishing areas—are of great potential value. Over time, many traditional communities have succeeded in developing satisfactory answers to the difficult problems of access control, wealth allocation, and reduction of conflicts between different gear and fishermen groups. Most solutions are based on the allocation of property rights to specific groups, as small as permit the rate of mixing and mobility of stocks, in order to alleviate the negative effects of open exploitation in common property conditions. With the expansion of long-distance fisheries, the prospects offered by those methods have progressively gone unheeded. Many fell into disuse with the intrusion of newcomers in traditional common fishing grounds or the acquisition by some members of the group of considerably more efficient fishing capacities. Japan is a notable exception of a country where the expansion of High Seas fisheries took place without customary regulations of inshore fisheries. In most continental fisheries such rules have also been maintained. The establishment of exclusive economic zones is a first step back to resource allocation practices: it gives national administrations the opportunity, whenever appropriate, to revive traditional schemes or to draw from them for the formulation of new ones.

The Large-Scale Fisheries

Among the methods proposed for a comprehensive management of commercial fisheries, the allocation of individual quantitative rights has sometimes been recommended. Such a system should, in principle, enable the fishermen, fishing companies, or processing plants individually to keep their fishing capacities and costs at the minimum required for harvesting their quota (Christy, 1973). The conditions largely prevailing in agriculture, for example, where the individual users assume the responsibility and costs of management, could thus be recreated (Pearse, 1979). Unfortunately, the application of such a method, though very attractive as a concept, faces very serious enforcement difficulties. The first reason is that catches are appreciably more difficult to control than vessels. Evidence of systematic underreporting has been noted on various occasion for the CECAF (FAO, 1980a) as well as other fishing regions when catches become explicitly or implicitly controlled. This observation is even more relevant considering the present state of the statistical systems in several Northwest African countries. In addition, the use of catch quotas implies that fluctuations of stock biomass can be monitored in real time allowing for the regular adjustment of the quotas to maintain the desired exploitation rate. This requires research capabilities which few Northwest African countries possess. Its implementation would be particularly difficult for stocks where the biomass undergoes important natural fluctuations, either seasonal (shrimps, cephalopods, and other short-lived species) or annual (small pelagics). In addition, its application to multispecies fisheries faces well-known operational difficulties: either a series of catch quotas will be individually fixed for the major stocks but their harvesting during the same fishing operations will be operationally incompatible, or a single multispecies quota will be adopted, but it will be difficult to prevent an excessive and potentially detrimental concentration of the allowed total take on the most attractive species. These considerations are particularly relevant to the area for, in addition to the high number of species available—especially in its southern part, the limited administrative capabilities of many coastal countries will oblige them to manage their fisheries by large composite units.

Licensing systems setting the number and controlling the fishing power of vessels are likely to be more robust and, therefore, better adapted to Northwest African fisheries conditions (Doucet *et al.,* 1981: Troadec and Garcia, 1979). As compared to quotas, the same fishing capacity could keep fishing mortality more stable for a given amplitude of biomass fluctuations (Walter, 1976). This feature should permit reductions in fishery research and stock monitoring programs. However, fishery administrations should have the authority to maintain the fishing power of the fleets effectively and be able to overcome technical and political difficulties in implementation. The main difficulty will be the control over the number of boats, while also avoiding the creation of a monopoly situation for the licensed fishermen. Conventionally, recourse to market mechanisms for the allocation and transfer of licences should permit the desired equilibrium in fleet size to be dynamically maintained. Thus, potential newcomers may offer the less efficient fishermen the possibility to buy out their licences at prices attractive to both. However, the frequently observed low mobility of fishermen and fishermen communities which arises from the specialization of their work will sometimes make necessary to recognize preferential rights to existing fishermen. At the same time, this low mobility may preclude rapid—e.g., yearly —adjustments in fishing capacity and labor forces. For fisheries based on higher fluctuating stocks lower rates of exploitation may, for that reason, be desirable.

Gains in fishing efficiency would then have to be offset. These will result from:

1. The adoption of technological innovations and developments in the noncontrolled characteristics of the fishing units.
2. The better distribution of fishing operations relative to the densest and more accessible fish concentrations as the fishermen's knowledge of the stock distribution, migration, and behavior develops.

The capacity to counterbalance these seepage effects in fishing efficiency will depend on how the administration will succeed in capturing the economic rent and/or preventing its reinvestment in excessive fishing capacities. When, as is frequently the case, licensing schemes are introduced late, i.e., once the rent is totally dissipated, experience shows that it becomes psychologically and politically difficult to tax fishermen whose incomes have for a long time been marginal. The involvement of fishing communities and unions in the development of measures (e.g., buy-back schemes) to offset seepage effects will often be a condition to the successful regulation of the overall fishing capacity. The example of the Japanese offshore fisheries shows that the industry itself may take the responsibility of implementing such regulations (Kasahara, 1964).

Such difficulties should, in principle, be more easily overcome with foreign fleets as the coastal state has, in principle, a greater flexibility to adjust, year after year, the fishing effort at the level it desires and to extract the economic rent. In that respect it should be noted that, for fisheries which are to be temporarily exploited with foreign fleets, a priority objective will often be to maximize the net economic revenue. Under such circumstances, coastal countries may envisage the periodic auctioning of a limited number of fishing licences corresponding to a preset target of fishing effort. The greater flexibility offered by such a scheme, notably with respect to the adjustment in fleet size and composition, should appreciably facilitate the regulation of effort and the maximization of the rent. Comparatively higher rents could be extracted inasmuch as the scheme facilitates the selection of more efficient boats. Finally, the scheme should be less demanding in administrative and research capabilities (Doucet *et al.*, 1981).

However, its implementation is likely to also encounter a number of operational difficulties. As previously noted, countries and national fisheries administrations are seldom free to select their foreign partners on purely economic considerations. Bidding may not operate as expected owing to collusion among certain bidders, as well as competition from other countries likely to offer access to similar resources at better conditions. These may be the reasons why Northwest African countries have so far been applying mostly fixed licence fees (e.g., by gross registered tonnage) and have given scant attention to the opportunity to maximize net economic rent by adjusting the total number of licences back to a level close to that ultimately producing the maximum economic yield and simultaneously increasing the value of the individual fees.

Whatever scheme is finally adopted, its successful implementation will depend on the coastal states' capability in effectively undertaking control and surveillance measures. This is one of the most critical issues affecting Northwest African countries today (FAO, 1981b). Even if foreign poaching can be expected to decline progressively as costs of fishing increase as a result of the introduction of surveillance schemes, finely timed management schemes (e.g., implying that the location of fishing operations and the catch composition of different foreign and domestic fleets are progressively controlled) will be required in the long run to optimally utilize fish resources and to reduce conflicts among fleets. An appreciable strengthening of means and capabilities among the coastal countries will be needed for that purpose.

Conclusions

There are now relatively few, easy to develop, unexploited stocks. This situation has not only brought an end to the phase of extremely flourishing expansion which, in the course of the present century, has above all benefitted industrial fisheries; it has also, thanks to the change provoked in the legal regime of the sea, involved alterations in the rules of access to the resources, in the methods of exploitation, in international trade patterns, in development strategies, in management approaches, in the relative importance of administration and of fisheries research, and, in response to these alterations, of the balance of studies devoted to the biologic, economic, and social aspects of fisheries. The consequences of these changes are considerable.

International fisheries, whose particular requirements have dominated the science and practice of fisheries development and management over recent decades, will in the final analysis have played only a limited role in the history and geography of world fisheries. With the change of emphasis to national fisheries, the unique features and strengths of local fisheries—and notably the artisanal ones—will come to be more highly appreciated and valued. The applies particularly to the cultural experiences of traditional communities, especially those useful solutions, worked out in the course of centuries, to the difficult problems raised by the coherent utilization of mobile and variable natural resources.

In comparison with other agricultural sectors, with industry, education, health, etc., the importance of international fisheries has finally delayed the recognition of the limitations of western approaches for promoting the development of Third World countries and of the value of a nationally determined development. In this respect, the new legal regime should encourage the proper consideration of local factors and of the constraints which they impose upon a coherent improvement of the fisheries in each country. This applies particularly to the fisheries of the Third World where conditions differ–in varied but distinct ways–from those to be found in developed countries. The review of the past practices and performances of fishery development and technical assistance programs in Northwest Africa demonstrates how the pressure of international fisheries in this region has hindered the expansion of local fisheries. In an area where the stocks were sufficiently abundant to attract long-distance fleets, the effects upon yields of competition for the resources and of disparities in access to markets have heavily contributed to the maintenance of the margin of advantage of the most competitive countries. In these countries, as in many other developing countries, the heavy influence of the model for expanding industrial fisheries in advanced environments appears to have favored the adoption of uniform intervention schemes which place emphasis upon the straightforward transfer of foreign solutions rather than upon the analysis of local conditions and the formulation of development schemes which conform to such contexts.

Progress in this connection will be made first of all through the strengthening of quantitative analyses of unit fisheries, of the options and priorities available for their improvement. Such studies must serve as the basis for the formulation of strategies for long-term development. Without such studies it undoubtedly will be dificult to avoid the inevitable gaps created by a series of unintegrated, isolated actions, in particular the frequently observed tendency toward overinvestment and choice of ill-adapted solutions. For example, the utilization of by-catch from specialized shrimp fisheries depends much more upon the clarification of optimum regimes of exploitation in relation to the mix of simultaneously capturable shrimp and finfish resources than upon the simple reduction or utilization of the by-catch fish from a fishery conceived essentially for the export of shrimp alone.

By starting with an analysis of the local situation, it should also be possible to achieve a better mobilization of national concerns and means. Thus, in such areas as fisheries research including the collection of statistics, training, technical assistance, etc., one would expect that programs based upon an adequate analysis of prospects for the national fishery would generate better support of national administrations for actions whose relevance appears to them most evident. This was not always the case under the old regime when countries were often urged to adopt uniform systems–for example, for the collection of statistics–which did not necessarily reflect their participation in the fisheries or the level of their development and their research priorities.

Such a process should lead equally to a revision of development and technical assistance strategies. In particular, greater importance must be given to actions upon the environment of fisheries and to the external stimulation of their expansion rather than to the direct promotion of production, notably through the provision of public aids for the purchase of additional catching capacity or the substitution of the state for private initiatives to acquire new production facilities. Over the last century, the evolution of artisanal fisheries in technologically advanced countries, from a subsistence to a market economy, has generally occurred without direct intervention upon the means of production by national administrations, other than the extension of basic infrastructures (such as railways, harbors, preservation and storage systems, markets) to the coastal areas (Cadoret *et al.,* 1978). It would seem opportune in this respect to deepen and extend the study of this process as such a study is likely to provide models for the development of the fisheries of Northwest Africa and the Third World generally which would be more relevant than the model provided by the expansion of industrial fisheries in developed countries.

The role of public administrations has long been recognized in such activities as the financing of heavy investments (port infrastructures, transport systems, cold storage, etc), the promotion of export markets, and the encouragement of professional organizations. On the other hand, the actions they should take regarding fisheries management and assurances of long-term security and profitability to operators and potential investors are much less well recognized. Thus, control over the level of exploitation, by regulating the number and fishing power of authorized vessels, has been rarely used to promote development. Nevertheless, if one considers that expansion of self-productive—that is to say through the sole initiative of the operators—once and until a fishery produces benefits, it can be stated that national administrations have acquired, with their authority over the resources, a powerful instrument to stimulate development. One such possibility is confirmed by reference to fisheries to which this principle has been applied, e.g., the Japanese High Seas fisheries (Kasahara, 1964); or the West Australian shrimp fishery (Meany, 1978). These experiences have demonstrated that, by establishing very early in the expansion of these fisheries a licensing system to control the rate of exploitation, development as a result accelerated: private investors were encouraged by the reduction in long-term risks provided by regulation of the fishery.

In effect, development and management must not be considered to be two consecutive functions but to be complementary tasks affecting all stages of the expansion of a fishery. Schematically, the steps which should be taken to improve a fishery must include the following:

1. Control over the rate of fishing effort, including steps regarding the distribution of the activities of different fleets among the various available stocks, in such a way as to reduce conflicts between the different fisheries—notably those which concentrate their operations on the successive stages of the biologic cycle of the exploited stocks (gauntlet fisheries)—and to avoid the danger of overexploiting the most attractive stocks and of economically underusing stocks of lower density and market value.
2. Reduction of aggregate fishing costs through introducing improved technological methods which will simultaneously facilitate the development of stocks previously economically unattractive.
3. Increasing the value of the basic raw material through better handling and processing and, even more, by putting into operation new strategies for marketing, especially for export.

In the past, there has been a tendency to be concerned first with development and only afterwards with management; in other words, to think about management only once the stocks have become fully exploited. According to this concept, the countries of the region should now only be concerned with regulating the effort of foreign fleets, without worrying excessively about possible overinvestment in national fleets, hoping such investments are expedient and temporary. The experiences observed of the past evolution of the fisheries of the region do not justify such optimism. Despite abundant resources and substantial public aid, production from these fisheries has grown neither as quickly nor in the ways anticipated by the planners.

The taking into consideration of the fundamental concepts of management, especially at the time of formulating developmental strategies, clearly seems to be a precondition for fisheries expansion and for the success of action plans. Management is, as one has seen, a factor in encouraging private investment by offering the necessary assurances of profit and security. It is also a factor in improving the economic feasibility of national fisheries upon stocks of low market value (e.g., offshore fishing for small pelagic species), in assuring the success of projects designed to promote the social and economic conditions of artisanal fisheries, etc. Management can also be instrumental in helping to reduce conflicts between fleets and communities of national fishermen, when their catches affect different age strata of the stocks.

On the other hand, management has little chance of being acceptable if the schemes introduced do not take account of development needs. Thus, there will be a good reason to create alternative economic activities, to push forward general education programs, etc., in order to facilitate the transfer, under acceptable conditions, of the surplus manpower of an artisanal fishery upon which excessive population prohibits the achievement of all its potentials.

In all these tasks, the most important and the most difficult is certainly that of regulating the fishing effort and of controlling its distrubution over the major resources which will be desirable to manage separately. The new legal regime permits and necessitates the use of new methods capable of going beyond the simple conservation of the resource. The allocation or lease of parts of the resource, or of fishing rights in demarcated areas, to specific artisanal fishing communities, the control over fishing capacities by allocating licenses in the commercial fishing sector, all offer new perspectives compared with the old fixation of total allowable catches.

Through these various actions, there will be the benefits of being able to resort to schemes involving the fishermen and the fishing industry in that such schemes provide them with a direct interest in the success of management and in the regulations involved. The area of application of such measures—such as the allocation of parts of resources which are only slightly mobile or not mobile at all, or of individual catch quotas, or of fishing rights in defined sectors, to the smallest possible groups of fishermen—certainly has its limits. However, their implication is considerable. These methods all aim at breaking away from the difficulties inherent in the rational exploitation of open-access resources. In order to interest the fishermen in their management, it is vital that they know what resources are accessible to them and that there will be no intruders to be part of the longterm benefits which will arise after the constraints and losses they are prepared to accept immediately. The creation of exclusive economic zones is a first step in this respect. After the extension of national responsibility for fisheries, a better association of various fishermen groups with the development and management of the various domestic fisheries in which they are specifically engaged must follow.

References

Belvèze, H. and J. Bravo de Laguna. 1980. Les ressources halieutiques de l'-Atlantique centre-est. Deuxième partie. Les ressources de la côte ouest-africaine entre 24°N et le détroit de Gibraltar. FAO Doc. Tech. Pêches 186.2; pp. 64.

Cadoret, B., D. Duviard, J. Guillet, and H. Kerisit. 1978. Ar vag, voiles au travail en Bretagne atlantique. Tome 1. Grenoble, France, Editions des 4 Seigneurs, 369 pp.

Christy, F. T. Jr., 1973. Fishermen's quota: a tentative suggestion for domestic management. Occas. Pap. Law Sea Inst. Univ. R.I., 19, p. 7.

Christy, R. T. Jr., 1979. Economic benefits and arrangements with foreign fishing countries in the northern sub-region of CECAF: A preliminary assessment. CECAF/ECAF Ser. 79/19, 33 pp. (Issued also in French.)

Christy, R. T. Jr., and A. Scott. 1965. The Common Wealth in Ocean Fisheries. Some Problems of Growth and Economic Allocation. Johns Hopkins Press, Baltimore, 219 pp.

Doucet, F. J., P. H. Pearse, and J.-P. Troadec. 1981. Mauritanie. Politique de développement et d'âménagement des pèches dans la zone économique exclusive. Rapport rédigé pour le gouvernement de la République islamique de Mauritanie. Rome, FAO, FI:TCP/MAU/001, p. 135.

FAO. 1971. Fishery Committee for the Eastern Central Atlantic (CECAF). Report of the Second session of the Fishery Committee for the Eastern Central Atlantic (CECAF). Casablanca, Morocco, 13-19 May 1971. FAO Fish. Rep. 197, 25 pp. (Issued also in French.)

FAO. 1979. Report of the *ad hoc* working group on fishery planning. Dakar, Senegal, Fishery Committee for the Eastern Central Atlantic, CECAF/TECH 79/14, p. 63.

FAO. 1980a. Report of the Sixth session of the Fishery Committee for the Eastern Central Atlantic (CECAF), Agadir, Morocco, 11-14 December 1979. FAO Fish. Rep. 229, p. 62. (Issued also in French.)

FAO. 1980b. Marine fisheries in the new era of national jurisdiction. *In:* State of Food and Agriculture, 1980. FAO, Rome, pp. 83-181.

FAO. 1981a. Fishery Committee for the Eastern Central Atlantic/Comité des pêches pour l'Atlantique centre-est. CECAF statistical bulletin No. 3: nominal catches, 1969-1979/Bulletin statistique du Copace No. 3: captures nominales, 1969-1979. CECAF Stat. Bull./Bull. Stat. COPACE p. 211.

FAO. 1981b. Fishery Committee for the Eastern Central Atlantic, Report of the Consultation on Monitoring, Control and Surveillance. Freetown, Sierra Leone, 30 June to 3 July 1981. Dakar, Senegal, Fishery Committee for the Eastern Central Atlantic, CECAF/TECH 81/35, p. 30 (E).

FAO. 1981c. Yearbook of fishery statistics: catches and landings. Annuaire statistique des pêches: captures et quantitiés. Annuario estadistico de pesca: capturas y desembarques. FAO, Rome, Vol. 50, p. 396.

Fishery Research Unit, Tema/ORSTOM. 1976. Rapport du Groupe de travail sur la sardinelle, *S. aurita*, des côtes ivoiro-ghanéennes. Paris, ORSTOM, p. 40.

Fréon, P., and J. Weber. Djifère au Sénégal: la pêche artisanale en mutation dans un contexte industriel. lére partie: Le milieu, la senne tournante, son impact naturel (in preparation).

Grasset, G. 1972. Essais-démonstrations comparatifs d'emploi d'une senne tournante et coulissante adaptée à la pêche piroguière. Rapp. PNUD (FS)/FAO Proj. SEN/66/508 "Etude et mise en valeur des ressources en poissons pélagiques", Dakar, 4/72, p. 22.

Gréboval, D. Analyse bio-économique de l'aménagement des principales pêcheries démersales de la région nord COPACE (in preparation).

Griffin, W. L., and W. E. Grant. 1981. A bioeconomic analysis of the Ivory Coast shrimp fishery. Paper presented at the Workshop on the scientific basis for the management of penaeid shrimp. Key West, Florida, 18-24 November 1981. Sponsored by Southeast Fisheries Center, U.S. NMFS, the Gulf States Marine Fisheries Commission, in collaboration with FAO, 48 pp. (mimeo).

Griffin, W. L., J. P. Warren, and W. E. Grant. 1979. A bioeconomic model for fish stock management; the cephalopod fishery of Northwest Africa. Dakar, Senegal, Fishery Committee for the Eastern Central Atlantic, CECAF/TECH 79/16, p. 42. (Issued also in French.)

Gulland, J. A. (Ed.). 1971. The fish resources of the ocean. West Byfleet, Surrey, Fishing News (Books) Ltd., 255 pp. Rev. ed. of FAO Fish. Tech. Pap. 97, p. 425. (1970).

Gulland J. A., J.-P. Troadec, and E. O. Bayagbona. 1973. Management and development of fisheries in the Eastern Central Atlantic. J. Fish. Res. Board Can. 30(12) 2:2264-2275.

Indo-Pacific Fishery Commission (IPFC). 1980. Summary report of the Symposium on development and management of small-scale fisheries. Kyoto, Japan, 21-23 May 1980. Proc. IPFC 19(1+2):72-96.

Jarrold, R. M., and G. V. Everett. 1981. Some observations on formulation of alternative strategies for development of marine fisheries. Dakar, Fishery Committee for the Estern Central Atlantic, CECAF/TECH 81/38, p. 56.

Kasahara, H. 1964. Japanese fisheries and fisheries regulations. *In:* California and the World Ocean. Museum of Science and Industry, Los Angeles, California, pp. 57-61.

Larkin, P. A. 1977. An epitaph for the concept of maximum sustainable yield. Trans. Am. Fish Soc. 196(1):1-11.

Meany, T. F. 1978. Restricted entry in Australian fisheries. *In*: Limited Entry as a Fishery Management Tool, edited by R. B. Rettig and J. J. C. Ginter. Proceedings of a National Conference to consider limited entry as a tool in the fishery management. Denver, 17-19 July 1978. Seattle, University of Washington Press, Washington Sea Grant Publication, pp. 391-415.

Moiseev, P. A. 1976. The living resources of the World Ocean. Jerusalem, Israel Programme for Scientific Translation, Jerusalem, IPST Cat. No. 5954, 334 pp.

Panayoutou, T. The development and management of small-scale fisheries (in preparation).

Pearse, P. H. 1979. Property rights and the regulation of commercial fisheries. *Journal of Business Administration* 11(1+2), 185-209.

Sissenwine, M. P. 1978. Is MSY an adequate function for optimum yeild? *Fisheries* 3(6), 22-4, pp. 27-42.

Smith, I. R. 1979. A research framework for traditional fisheries. *ICLARM Stud. Rev.* 2, p. 40.

Troadec, J.-P. 1976. Développement de la pêche dans le monde; répartition de l'industrie des pêches; nature et répartition des ressources halieutiques; potentiel des ressources halieutiques mondiales; perspectives dans l'exploitation des ressources vivantes de l'océan. *In* Océanographie Biologique Appliquée-l'Exploitation de la Vie Marine, par P. Bougis *et al.* Paris, Masson.

Troadec, J.-P., W. G. Clark, and J. A. Gulland. 1980. A review of some pelagic fish stocks in other areas. *Rapp. P.-V. Réun. CIEM* 177, 252-77

Troadec, J.-P., and S. Garcia (Eds.). 1979. The fish resources of the Eastern Central Atlantic. Part 1. The resources of the Gulf of Guinea from Angola to Mauritania. *FAO Fish. Tech. Pap.* 186.1, p. 166. (issued also in French).

Unar, M. and N. Naamin. 1981. A review of the Indonesian shrimp fisheries and its management. Paper presented at the Workshop on the scientific basis for the management of penaeid shrimp. Key West, Florida, 18-24 November 1981. Sponsored by Southeast Fisheries Center, U.S. NMFS, the Gulf State Marine Fisheries Commission, in collaboration with FAO, 25 p. (mimeo).

Walter, G. G. 1976. Non-equilibrium regulation of fisheries. *Sel. Pap. ICNAF* 1, 129-40.

Weber, J. and P. Fréon, Djifère au Sénégal: la pêche artisanale en mutation dans un contexte industriel. 2ème partie. Conséquences d'une innovation technique sur l'organisation économique artisanâle, ou les excès d'une réussite (in preparation).

6
International Tuna Management Revisited

James Joseph

Introduction

Ten years ago international arrangements for the scientific study and management of tuna were reviewed in a predecessor to this volume (Joseph, 1972). The main thrust of that contribution was to stress the necessity for international cooperation on a global basis to insure rational future utilization of tuna resources. The need for international cooperation was emphasized because of the highly migratory nature of the numerous tuna and tuna-like species, the mobility of the fleets that harvest these species, and the international character of the tuna processing and marketing system.

During the decade since the original article was written nothing has taken place to alter the conclusion that international cooperation is necessary for tuna conservation and rational management; tuna are still migratory; fleets continue to be highly mobile; tuna processing and marketing has become even more internationalized. However, a number of important things have taken place with profound impacts upon the chances for realizing the goal of effective international management. First, and most important, has been the general move by most

---

James Joseph has engaged in tuna research since 1958, and for the last ten years has served as Director of Investigations for the IATTC, an international fishery commission with headquarters in LaJolla, California. Concurrent with that position, he holds appointments as Affiliate Professor of the University of Washington and as Research Associate at the Scripps Institution of Oceanography. He is the author of many publications on the biology and management of tuna and billfish resources and has recently coauthored a book entitled "International Management of Tuna, Porpoise, and Billfish—Biological, Legal, and Political Aspects."

coastal states to extend their jurisdiction over fisheries resources to 200 nautical miles off their coast. Second, the nations of the world under the auspices of the United Nations have been struggling to codify an international Law of the Sea (Third United Nations Conference on the Law of the Sea-UNCLOS III), and highly migratory species, including tunas, have been given special consideration in these negotiations. Third, regional negotiations to formulate institutional arrangements for tuna management and conservation that are more in concert with recent trends in the Law of Sea have been taking place, particularly in the eastern Pacific Ocean. Fourth, differing national policies have resulted in serious conflicts regarding tuna management.

It is the purpose of this chapter to reevaluate the requirements for rational tuna management in light of both the behavior of the animals themselves and the nature of the tuna industry. Then particular emphasis is focused on the current political issues and their impact on present and future tuna management. Finally, alternative formats for resolution of these issues are discussed.

Biological Factors

The tuna and tuna-like fishes, including their close relations the billfishes, have evolved to form the suborder Scombroidei, a group of uniquely adapted species whose migratory mode of existence sets them apart from most other fishes. They are nearly all characterized by rapid growth and most species attain a large size. The bluefin tunas and the black marlin are among the world's largest fishes. Other species, such as yellowfin and bigeye, reach sizes in excess of 200 pounds within 4 or 5 years. They spawn large numbers of eggs over vast areas of the ocean. The eggs drift after fertilization in the upper strata of the ocean where they are subject to the vagaries of ocean currents and are preyed upon by a multitude of animals. The eggs hatch quickly and the small fish grow rapidly, maintaining the populations of these species at high levels of abundance. It is this rapid growth that results in large biomasses of fish that support important fisheries worldwide. Because of their high fecundity, rapid growth, and worldwide distribution, it would be virtually impossible to overfish the tunas to a point that would threaten biological extinction. However it is possible to fish them to such a low population level that production is reduced substantially, resulting in economic chaos in world fisheries for tuna.

The anatomical and physiological adaptations which tunas and billfish have undergone have resulted in a group of fishes that are capable of leading a nomadic existence, throughout the pelagic zone of the tropical and temperate region of the world oceans. They are characterized by a fusiform shape which is adapted hydrodynamically for sustained high speed. All tuna have a highly developed circulatory system that can transfer energy and oxygen to all parts of the body very efficiently and very quickly. They have evolved anatomical "heat exchangers" which permit them to release or conserve heat which in turn allows them to adjust their internal temperature to maintain maximum metabolic efficiency.

This capability makes these fish unique among "cold-blooded" species in that they are able to maintain their internal body temperatures as much as 15°C above the ambient water temperature. Tunas and billfishes are among the fastest swimming species in the ocean and, in fact, cannot stop swimming, for because of their high density they would sink if they did. They would also suffocate if they stopped swimming because they must move large volumes of water over their gills to meet their high oxygen demands. Their minimum swimming speed moves them through the ocean at more than one body length a second.

Their unique characteristics have resulted in tuna and billfish being distributed worldwide, often undertaking transoceanic migrations in a matter of months. For example, North Pacific bluefin spawn in a very restricted region in the western Pacific east of Formosa. Their young drift north with the Kuroshio current where they are first captured by Japanese fishermen off Kyshu and Honshu when about a foot long. During their life they may transit the Pacific ocean as many as three of four times being exposed to fisheries off Mexico, Canada, the United States, Japan, and Taiwan (Bayliff, 1980). The southern bluefin as well as the albacore are equally nomadic. Tagged skipjack tunas migrate extensively throughout the Pacific Ocean (Forsbergh, 1980; Kearney, 1981). In fact, in the eastern Pacific, where a major skipjack fishery is carried out, skipjack tuna do not regularly spawn–their migrations carry them from the central Pacific into the eastern Pacific on feeding forays where they stay for only a few months of their life. Yellowfin tuna are less migratory but still undergo extensive movements. Tagging studies in the eastern Pacific show them to travel up to about 3000 miles from where they have been released, although on the average most tagged yellowfin have been recovered within 600 miles of where they have been released.

The migrations of tunas and billfishes carry them through the national zones of jurisdiction of many countries during their lifetime, as well as to high seas waters beyond the jurisdiction of any single nation. What happens to them on the high seas, or in any nation's zone, affects what will become of them in other areas. It is this characteristic that sets tunas and billfish apart from other fishes, such as the snapper, herrings, anchovies, and flatfishes, and which requires that they be given special treatment to insure their proper conservation and management. Subsequently I will discuss the approaches that have been taken in UNCLOS III in formulating guidelines for rational tuna management as well as specific approaches that certain groups of nations have taken.

The Tuna Industry Compared–1970 and 1980

There are three principal ways in which tuna are caught–longlining, baitfishing, and purse seining. These fishing methods have been discussed extensively elsewhere (Joseph and Greenough, 1979) and are not reviewed again here. Other methods of fishing for tuna have been employed, including traps, gillnetting, handlining, and harpooning, but they account for only a very small share of the catch and in the main are mostly subsistance fisheries.

Baitfishing has long accounted for the major share of the world catch of tunas. Prior to 1970 more than one-half the total world catch was taken by this method, and it still accounts for the major share of the catch. It has been kept in the number one spot by the Japanese development of the skipjack fishery in the western Pacific, where it now is the single largest tuna fishery in the world. The Japanese accomplished this by being able to carry live bait from the cold waters of Japan into the warm tropical waters of the Pacific by refrigerating the circulating seawater in the bait wells.

Prior to 1970 the second most important tuna fishing method was longlining. However, this form of fishing has subsided in importance because of the high costs of catching fish in this way. Fuel costs have increased steadily, crew wages have gone up dramatically, and longlining is labor intensive compared to the other major forms of tuna fishing. Japan first developed the longline fishery and dominated it up until the mid-1970s, when those consequent increasing labor costs resulted in the transfer of investment capital and ownership to South Korea and the Republic of China where labor was cheaper. In recent years the Japanese longline industry has directed the major share of its effort to supplying the high priced sashimi (raw fish) market.

Purse seine fishing, which was in last place among the major methods of tuna fishing to 1970, has now replaced longlining in second place. Purse seine fleets have expanded throughout the Pacific and Atlantic Oceans. So far there has been little development in the Indian Ocean. Historically this form of fishing was confined primarily to the eastern Atlantic and eastern Pacific Oceans where boundary curents and shallow upper mixed layers made purse seines more efficient. However, in the late 1970s fishermen began to experiment with deeper, faster sinking nets and more powerful winches which made purse seines more effective in the western Pacific. This has contributed to a major increase in construction of purse seine vessels. Another important factor in the expansion of purse seining has been the development of "raft fishing." In this technique artificial rafts are deployed in tuna areas to attract tuna schools and make them more vulnerable and available to the fishing gear. There has been considerable success in this form of fishing. Purse seining will continue to grow in importance in the future and may replace bait fishing as the most important method of catching tuna.

Since 1970 the tuna fishing fleets of the world have continued to grow. As an index to this growth we can look at the eastern Pacific Ocean. At the end of 1960 the international fleet operating in the eastern Pacific comprised about 40,000 short tons of carrying capacity. By the early 1970s there had been nearly a fourfold increase to about 150,000 tons. During the 1970s growth continued and at the present time the international fleet comprises 190,000 tons. Presently and additional 75,000 thousand tons of vessel carrying capacity are being constructed, much of which will fish in the eastern Pacific. Fleet growth is also being experienced in other areas of the world. In 1960 world tuna fleets were estimated to comprise 350,000 capacity tons and by 1970 about 700,000 capacity tons (Joseph and Greenough, 1979). At present, the world tuna fleet is estimated to have about 1,000,000 short tons of carrying capacity.

World catches of tuna and tuna-like species are generally divided into three categories. The principal market species in 1979 in order of importance in terms of landed weight are the skipjack, yellowfin, albacore, bigeye, northern bluefin, and southern bluefin (Table 6.1). They are the most sought after and valuable species of tuna and are traded internationally throughout the world. In terms of tonnage the principal market species account for just over 70% of the world catch of tuna and tuna-like fishes.

The secondary market species include the smaller tunas and tuna-like fishes. There are more than 40 different species in this group. They are generally much less valuable, are not canned to a large degree, and in most cases do not enter into international trade. This category makes up over 25% of the world catch.

Billfishes make up the third category. Most of the commercial catch of billfish is taken by longlining. Billfish are generally sold, particularly in the case of marlin, at a premium price to the sashimi market. The importance of billfish to the sportfishing industry should be emphasized as throughout the world there are many resort centers that have been built up on the basis of big-game fish stocks. These are mostly billfish but in some areas include tunas–particularly northern bluefin tuna in the western Atlantic Ocean.

Throughout the remainder of this chapter in the discussions of catch and marketing trends, etc., I will be referring to the principal market species unless otherwise indicated.

Prior to 1940 the world catch of tuna never exceeded 250,000 tons but after World War II catches began to increase, and by 1970 they reached over 1 million metric tons. This was an annual growth of about 4.6% per year. From 1970 to 1979 this steady increase in production continued, average annual growth being 5.5% per year. At present the world catch of tuna is 1.7 million metric tons.

Table 6.1 shows clearly that skipjack is the most important species, with catches making up 40% of the world total in 1979. This is quite different than the situation during the 1960s when it was second in importance to yellowfin. The primary reason for this change has been the rapid development of the skipjack fishery in the western Pacific. Since 1970 the catch of the other species, with the exception of albacore and bluefin, have also increased.

The number of nations fishing tuna on a commercial basis has increased since 1970 from about 40 to about 60. Japan and the United States are still the dominant tuna catching countries, accounting for about 52% of the catch. This percentage has remained relatively constant from 1970 through 1979. South Korea has replaced Taiwan as the number three producer of tunas. After Taiwan, France and Spain have retained their importance; however, the Philippines and Indonesia have rapidly developed their tuna fisheries to establish their places on the list of leading tuna producing nations. These eight nations catch over 80% of the tunas of the world. However, significant strides are being made by Mexico, Papua New Guinea, Fiji, the Solomons, and other countries of the Pacific basin. The steadily increasing value of tuna in the international market has made it a more sought after group of fishes. Using the FOB price at San Diego as an index, the price between 1970 and 1980 increased by about a factor of three. Although

Table 6.1. The 1979 World Catch in Metric Tons of Principal Tuna Species[a]

Area	Skipjack	Yellowfin	Albacore	Bigeye	Northern Bluefin	Southern Bluefin	Total Six Species
21	292	787	1,106	2,116	1,901	–	
27	2,644	905	38,315	640	1,970	–	
31	2,080	15,380	7,194	1,112	2,601	–	
34	77,532	113,680	4,285	20,780	1,985	–	
37	11	–	191	–	4,873	–	
41	1,439	3,451	8,608	2,358	–	10	
47	3,716	9,956	14,626	6,054	–	7,619	
Atlantic Ocean	87,714	144,159	74,325	33,060	13,330	7,629	360,217
51	27,022	44,034	5,335	14,071	–	2,685	
57	5,640	16,615	6,898	17,561	–	15,457	
Indian Ocean	32,662	60,649	12,233	31,632	–	18,142	155,318
61	143,045	38,962	62,009	12,531	14,325	–	
67	468	1,648	2,734	–	88	–	
71	316,898	153,473	10,239	26,590	218	–	
77	107,711	171,550	17,653	69,256	5,477	–	
81	8,465	7,022	20,468	4,148	–	7,109	
87	1,722	6,567	1,593	9,159	–	–	
Pacific Ocean	578,309	379,222	114,696	121,684	20,108	7,109	1,221,128
TOTAL	698,685	584,030	201,254	186,376	33,438	32,880	1,736,663
	40.2%	33.6%	11.6%	10.7%	2.0%	1.9%	

[a]Source: *FAO Yearbook of Fishery Statistics.*

this is a significant increase it is not comparable to the increases in the costs of vessel construction or diesel fuel on the international market.

Because it is such a high-priced commodity, tuna is mostly consumed in the more affluent nations. The United States and Japan still consume the most tuna, utilizing together nearly 70% of the world production. The United States consumes about 40% and Japan most likely more than 25%. This represents only a slight decrease in the proportion of world production consumed by these two nations since 1970. Western Europe accounts for about 17% of world consumption. A slight increase in tuna consumption by other nations is taking place, mostly in Latin America and the middle east where more people seem to be consuming tuna.

Along with these slight changes in the distribution of the catch and consumption of tuna there have been changes in the international tuna industry itself. In 1970 the majority of tuna boats were owned by individuals. Since then, a steadily increasing number have transferred to ownership by the tuna processors, who seem to be vertically integrating the high costs of vessel operation with processing and marketing activities.

The trend toward extended jurisdiction has made it increasingly difficult for foreign flag vessels to obtain access to coastal zones. In fact, of the present world catch of tuna, roughly about 60% has been captured within economic zones that commonly extend to 200 nautical miles. This is significantly more than 10 years ago when the commonly accepted width for economic zones with jurisdiction over fisheries was from 3 to 12 nautical miles. At that time nearly all of the catch of tunas was taken on the high seas beyond the juridical zones of any nation.

Because access to many major tuna fishing grounds has become uncertain and difficult to obtain, and because the infrastructure, investment capital, and experience needed to develop a domestic tuna industry is lacking in many resource adjacent nations (RANs), there has been an increasing tendency on the part of many industrialists to enter into joint ventures with these RANs. So far, the results of these joint ventures has been extremely variable, with some successes and some failures. Judging from current events, the future trend will be toward an increasing number of joint ventures.

Tuna Management Prior to 1970

From the Seventeenth Century through the middle of the Twentieth Century, common international practice recognized coastal states' sovereignty over the seas as extending along a narrow coastal strip, generally from 3 to 13 nautical miles wide. The commercial exploitation of tuna developed under this policy of a coastal state's jurisdiction over living marine resources applying only with a narrow zone. Tuna fishing vessels were free to roam the oceans of the world in quest of their prey. During the post-World War II years to 1970 more the 90% of the catch was taken in what was termed international waters. The tunas that inhabited these international waters were considered to belong to whomever first rendered them to his own use. No one owned them, nor did any state have jurisdiction over them while they were beyond the 3 to 12 nautical mile coastal zone of national jurisdiction.

It was under this international "regime" that the great tuna fishing nations developed their far-ranging fisheries. American fishermen operating out of southern California developed fisheries that expanded southerly to off Peru and westward to about 120°W by the mid-1960s. During this same period the French and Spanish developed surface tuna fishing off West Africa. French tuna vessels operating from the west coast of France fished as far south as 20°S latitude off West Africa and seaward more than 300 miles offshore. Spanish boats operated in nearly the same areas as did the French. In the western Pacific the Japanese began development of the largest single tuna fishery in the world by the mid-1960s they were fishing skipjack tuna west of the international date-line and south to about 10°N latitude. When the McArthur Line (set by the United States military occupation forces in Japan after World War II) was lifted in 1952, the Japanese longline tuna fleet expanded throughout all of the tropical and temperate oceans of the world within a matter of 10 years. By 1979 these four nations were catching more than 60% of the world catch of tuna. Nearly all of this catch was taken on the High Seas beyond the jurisdiction of coastal states and most was captured at great distances from home port.

During this postwar period of rapid expansion, fishing vessels were basically free to roam the oceans of the world in search of tuna. They generally were not required to request permission from RANs to fish in waters off their shores, nor were permits or licenses a necessity. An exception was the fishery of baitfish, which generally were taken within a mile of shore. To fish for baitfish coastal state permission was required, and licenses were usually purchased because the baitfish clearly inhabited territorial waters.

During this period of rapid expansion in tuna fisheries all over the world, a great deal of concern was generated over the potential production that might be derived from the sea. There was also a growing concern on the part of both nations and industries over the continued viability of the tuna stocks. Out of this concern international organizations for tuna research and management were created. These organizations were delegated various degrees of responsibility for the scientific study and management of tuna.

The first of these organizations to have sole responsibility for tuna was the Inter-American Tropical Tuna Commission (IATTC). This organization was given specific responsibility for the study and management of tuna and tuna-like species, baitfishes, and, quite recently, marine mammals taken in association with tuna fishing activities. A second and very similar organization was formed among countries having an interest in the tuna of the Atlantic Ocean. This organization was the International Commission for the Conservation of Atlantic Tunas (ICCAT).

Two other organizations, the Indian Ocean Fisheries Commission and the Indo-Pacific Fisheries Council, have broad responsibilities for all types of fisheries in their respective regions, including responsibility for conducting some aspects of tuna research, but they do not have direct regulatory authority. These two organizations fall within the framework of the Food and Agriculture Organization (FAO) of the United Nations.

The objectives of all of these organizations have to do with the conduct of research on the tuna stocks of the world and on the fisheries that exploit them. The results of this research are used in assessing the effect of exploitation on the condition of the stocks being exploited. Only two of the organizations, IATTC, and ICCAT, have actually recommended and implemented management measures for the species within their areas of responsibility.

In the eastern Pacific the IATTC has instituted a management program for yellowfin tuna. Based on scientific studies conducted by the Comission's staff, a quota on the total catch of yellowfin tuna was recommended. This was done so as to prevent further overfishing of the yellowfin stock, the catch of which had been declining over a series of years in the face of increasing fishing effort. The conservation quota was placed into effect for the first time in 1966 and was continued each year through the 1970s. In its initial stages the conservation quota was established as an overall quota on the catch of yellowfin, with the flag vessels of all nations fishing in the eastern Pacific being subjected to the same controls. The season opened to unrestricted fishing in the Commission's Yellowfin Regulatory Area (CYRA) on 1 January of each year. It was closed at a time during the year when the Commission staff determined that the yellowfin catch taken to date, plus what would be taken by vessels that put to sea before the closure date (they were allowed to fish for yellowfin in the CYRA unrestricted until the completion of this last open trip), plus the incidental yellowfin catch taken by vessels fishing for skipjack after closure (but not to exceed 15% of the total landed) equal the recommended quota. As time progressed an increasingly larger share of the overall quota was set aside for vessels suffering undue economic hardship as well as for certain RANs. By 1979 almost one-half of the over all quota was set aside for special country allocations.

For the Atlantic Ocean and the Mediterranean Sea ICCAT recommended regulations on the harvest of bluefin tuna in 1974. A minimum size limit of 6.4 kg (14 pounds) was set. In addition, all nations fishing bluefin for either sport of commercial purposes were requested to limit their exploitation of the bluefin tuna stocks to recent levels. In 1972 ICCAT recommended a minimum size limit on yellowfin tuna of 3.2 kg (7 pounds). This was followed in 1979 by a recommendation for the same minimum size limit on bigeye tuna. The latter limit was set because of confusion in differentiating small yellowfin from small bigeye.

No other international conservation regulations have been in effect for tuna. It is noteworthy, however, to mention Japanese regulations on southern bluefin tunas. Size limits and closed areas were instituted by the Japanese longline industry and apply only to Japanese vessels fishing for bluefin tuna in the southern hemisphere.

Extension of Jurisdiction

The jurisdiction of coastal states over the oceans of the world has ranged between very broad and very narrow zones. In the first case Pope Alexander VI in the

Fifteenth Century divided the oceans of the world between two powerful maritime nations, Spain and Portugal. Their ownership of the oceans went unchallenged until near the end of the Sixteenth Century, at which time some of the northern European nations, notably the Dutch, had become powerful maritime nations. These nations, under the leadership of the Dutch statesman Hugo Grotius, expressed a position opposite to that of Pope Alexander VI–the concept that a coastal state's jurisdiction over the use of the sea for transportation and resource extraction should not extend beyond a narrow coastal zone. This was the basis for the establishment of the 3-nautical mile limit, which was the "common practice" from the time of Grotius until the end of World War II.

Shortly after World War II interest in living and nonliving resource extraction increased markedly–particularly in the case of oil and fish stocks (including whales). Although seaward extension of jurisdiction seemed to be encouraged by a proclamation by U.S. President Harry S. Truman in 1945 dealing with resources of the continental shelf, the forefront of the battle was led by Chile, Ecuador, and Peru (the CEP countries). These CEP countries, in the Declaration of Santiago in 1952, established territorial limits out to 200 nautical miles offshore. This extension included a substantial portion of the most important tuna fishing grounds in the eastern Pacific Ocean and set the stage for the seizure of unlicensed tuna vessels fishing inside the established limits. Other nations, including several in Latin America, began following the lead of the CEP nations, although they generally established 200-mile extended economic zones (EEZs) rather than 200-mile territorial seas.

This situation resulted in the United Nations convening two world conferences which attempted to codify a series of laws pertaining to conduct of nations with respect to the sea. The first United Nations Conference of the Law of the Sea (UNCLOS I) was convened in 1958, the second (UNCLOS II) in 1960. Although the conferences succeeded in establishing numerous prinicpals upon which nations could agree, they were unable to agree on the breadth of the territorial sea (UNCLOS II failed by one vote to agree upon a 6-mile territorial sea and a 6-mile contiguous zone in which the coastal state had jurisdiction over resources).

Following the failure of UNCLOS II to reach agreement on the breadth of the territorial sea and contiguous zone, nations began to rapidly extend their jurisdiction to 200-miles–some establishing 3-mile territorial seas with 197-mile EEZs in which they exercised jurisdiction over living resources, and others establishing 12-mile territorial seas with EEZs of 188 miles.

By 1970 there were approximately 11 nations claiming in one form or another jurisdiction over fisheries resources to 200 nautical miles, whereas the comparable number in 1960 was only five nations. With the increasing number of nations extending their jurisdiction seaward, the rapidly increasing incidence of seizures of foreign flag vessels in these extended juridical zones, and the popular proposal of Ambassador Pardo of Malta that the resources of the sea and seabed are the "common heritage of mankind" a movement among the majority of the major coastal states of the world to convene a third United Nations Conference in the Law of the Sea gained popular support. UNCLOS III was convened in 1972 and has been underway ever since. Though no final agreement has been reached there is virtual agreement on the establishment of a 200-mile EEZ. By 1980, 91 nations had claimed some form of 200-mile extended zone. This represents an eightfold increase during the last decade. Among those nations claiming EEZs are the traditional opponents of extended jurisdiction, notably the United States, the USSR, Japan, France, Spain, and the United Kingdom. Although UNCLOS III has failed to reach final agreement on a Law of the Sea there is, for all practical purposes, acceptance of a 200-mile extended zone. However, the types of claims differ and fall into three major categories.

The most common claim is for a narrow territorial sea of 3-12 nautical miles and an extended economic zone to 200 nautical miles (the "200 including tuna" nations). Most of the nations of the world make such claims. Their claims include jurisdiction over all living resources within the 200 nautical mile zone including highly migratory as well as less migratory and nonmigratory species. In the context of tuna, most of the RANs fall within this category (notably Mexico, Panama, and Costa Rica within the eastern Pacific).

The next most popular claim is that of the territorialist governments, who claim a 200-mile territorial sea, that is, an extension of their territorial limits to 200 nautical miles seaward. Most notable among these are the CEP countries (Chile, Ecuador, and Peru) who led the movement to extended jurisdiction. There are 14 nations claiming 200-mile territorial seas.

A third category of nations are those claiming 200-mile EEZs, but which exclude the highly migratory species from falling within the jurisdiction of these extended claims (the "200 excluding tuna" nations). Three major tuna fishing nations, the United States, Japan and the Republic of Korea, who maintain large fleets of distant water tuna vessels, exclude the migratory species from their jurisdiction. Their rationale for this is that tunas and related species are highly migratory and cannot be managed effectively over only a portion of their range.

UNCLOS III and Highly Migratory Species

Recognizing the unique characteristics of tunas and other highly migratory species, UNCLOS III drafted an article dealing specifically with them. Article 64 Part V of the Draft Convention on the Law of the Sea (Informal Text) Resumed Ninth Session, Geneva, 28 July to 29 August 1980 addresses the question of management of highly migratory species and mandates international cooperation:

64-1. The coastal State and other States whose nationals fish in the region for the highly migratory species listed in annex I, shall cooperate directly or through appropriate international organizations with a view to ensuring conservation and promoting the objective of optimum utilization of such species throughout the region, both within and beyond the exclusive economic zone. In regions where no appropriate international organization exists, the coastal State and other

States whose nationals harvest these species in the region shall cooperate to establish such an organization and participate in its work.

Obviously the drafters of this article recognize the unique characteristics of the highly migratory species and they call for nations to manage these resources on an international basis. Up to this point Article 64 is quite clear. However, in paragraph 2 it goes on to state:

64-2. The Provisions of paragraph 1 apply in addition to the other provisions of this Chapter.

Because of this language, the exact interpretation of Article 64 is in dispute. The "territorialist" and "200 including tuna" states interpret Article 64 to call for international cooperation but to vest in the coastal state management authority. Based on paragraph 64-2, they consider that Articles 61 and 62 apply to all species including tuna. In this case Article 61 would give authority for conservation of tuna in the EEZ to the coastal state as well as authority for determining the allowable catch. Article 62 would provide authority for the coastal state to oversee harvesting of migratory species while they are in the EEZ. Specifically the coastal state would determine its own harvesting capacity, license other states to harvest any surplus, and set regulations relating to licensing and harvesting.

The "200 excluding tuna" states, on the other hand, do not interpret paragraph 64-2 as mandating the application of Articles 61 and 62 to the highly migratory species defined in Article 64. They argue that the migratory species are recognized in the text as presenting unique problems requiring unique solutions for the proper conservation and management, and they maintain that Articles 61 and 62 do not apply to these species. These differing viewpoints are of course the essence of the problems between RANs and non-RANs concerning tunas.

These problems are exemplified in the eastern Pacific where confrontations between countries supporting the two viewpoints are commonplace. Taking for examples Mexico and the United States, the problems over tuna have reached such proportions that they strain the relations between two neighboring states. The United States law establishing their 200 EEZ, the Fisheries Conservation and Management Act of 1976 (FCMA), does not include the highly migratory species as falling within its management authority. The FCMA specifically mandates that the highly migratory species, defined as tuna, must be managed internationally. (Curiously the FCMA definition of highly migratory species does not include billfish, many of which are as migratory as some species of tuna.) The United States does not recognize the management authority of any state over tuna, and the FCMA specifically requires that an embargo be placed on the tuna products of any nations seizing United States flag vessels fishing in their national zones between 12 and 200 miles. In contrast, Mexico does not differentiate in its law between tuna and any other species and requires any nation wishing to fish tuna in Mexican waters to first obtain permission. As a result a number of United States flag tuna vessels have been seized and their owners fined by Mexico for violation of its 200-mile EEZ. In return, as they must do by their own law, the United States has embargoed Mexican tuna. For similar reasons embargoes also have been placed into effect against Ecuador, Peru, and Costa Rica.

Unless nations can reach some form of agreement on how tuna can be managed, hope seems dim for agreement on the *modus operandi* and subsequent codification of laws to govern the highly migratory species. Article 64, the issue of the highly migratory species, may indeed be one of the major stumbling blocks in arriving at a Law of the Sea text which has a reasonable probability of general acceptance.

Negotiations in the Eastern Pacific

Attention has already been called to the fact that the tuna fishery of the eastern Pacific was developed and for many years dominated by United States fishermen. During this period open access was the common practice. When fishing effort expanded beyond the ability of the resource to replace itself a conservation program was instituted by the IATTC. At the outset this program was based on the simple principle of first come, first served. Whichever nation had the greatest catching capacity caught the most fish. This meant that the United States with about 90% of the fleet took about 90% of the catch. However, two important changes were taking place. The nations of the world had begun to move rapidly toward extended jurisdiction, and many of the RANs from Chile to Mexico began to actively pursue their interests in developing their tuna industries.

With the expansion of Latin American tuna industries, the United States, share of the tuna catch from the eastern Pacific declined steadily, from more than 90% in the mid-1960s, to about 55% today. In the meantime, the Latin American nations have been slowly increasing their shares. They now take about 35% of the total eastern Pacific tuna catch and have aspirations to further increase this share. The remaining catch is taken by vessels of nations from outside the eastern Pacific region. The operating structure of the overall yellowfin tuna catch quota system, as discussed previously, was administered in the eastern Pacific during 1966 through 1968. During that period, however, many of the coastal developing states maintained that the form of the conservation program limited their ability to expand their tuna fisheries. They argued that because the United States had most of the fleet, they would continue to catch most of the fish under the existing conservation program, which allowed free competition for the yellowfin quota. They felt that special allocations or shares of the overall quota should be assigned to them because much of the catch was being made in their coastal waters, and they also noted that this share would increase with the extension of their coastal zones to 200 miles.

In 1969, pursuant to these arguments and to insure that the coastal developing states would continue to cooperate in the conservation program, the Commission for the first time set aside a share of the over all quota which could be taken during the closed season by a special class of vessels. A quota of 4000 tons was assigned to each country (including developed nations) to be taken by their flag vessels with a carrying capacity of 300 tons or less. The over all yellowfin quota was also increased substantially in 1969 under an experimental fishing program.

Although it was originally agreed that the 1969 provision for small vessels was being established to resolve a special problem and was to apply for that year only, it has been retained and expanded in various ways. In 1970 the 4000-ton allocation was increased to 6000 tons. Another important change in the regulations was that for the first time a "grace period" of 10 days was established during which vessels could put to sea for their last unregulated trip (or, as it is commonly referred to, their last open trip or LOT), provided they were in port at the time of the closure date. In 1971 the 6000-ton allocation was renewed, the LOT grace period was increased to 30 days, and a special allocation of 2000 tons was assigned to a certain class of vessels experiencing special economic problems. Although Mexico was not named in the paragraph of the resolution setting this latter allocation, the criteria were drawn so that Mexico was unique in meeting them. The overall quota was also increased in 1971.

In the following years it became increasingly difficult to reach agreement at the intergovernmental meetings held in conjunction with Commission meetings. Sometimes two, three, or four meetings were held before final agreement on a conservation regime was reached. Certain of the Latin American states led by Mexico continued to maintain that the concepts under which the conservation system for yellowfin in the eastern Pacific had evolved were no longer valid. They maintained that by virtue of their adjacency to the resources, they and all other RANs should be able to exercise special rights with respect to the tuna fishery. To maintain the conservation program and to secure the cooperation of Mexico, Mexico's special allocation continued to be increased by the Commission. In 1973 it went to 6000 tons, in 1974 to 8000 tons, in 1975 to 10,000 tons, and in 1976 to 13,000 tons. Over all conservation quotas were also increased during this period.

In 1977 part of all the 6000-ton small boat allowances for Costa Rica, Panama, and Nicaragua were applied to their flag vessels regardless of size. In 1978 Costa Rica's allocation was increased to 7500 tons and Mexico's to 26,500 tons. However, while special allocations had previously applied only during the closed season, Mexico's allocation of 26,500 tons for 1978 applied during the entire year.

Through the early 1970s, Mexico had consistently urged during intergovernmental meeting debates that new principles for tuna management should be reflected in the Commission's charter and programs. Thus, in 1975 they announced their intention to convene a meeting of plenipotentiary nations to draft a new tuna convention for the eastern Pacific region. In anticipation of this meeting they requested through the Commission that the Director of Investigations undertake a study of various management alternatives for the conservation of yellowfin tuna and that these alternatives should include the concepts on the utilization of ocean resources emerging from the ongoing Third United Nations Conference on the Law of the Sea. This study, entitled "Alternative for International Management of Tuna Resources," by James Joseph and Joseph W.

Greenough, was prepared and distributed in October 1976 to serve as a background paper for a plenipotentiary meeting to be convened in 1977.

Two preparatory meetings were also held in anticipation of the plenipotentiary meeting. The first was held in November 1975 at the United Nations in New York. In attendance were the coastal nations of the western Americas. There views were exchanged on the range of possible provisions to be considered for incorporation in a new treaty governing the conservation and management of tuna in the eastern Pacific. The second meeting was a follow-up meeting between Mexico and the United States held in Mexico City in January 1976 where a further exchange of views took place.

The first plenipotentiary conference was held in San Jose, Costa Rica, in September 1977 and was convened jointly by Costa Rica and Mexico. The coastal states of the eastern Pacific region and noncoastal states who were IATTC members were in attendance. Mexico and Costa Rica distributed a working document outlining a number of principles governing the conservation and management of tuna in the eastern Pacific intended to be incorporated in a treaty. The document included 38 articles dealing with all aspects of their proposal. Among the more important elements proposed were:

1. A new organization for tuna management would be established with membership open to all nations participating in the plenipotentiary meeting.
2. The treaty area would be the eastern Pacific Ocean, and the species to be included would be defined in an annex.
3. All member nations would enjoy fishing privileges throughout the eastern Pacific to within 12 miles of the coast of each cooperating RAN of the region.
4. Each coastal state would have a guaranteed annual allocation based on the concentration of the yellowfin resource within its 200-mile zone, and a RAN's allocation would be equivalent to the average harvest of all vessels, of the international fleet within its 200-mile zone over the preceding 5 years.
5. Shares of RAN allocations not utilized by the coastal states would be available to other states on a competitive basis.
6. RAN allocations would apply from the beginning of the year, and if a RAN took its full allocation during the open season it could fish freely during the remainder of the open season.
7. The sale or transfer of allocations would be prohibited.
8. Participant fees would be paid by all members for each ton of tuna caught with the fee being 5% of the dockside value of each member's catch.
9. Vessels paying participant fees would receive a regional license enabling them to fish throughout the treaty area.
10. Participant fee receipts would be divided into three categories and would be (a) used to cover all operating costs of the new management organization; (b) distributed to RANs on the basis of catches made in their 200-mile zones; and (c) distributed to all participants in the fishery on the basis of their catches.
11. All assets of the IATTC would be transferred to the new organization, and

the headquarters and personnel of the IATTC would carry over to the new body.

12. The conference would adopt measures to resolve the tuna/porpoise problem as soon as possible and to abolish the last open trip provisions contained in the annual resolution implementing the IATTC conservation program.

These principles proposed by Mexico and Costa Rica closely paralleled the partially allocated quota or PAQ management alternative presented in the background study by Joseph and Greenough.

Only limited agreement was reached on the Mexican-Costa Rican proposal. There was agreement on the concept of an international licensing system with fees, but no agreement on the details of how it might be established or fee level. No agreement was reached on establishing a RAN allocation scheme or on the criteria for determining such allocation. In the opinion expressed by the United States Delegation, the IATTC had always responded to RAN needs, and there was no apparent reason why it could not continue to do so in the future. The United States' view was that the present IATTC should continue to operate until such time as the RANs of the eastern Pacific were ready to join a new organization. Mexico's opinion was that a new treaty organization should be established as quickly as possible because the IATTC no longer reflected current trends in international ocean policy, and they indicated that their continued membership in IATTC was doubtful unless changes were made. The final statement emanating from the conference was short and cryptic, noting only agreements to hold a new conference as soon as possible in a place and date to be decided.

A regular annual meeting of the Commission was held in the following month (October 1977) in Mexico City. At this meeting both Mexico and Costa Rica announced their intention to withdraw from the IATTC. The appropriate instruments of withdrawal were subsequently filed with the depository government (the United States) with Mexico's withdrawal effective in November 1978, and Costa Rica's in April 1979.

Between the First Plenipotentiary Meeting in November of 1977 and the Second Plenipotentiary Meeting in January, 1979, a series of four informal meetings were held. Some progress was made and agreement on several issues was reached.

At the Second Plenipotentiary Meeting two documents outlining a draft convention were presented, one prepared jointly by Mexico and Costa Rica and the other by the United States. These were subsequently combined into a single document which formed the basis for discussion among the participants. During the meetings two general positions emerged, one supported by the territorialist governments subscribing to the concept of a 200-mile territorial sea and the other by those governments having 200-mile extended economic zones. The territorialist governments presented their position that coastal states should retain complete jurisdiction over the tuna resources whenever they were present in their territorial waters as well as over associated fishing activities. They defended the right of the coastal state to determine the total allowable catch in its jurisdictional zone but at the same time recognized the obligation of the organization to fix the total allowable catch in convention waters. They proposed that licenses to fish for any surplus resources in jurisdictional zones should be granted by the coastal states and that licenses to fish in convention waters beyond 200 miles should be granted by the organization.

The majority of the EEZ governments agreed on the need for an international organization to conserve and manage tuna resources of the eastern Pacific Ocean and felt that the responsibilities of the organization should extend throughout the convention waters including the waters of the EEZs; they also agreed that voting in a new organization should be based on the principle of unanimity; and they further agreed that all of the principal market species of tuna as well as several of the secondary market species should be included within the terms of the reference of the organization. Although the principle of international certificates of access or licenses and the principle of RAN allocations were agreed to, the details of how these would be handled were not settled.

The most significant results of the conference were that (1) for the first time a joint working text was prepared for a draft convention in which areas of agreement as well as disagreement were identified and (2) there was a general feeling that the views of the territorialist and the EEZ governments were so divergent that it would be very difficult for them to reach a common ground in the immediate future.

Subsequent to the Second Plenipotentiary Meeting in January 1979 and prior to June 1980, six additional informal negotiating sessions were held. The principal participants in these sessions were Costa Rica, Mexico, and the United States. There were four basic issues around which extensive discussions centered: (1) The level of RAN allocations, (2) international license fees, (3) the matter of the last unregulated fishing trip for yellowfin tuna during the open season, and (4) the possibility of a special allocation for the fleet of small, disadvantaged United States vessels.

Following the last meeting held in Washington, D.C. during May of 1980 the positions of Costa Rica, Mexico, and the United States on the four issues were:

1. Yellowfin allocations for RANs: All agreed that there should be allocations to RANs and that these should be based on the concept of concentration of resources within the 200-mile zone. Costa Rica and Mexico proposed that concentration be measured by historical catches made in the zone by the international fleet. Based on the preceding 5 years, such allocations for all RANs combined would amount to about 69% of the total catch of yellowfin from the CYRA. The United States proposed that concentration should be measured by the estimated abundance of fish within the 200-mile zone relative to abundance beyond 200 miles. Based on this approach and the same 5-year period, 45% of the total catch would be allocated. Although no final agreement was ever reached on this issue, at one point the United States was willing to accept 57% and Costa Rica and Mexico were willing to accept 62.5%.
2. Licenses: Agreement was reached on issuing international licenses to vessels for a fee of $100.00 per net registered ton. The license would be issued by the organization, would be valid for 1 year, and would apply throughout con-

vention waters beyond 12 miles of the coasts of all participating states. License fee proceeds would be distributed to RANs in proportion to the catches of all species of tuna made in their waters.

3. Last Open Trip: From the outset Mexico and Costa Rica proposed that there be no last unregulated trip and that all states without allocations would stop fishing for yellowfin when the unrestricted yellowfin season closes. On the other hand, the United States proposed that the present system used by the IATTC be continued. Under this system all vessels in port at the time of closure are allowed 30 days in which to return to sea for their last unrestricted yellowfin fishing trip. Throughout the negotiations the United States held firm to this position. Mexico and Costa Rica altered their position to allow a 20-day period in which to set sail on the last voyage and a maximum of 60 days for the last trip. No final agreement was reached on this issue.
4. Disadvantaged vessels: Costa Rica and Mexico proposed originally that no special provisions be made for disadvantaged vessels (vessels with less that 500 tons of carrying capacity). The United States proposed that a 6000-ton yellowfin allocation be established that could be taken after closure by the disadvantaged vessels of their fleet. Subsequently Mexico and Costa Rica agreed to the 600-ton allocation for disadvantaged United States vessels provided that there be certain restrictions on the number and kind of vessels, that there be no new vessels added to the list, and that there be certain restrictions on where they could fish off the Mexican coast.

In a final attempt to reach an agreement Costa Rica, Mexico, and the United States discussed the possibility of a 3-year interim agreement which would permit the continuation of the IATTC conservation program and allow additional time for drafting a formal permanent agreement that would be acceptable to all states concerned. Agreement was not reached on the interim agreement proposal because no compromise could be reached to resolve the last open trip dilemma, even though tentative agreement was reached on the other three key issues.

Confrontation in the Eastern Pacific

After failing to reach agreement on a new tuna treaty during the numerous meetings held between 1977 and 1980, Mexico and Costa Rica announced their intent to enforce their 200-mile EEZs against tuna fishing vessels. Costa Rica began seizing foreign flag vessels in February 1979 and Mexico began seizing vessels in July of 1980, less than 2 months after the last informal negotiating session. In response to these seizures, the United States embargoed all tuna products from both Costa Rica and Mexico, as required by law under the U.S. Fisheries Conservation and Management Act (FCMA). As a result of these actions the tuna fishing nations of the eastern Pacific face a number of serious problems.

Under Mexico's Plan for Fisheries Development its national tuna fleet has grown from 15,000 tons of carrying capacity in 1979 to more than 50,000 tons in 1981. This fleet is expected to increase to at least 85,000 tons by 1983. During 1979 Mexico landed 30,000 tons of tuna while their domestic consumption was about 26,000 tons. The 4000 tons balance was exported to the United States in addition to 40,000 tons of tunas landed in Mexican ports by non-Mexican flag tuna vessels. A large infrastructure has built up supporting these unloadings. In 1981 the Mexican fleet landed about 64,000 tons of tunas, but no non-Mexican flag vessels were unloaded. Development of Mexico's internal and export markets since her fleet expansion program has been extremely slow and internal consumption during 1981 was still less than 30,000 tons. The tuna not utilized in Mexican markets (about 35,000 tons) was to be exported to the United States where about 85-90% of all of the tuna taken from the eastern Pacific is consumed. The United States embargo upset Mexico's export plans and during 1981 Mexico has had to carry large inventories of both canned and frozen tuna. The turn-around time for unloading Mexican boats has also been extremely slow. This situation may continue to worsen unless either the United States embargo is lifted, Mexico's internal consumption nearly triples, or new export markets for Mexican tuna are developed. If it continues it will most likely have a negative effect on Mexico's development plans. A similar situation has existed in Costa Rica, but because their tuna industry is smaller and less government financing has been available for aiding its growth, problems there have not been as severe as in Mexico. Costa Rica tuna development plans have not yet been realized. In fact, Costa Rica production of tuna has dropped sharply in recent years.

In the case of the United States, their confrontation with Costa Rica, Mexico, Ecuador, and some of the other RANs has caused serious disruptions in the tuna industry. First, processors who have normally depended on imports for more than 50% of their raw product, and who have invested in the producing sector in Mexico and other Latin American nations, are unable to import raw tuna due to to the embargoes that have been imposed. Second, United States producers have been excluded from major fishing grounds within national 200-mile zones where they have historically taken a significant portion of their catch. They are presently unable to fish off Mexico and must purchase expensive licenses to fish off most of the other Latin states. Historically nearly one-half their catch has been taken inside 200 miles. Since being excluded they must fish longer and harder outside 200 miles to try to make up the difference.

As already noted, yellowfin tuna in the eastern Pacific Ocean have been under a conservation program that was started in 1966. Due to the unresolved differences in attempting to reach an accord among Costa Rica, Mexico, and the United States no conservation agreement was reached during 1980 and 1981. Furthermore, there was no effective conservation program in 1979 because by the time the countries agreed to an interim arrangement it was so late in the year that the closure had no effect on curtailing fishing effort. Thus, the international fleet has fished on an unrestricted basis for the last 3 years. The result has been overfishing of the yellowfin resource as evidenced by declining average sizes, declining indices of abundance, and declining catches—from about 180,000 tons to about 150,000 tons on the average. If fishing effort continues unrestricted or increases with fleet growth, catches will in all probability continue to decline.

The failure of the long-enduring and effective conservation program in the eastern Pacific results from political differences in how nations perceive their jurisdiction over tuna resources and their conservation responsibilities both within and beyond their EEZs. Coastal developing states or RANs in whose EEZs tuna occur in abundance generally do not have developed tuna fisheries but wish to develop such. These states claim jurisdiction over tunas in their EEZs and insist on allocating their yellowfin resource on the basis of this claim. Such allocations only make sense if there is an over all limit on the total allowable catch as is the case with yellowfin. No allocations are proposed for skipjack because there is presently no over all limit on the skipjack catch. The non-RANs fishing in the eastern Pacific have a different view. In the main, they are highly developed tuna fishing nations harvesting a large part of their catch inside RAN 200-mile zones. The most important of these non-RANs, the United States, does not recognize the jurisdiction of states either to manage tuna inside their EEZs or to control access to these resources. Obviously these two differing positions create a dilemma that must be resolved because effective conservation and management of tuna cannot be attained without cooperation among the RANs and non-RANs.

Problems between RANs and non-RANs are not unique to the eastern Pacific. The same situation exists in the Atlantic where most RANs wish to develop their tuna fisheries. As in the eastern Pacific most of the eastern Atlantic catch is taken by non-RANs. These highly developed tuna fishing nations, represented mainly by Japan, Korea, France, and Spain, must have access to the rich tuna fishing grounds to maintain viable fisheries, and they must therefore cooperate in any conservation program if that program is to be effective. Discord over access to fishing grounds in the eastern Atlantic is already commonplace, and when conservation quotas are established, discord over catch allocations probably will occur. In view of the expressed need for limiting the harvest of certain tuna stocks in the eastern Atlantic, the lack of international regulatory measures is somewhat surprising. One cannot help but surmise that the problems in the eastern Pacific surrounding the allocation of the yellowfin quota account, at least in part, for the lack of over all quotas in the eastern Atlantic.

The IATTC has conducted research on the tunas of the eastern Pacific for 30 years. Based on this research they have for many years recommended conservation measures designed to maintian tuna stocks at levels that can support maximum yields on a continuing basis. The program was very successful in terms of its objectives. The science was good and the resulting management program was effective in maintaining tuna stocks at optimum levels. However, some member nations were dissatisfied with the way in which the recommendations of the Commission were implemented, even though the Commission itself recommended only over all harvest levels. The detailed recommendations on special country allocations and other matters emanated from an intergovernmental ad hoc meeting outside the authority of the Commission, and these recommendations were transmitted to the Comission in the form of a resolution. The reasons for doing this were twofold: (1) it provided a means for eliciting the cooperation and participation of nonmember governments in the conservation program, and (2) it provided a way of partitioning part of the over all catch quota into special allocations based on various economic and political criteria that the Commission is not empowered to consider under its convention. Because of dissatisfaction with the way these allocations and special provisions were made, some of the member countries have withdrawn from the Commission, the conservation program of 16 years has been discontinued, and even the continued existence of the IATTC may be in jeopardy.

It is reemphasized that all of this has resulted from the way in which the Commission's conservation recommendations have been applied politically; it is not a reflection of the scientific work of the Commission, which has been more than exemplary. It would be shameful for research on tunas in the eastern Pacific (or elsewhere) to be curtailed. For the eastern Pacific tuna fishery a data base on catch, effort, and length-frequency information extends from the beginning of the fishery. To terminate the collection of such data because of opposing political viewpoints would not be in the best interest of the nations involved in the fishery and it would be irresponsible toward future generations, for it is fundamental to monitoring the effects of man's activities on tuna stocks and once it is lost it cannot be retrieved. Regardless of politics, we must continue to strive to to maintain it.

One might consider whether conservation programs which involve catch quotas should be instituted. If they are, then arguments on quota allocations will tend to cause the demise of the programs with the likely result being unrestricted exploitation and consequent overfishing. If they are not, the result will be the same. However, the control of harvesting through overall quotas and some form of allocations is, in my opinion, necessary for some stocks of tuna and will be required even more so in the future. The only way to accomplish this effectively is through international cooperation that allays the discords that can arise all too easily between RANs and non-RANs. The remainder of this chapter discusses how this may be accomplished.

A Format for Tuna Management Beyond 1980

Throughout this chapter and in many other reports (Kearney, 1979; Saila and Norton, 1974; Joseph and Greenough, 1979) the need for conservation and rational management of tuna resources has been demonstrated. As discussed earlier a conservation program operated effectively in the eastern Pacific from 1966 through 1978. In certain other ocean areas conservation programs have been proposed, but effective ones for whatever reason have not been instituted.

The program in the eastern Pacific was undertaken during a time when most nations claimed only narrow juridical zones over living resources and tunas were considered to be the property of whoever caught them In the predecessor to this volume and elsewhere I have repeatedly stated that this system of open-access management would fail because many RANs were extending their jurisdiction to 200 miles. Based on this the RANs maintain that their adjacency to the resource should be recognized in the form of special considerations for harvesting the tuna resources found in their coastal zones. They also maintain that control of access to the living resources of the EEZ is retained by the RANs. This prognosis has proved true–the tuna conservation program in the eastern Pacific has failed for the reasons given 10 years ago, and in other areas apparently needed conservation programs have not been instituted. It seems quite clear that management of tuna under the system that existed during the 1960s and 1970s is unacceptable to many, if not most, RANs. Some other approach to conservation and management is required.

A currently popular alternative for the management of highly migratory tuna is unilateral action by the RANs. Each RAN would be responsible for the scientific study of the resource, establishment of harvest, supervision of exploitation, and control of access within its own coastal zone. Unfortunately, because of the nature of the animals and the requirements of the RAN fleets and markets, this unilateral approach will not work because tuna frequently cross man-made boundaries, and the fishing fleets must be able to fish wherever tunas concentrate. These concentrations fluctuate in time and space. For example, in one year or season the fish may be concentrated heavily in the zonal water of country A. In other years or seasons they may be concentrated in the waters of country B or C. In still other years or seasons they may be concentrated beyond the juridical zone of any nation. If controls are set on the level of harvest within only a particular coastal zone, when the fish leave that coastal zone they are subject to uncontrolled fishing and potential overexploitation. Further, for tuna industries to be economically viable they must have access to markets. To maintain access to markets there must be harmony among producing nations, processing nations, and consuming nations. Unilateral management of tuna will not work.

If the approach taken to tuna management prior to 1980 is no longer acceptable, and if unilateral action of RANs will not work, then it is essential to find an approach that will be acceptable to both the RANs and the non-RANs, as well as being effective in conserving and managing tuna. The ongoing UNCLOS III has addressed this issue and concluded that nations involved must cooperate through international channels if the highly migratory species, particularly tunas, are to be adequately conserved and properly managed. If we accept the recommendations of UNCLOS III, and if such international cooperation is to result in effective conservation and management of the world's tuna stocks, then several problem areas must be addressed.

The first major step is to assess how much tuna is available for harvest and what impact harvesting has on future abundance. Fundamental to such assessments is the acquisition and maintenance of a data base that includes catch statistics as well as biological information. All nations involved in a given fishery will have to cooperate to ensure that the necessary data are collected in a timely and efficient manner. In addition to the overall national obligation, the tuna industries of each nation must also cooperate and provide basic fishery production data as may be required. Only through such cooperation can there be made available the scientific data base needed for recommending conservation and management measures.

After determining how much fish is available for harvest, the more difficult task of determining who gets what share must be addressed and resolved. From earlier discussion it is noted that on a world basis non-RANs catch most of the tuna. Recent extensions to 200-mile EEZs in most cases permit RANs to reserve migratory species for their own use when they occur in their economic zones, implying that non-RANs must reduce their catches in these zones. However, because of the highly variable catches made in EEZs it is clear that such unilateral action will not benefit the RANs to any great degree since they cannot be assured of high concentrations of tuna every year.

If the fleets of coastal states are to be assured of consistently high catches, they must be able to follow the fish wherever they go. Whatever constraints RANs place on foreign vessels in their waters, they must expect similar constraints when they are the foreigners. Coalitions of RANs might tend to diminish this problem, but they would not solve it, because non-RANs take a major share of the world tuna catch. Excluding the non-RANs from the EEZ will force them to fish in the area beyond 200 miles. The result will most likely be unlimited fishing beyond 200 miles because, if excluded from the inshore fishing grounds, it seems highly unlikely that the non-RANs will want to control their activities on the high seas. As a consequence stocks of tuna could be very seriously overexploited after leaving the coastal zone and concentrating on the high seas.

One system for future management regimes that may overcome some of these problems is the partially allocated quota (PAQ) system discussed by Joseph and Greenough (1979) and incorporated in concept by Mexico, Costa Rica, and the United States in their negotiations for a new management regime in the eastern Pacific. An important component of such a system is the recognition of RAN claims that they are entitled by virtue of their adjacency to preferential shares of the resource. The most straightforward form of recognition would be for each RAN to be allocated unrestricted fishing within its own coastal zone, just as is the case for nonmigratory species. However, because a RAN cannot be assured of consistent catches, under the PAQ system quotas are assigned to RANs in proportion to the concentration of the resource in their juridical zones. Resource adjacent nations could then be allowed to harvest their allocations by fishing throughout the area of distribution of the stock. The proportion of the overall quota that is unallocated would be taken on a competitive basis.

Another key element of PAQ management includes provisions for an international licensing system. This is in recognition of the need to allow access to fishing areas wherever tuna concentrate as well as to recognize the sovereignty of the coastal states. Payment for a license should ideally be in the form of a participants' fee paid by all participants, RANs as well as non-RANs. This participant fee could be based on quantity of fish harvested, on vessel capacity, or on some other index. In turn the fee would be redistributed to the RANs as a payment for their permitting licenses to be issued that allow fishing in their EEZs.

Some form of international licensing system will be mandatory if fisheries for highly migratory species are to be economically viable, since a very large part of the catch is made inside 200 miles. In many of the major tuna fisheries, a single vessel may fish in the coastal waters of up to four or five different nations during a single voyage. If vessels are required to purchase expensive licenses from each of these nations for each trip, the cost of licenses quickly exceeds any potential profit from the sale of fish. It would quickly become uneconomical to fish under such a scheme.

Another important problem area that must be resolved under PAQ management is that of maintaining some balance between fleet capacity and the resource base. The management organization of the future must be delegated the responsibility to deal with the problem of unrestricted fleet growth. Provision will have to be made wherein international fleets will be maintained at levels consistent with the ability of the stocks to produce raw material. If this is to be accomplished a high degree of international cooperation and coordination will be required. Because of the great mobility of tuna fleets, the relationship between total fleet capacity and resource base must be given consideration in all areas where the fleets may operate. In the PAQ system proposed by Joseph and Greenough (1979) a detailed discussion of possible approaches to solving this problem is given in Appendix IV.

A final problem is that of equitable and uniform enforcement of conservation recommendations and regulations. An elaborate scheme for enforcement was included in the draft treaty prepared by Mexico, Costa Rica, and the United States which emanated from the Second Plenipotentiary Meeting. The scheme called for a uniform system of fines and sanctions for violations of conservation regulations. It also established an international enforcement arm which allowed for a reciprocal port inspection provision as well as a system of sanctions against nontreaty governments that ignored the recommendations of the organization.

Structure of a Global Management Body

None of the currently existing organizations for tuna research and management is dealing effectively with all of the problems just discussed, either through a lack of coordinated initiative or through lack of sufficient authority. The lack of adequate legal authority does not seem an intractable problem since most treaties are quite general and subject to broad interpretation. The major obstacle to effective conservation and rational management is the lack of political will to get the job done. Once the will is there, the job will get done.

In my earlier report in this series I proposed that tuna research and management could be most efficiently accomplished on a global basis. The rationale given was based on the migratory nature of tuna and tuna-like species in general, the mobility of the vessels that harvest these species, and the international nature of the market. The ideas I proposed were not unique, having been suggested earlier by Anonymous (1968) and Kask (1969). At present, some 10 years later, an even more compelling case can be made. Some tuna have been found to be even more highly migratory than believed earlier, and others less migratory. The fleets that harvest these species are made up of vessels that have, on the average, grown larger and more mobile. These vessels must and do travel the world's oceans in search of high concentrations of their prey. The markets are worldwide and expanding. After 10 years it is still my opinion that the collection and analysis of data, the establishment of closing and opening dates, size limits, catch quotas, the allocation of funds, and the identification and designation of duties and responsibilities can be accomplished most effectively and at the minimum cost through a broad, global approach.

A global body would be able to serve as a coordinator among states in numerous different ways. It could make certain that management actions taken in one fishery would have a minimum impact on fleets and markets involved in fisheries in other parts of the world. In situations where conservation regulations are in effect for different stocks in different areas, allocations among nations operating in both areas could be made on a rational and equitable basis by a body having responsibility for all areas. Such a global body could also act as an effective forum for sorting out problems of access among the RANs and non-RANs on a global basis. Almost all of the problems associated with tuna management, including those of excessive fleet capacity and enforcement and surveillance, could be handled most effectively and efficiently on a global basis.

How should a global tuna body be structured and what should be its duties and responsibilities? In answering the first part of this question one should take advantage of experience. There are a multitude of international organizations whose responsibilities deal in resources, research, assessment, and management. A number of these are fisheries organizations. Drawing on this experience, it seems logical to structure a body within the framework of a treaty which establishes a plenary body representing all high-contracting parties. Such a treaty should vest in the plenary body all responsibility for the functioning of the organization. It should also define explicitly the legal authority of the organization in terms of its capacity to enter into contracts, hold real property, sue and be sued, and its status with respect to member governments. The treaty should be especially specific in identifying the areas of scientific research to be conducted by the organization. Because fishing is an economic enterprise and fisheries management implies management of fishermen, research of an economic nature should be considered as an integral part of the responsibilities of the organization. The treaty must also designate a directorate or directorates within the organization to execute the technical functions of the plenary body.

An important question that arises is whether or not the purely scientific functions of an organization should be kept separate from the management functions. There are strong views, pro and con, as to whether these two functions should be kept totally separate or be included in the same organizational entity. Opinion on this issue has vacillated between the two extremes. The rationale for including them in the same structure is rather clear. The two are part and parcel of the same thing. Management is based on scientific advice and the two functions can be accomplished jointly in the most cost-effective manner. The rationale for not including them within the same organizational framework is less clear but has to do with the participants being willing to support the scientific function even if they may not like the consequences of the management.

With the distinction between scientific and management functions in mind, two approaches appear most straightforward: (1) two entirely separate organizations which are created by different treaties, with no overlap in plenary bodies, and located in different places, (2) a single body established under the same treaty, administered by the same plenary body but broken into a research and management arms with no line authority between the two. The creation of two entirely separate entities to accomplish effective tuna research and management seems to me to be inefficient and redundant. However, it may be necessary unless a strong political commitment is made by member nations to support such research and management. If a single body is to be created for research and management then it should perhaps be structured so as to allow continued participation by a nation in the research activities of the body even though that nation may choose not to participate in the management part of the organization.

Regardless of which organizational approach is taken there are certain common features to both. The species covered should be those tuna and tuna-like species that cannot be effectively researched and managed through action by a single coastal state, or by a small group of coastal states. This includes all species whose migrations carry them onto the High Seas in large enough numbers so that uncontrolled fishing there can have a measurable effect on abundance both inside and outside of juridical zones. Obviously the convention waters of such a global organization would be the oceans of the world in which the species included in the treaty occur. In defining the species to be included care should be taken to leave the list open ended. In this way as research reveals the migratory nature of the species under review, they can be deleted or added, depending upon the scientific findings.

Membership in the body should be open to any nation that is a RAN or whose nationals participate in the fishery on a significant scale. Significance in this context should be defined in either an economic or biological sense.

Obviously voting in such a body will be a consideration of great importance and therefore will have to be defined clearly. Several different voting schemes have been used by international fishery organizations. These include the concepts of unanimity, majority, and combinations of these for different issues. There probably will have to be unanimity in order to accomplish anything of major importance because it is unlikely that nations will agree to majority decisions being binding on dissenting voters.

Funding for the organization can be accomplished in several ways and really offers no major problems. It can be based on equal shares, weighted according to participation in the fishery, or degree of adjacency to the resource, or any combination of these. One point that should be considered however, is that many of the smaller, newly emerging nations are resource poor and have very limited free capital. This should be considered in assessing contributions.

No matter how an organization is structured and no matter what responsibilities are assigned it, staffing is a matter requiring special consideration. There are two major alternatives regarding staffing. One is that the secretariat only is funded by the organization. In this instance all data collection and research would be accomplished by the member states. The other situation would be that in which the directorates would be fully and independently staffed. Rather than enter into a detailed discussion of which approach is best I refer the reader to Joseph (1977), Gulland (1978), and Koers (1973).

Because of the large number of potential members who might form a global tuna body there is certainly room for concern as to its potential functional unwieldliness. One approach to overcoming this might be to partition the organization into divisions based on the three major oceans. Membership in the divisions would be handled at the division level. Global matters would be dealt with by the plenary body.

Because of the complexity of trying to establish a global body and the rather dynamic political state of the world today, it is unlikely that the formation of such an all-encompassing body can be accomplished in today's international political climate. Joseph (1977) has suggested a possible approach to this:

> It might be more realistically expected that such a global organization could result from the regional organizations that currently exist for the management of tuna and other highly migratory species. Since none of the regional organizations seem to be fully effective, a first logical step would be to strengthen them along the lines already suggested by modifying or rewriting their conventions.
>
> Once the regional bodies were empowered with the authority to carry out their functions effectively they could coordinate their activities through a central coordinating council composed, for example, of key members, of the regional organizations. Although this council would not be a policy making body it could serve in several important capacities, including the coordination of data collection, vessel transfers, catch allocations and enforcement. Once this system was operational, and if it were functioning effectively, the advantages of global management would be evident and political will for such a system attainable. The central council could than evolve into a governing rather than a coordinating body. The autonomous regional organizations would be grouped into divisions and staffing would be centralized.
>
> In summary, if the highly migratory species are to be managed effectively and their continued abundance insured for the benefit of future generations, a series of compromises on the part of all nations is essential. These compromises will need to be based not only on customary international law but also on the law of nature.

References

Anonymous. 1968. Report of the Meeting of a Group of Experts on Tuna Stock Assessment. Miami, August 1968. FAO Fish. Report, No. 61,45 pp.

Bayliff, W. H. 1980. Synopsis of Biological Data on the Northern Bluefin Tuna *Thunnus thynnus* (Linnaeus, 1758), in the Pacific Ocean. *In:* W. H. Bayliff (ed.), Synopsis of Biological Data on Eight Species of Scombroids. Inter-Amer. Trop. Tuna Comm. Spec. Rept. No. 2, pp. 261-294.

Forsbergh, E. D. 1980. Synopsis of biological data on the skipjack tuna, *Katsu-*

wonus pelamis (Linnaeus, 1758), in the Pacific Ocean. *In:* W. H. Bayliff (ed.), Synopsis of Biological Data on Eight Species of Scombroids. Inter-Amer. Trop. Tuna Comm., Spec. Rept. No. 2, pp. 295-360.

Gulland, J. 1978. Letters to the Editor. Alternatives for the Organization of Tuna Management. Mar. Policy., Vol. 2, No. 2. pp. 161-163.

Joseph, J. 1972. International arrangements for the management of tuna. A world resource. *In:* Brian J. Rothschild (ed.), World Fisheries Policy.

Joseph, J. 1977. The management of highly migratory species. Mar. Policy. Vol. 1, No. 4. pp. 275-288.

Joseph, J., and J. W. Greenough. 1979. International Management of Tuna, Porpoise, and Billfish. Univ. Washington Press, Seattle and London, 252 pp.

Kask, J. L. 1969. Tuna–A World Resource. Univ. Rhode Island Law of the Sea Inst. Occasional Paper No. 2, Kingston, R.I., pp. 32 + ix.

Kearney, R. E. 1979. An overview of recent changes in the fisheries for highly migratory species in the western Pacific ocean and projections for future developments. Suva, Fuji, South Pacific Bur. Econ. Co-op. SPEC (79)17 p. 99.

Kearney, R. E. 1981. Skipjack survey and assessment programme annual report for the year ending 31 December 1980. New Noumba, New Calidonia pp. i + 16. South Pacific Comm.

Koers, A. W. 1973. International Regulation of Marine Fisheries. Fisheries News (Books) Ltd., West Bayfleet. London, 366 pp.

Saila, S. B., and V. J. Norton. 1974. Tuna: status, trends and alternative arrangements. Paper No. 6 in a series prepared for the Program of International Studies of Fishery Arrangements. Resources for the Future, Inc., Washington, D.C. x + 55 pp.

7

Achievement of Fisheries Management Goals in the 1980s[1]

Brian J. Rothschild

Introduction

World fisheries have changed dramatically since the 1960s, when annual landings and fishing industries were growing rapidly. Now, growth appears to be virtually stagnant, despite dramatic changes in coastal state jurisdiction (FAO, 1981).

In the 1960s, recorded global landings were increasing at 7% per year; it was forecast that the global protein deficit would be significantly reduced as a result of increased fish catches. The construction of fishing vessels and gear was booming, and to most observers it was generally inconceivable that there would be a time, just two decades in the future, when fisheries production would stagnate.

Despite the optimistic forecasts of the 1960s, serious concerns were voiced regarding the stewardship exerted by the community of individuals who managed, harvested, processed, or otherwise utilized the world's fishery resources.

Brian J. Rothschild is a Professor at the University of Maryland's Center for Environmental and Estuarine Studies. He has been Senior Policy Advisor in the Office of the Administrator, NOAA; Director of the Office Policy Development and Long Range Planning, National Marine Fishery Service; and Director of the Southwest Fishery Center, National Marine Fishery Service, LaJolla, California; Deputy Director, Northwest Fisheries Center, Seattle, Washington. He has worked for various fishery organizations in New Jersey, Maine, Hawaii, Alaska, and the Gulf of Mexico. He has held academic appointments at the Universities of Washington and Hawaii, Scripps Institute of Oceanography, and the University of Miami. He has served as a consultant working on fishery problems of Korea, Egypt, Peru, Namibia, and the Indian Ocean.

[1]Contribution No. 1242, Center for Environmental and Estuarine Studies of the University of Maryland.

The concerns involved the fact that many stocks were overfished or becoming overfished; that many stocks were apparently fished in an economically wasteful manner, dissipating large amounts of economic rent; that scientific and technical information, while providing useful information on the long-term behavior of fish stocks under various intensities of fishing effort, failed under critical circumstances when information was required to provide management guidance for "next year's" fishery; and that, in a number of instances, managers and decision makers showed a lack of will to make the decisions and implement the actions that would change these undesirable and negative characteristics of global fishing institutions (see, e.g., Saetersdal, 1980). These undesirable characteristics were generally attributed to a common cause—the open-access regime under which most fisheries operated. Under the open-access regime, anyone could fish the resources whenever they wanted. Unrestricted entry into fisheries meant that the profitability of established fishery operations could be substantially reduced or eliminated as additional vessels entered the fishery. The spectre of additional fleets or vessels entering established fisheries, without any restrictions, heightened the perception of the high-risk of fisheries investment. The heightened risks induced fisheries to operate opportunistically, stimulating intensive fishing on individual stocks and reducing the biomass of each to historically low levels. After each stock was depleted, the fishery would move on to other stocks so that while global catches continued to increase, the total fishery productivity of the world's oceans decreased.

In the early 1960s global fishing effort was at a relatively low level and hence these problems of overfishing individual stocks, of economic waste, and performance of fishery management institutions were not nearly as evident and politically contentious as they were after substantial increases in global fishing effort occurred in the late 1960s and 1970s. Before the increases in effort, the results of bad management tended to appear rather inconsequential as fisheries (particularly, the highly publicized distant water fleets) had options to move from heavily fished stocks to lightly fished or unfished stocks, but as the total global fishing effort increased these options were foreclosed, seemingly amplifying the negative characteristics of fishery management.

The increase in fishing effort in the late 1960s and the early 1970s thus intensified the malaise associated with fishery management performance as either intentional or *de facto* allocations of fish stocks to various States and user groups often became the subject of strident debate and political contention. A common view of the root problem, however, was that it did not devolve from the open-access regime per se, the quality or quantity of information available to managers, management decisions, or the implementation of management decisions, but rather from essentially uncontrolled fishing by distant water fishery States, fishing in waters off the coastal States.

Thus, the political and public policy decision-making issues generally focused not on the general nature of access to the stocks or the careful evaluation of fishery management implementation, but simply on the more superficial issue, the elimination of distant water fishing operations in the waters of the coastal State.

In the mid-1940s and early 1950s, almost if by prescience, the stage began to be set by a few coastal States to eliminate or control distant water fishing by claiming coastal State jurisdiction over the fishing stocks which abounded in the waters off the coastal States. In 1945, President Truman proclaimed that the United States had authority to establish conservation zones on the High Seas contiguous to the United States coastline and in 1952, Chile, Ecuador, and Peru made territorial claims out to 200 miles from their coastlines. These actions eroded the freedom of the seas principle and along with other events stimulated a series of United Nations conferences on the Law of the Sea. These conferences tortuously proceeded to articulate a Draft Law of the Sea Treaty. Over the years, as the process leading to an acceptable Draft Treaty became more and more drawn out, those coastal States that perceived high political or economic stakes in excluding foreign fishing operations operations from their 200-mile zones, or otherwise perceived a need to control fishing in their zone, unilaterally extended their fisheries jurisdiction. By the late 1970s, most coastal States had some form of extended fishery jurisdiction.

The principal purpose of the extension of jurisdiction, the control of distant water fishermen, was thus attained. The process of controlling foreign fisheries often implied increased control of domestic fishermen as well, and thus most States explicitly or implicitly eliminated the open-access regime in the sense that they now had the authority to control access.

The attainment by the coastal States of exclusive authority to manage the stocks of fish in their coastal waters has totally changed the institutional framework for fisheries management. Through extended jurisdiction most States have increased and tightened authority to manage fish stocks in their fishery zones. The increased authority implied more effective management and heightened expectations for a greater return to the coastal States from the fishery resources in their fishery zones. There is no question that the imputed value of global fish stocks has increased substantially because coastal States now have the authority to control or manage fishing by both distant water fishermen and their own nationals. Economic waste can be minimized and increased returns to society (without necessarily increasing the catch) could be substantial. Improved management and concomitant reduction in investor risk would increase and direct investment, where appropriate, toward new opportunities that would result in further increased benefits including employment. In addition to increasing the absolute wealth generated by the fishery resources, opportunities would be enhanced for a more efficient and equitable distribution of fisheries wealth both among the industries and between the rich and the poor.

However, these great expectations for extended jurisdiction in the late 1970s have not yet materialized, some 5 years later, after most coastal States have extended their fishery jurisdiction. We now recognize that simply claiming authority over distant water fishing by coastal States was a piecemeal and simplistic solution to the problems of fishery management. We further realize that the simple declaration of management authority without a philosophical commitment to develop the underlying management system and the concomitant administrative and implementational architecture that are implied by the authority will be of little benefit and perhaps result in significant costs to the coastal States.

While some advances have been made as a result of extended jurisdiction, many of the old problems (e.g., overfishing) still exist, and new problems abound (e.g., McHugh, this volume, and Székely, this volume, relate aspects of the United States and Mexican experience respectively; Driscoll and McKellar, 1979, recite incredible difficulties involved in establishing a management regime in the North Sea; and Pearse, 1981, analyzes the contemporary management setting in British Columbia, Canada). Extended jurisdiction, while eliminating distant water fishing from the waters of the coastal State, has not provided in itself a mechanism for maintaining catch levels in those areas where the coastal States cannot replace distant water fishing effort with domestic fishing effort. Extended jurisdiction has made a reality out of management because before extended jurisdiction, management tended to be passive in the sense that quotas were frequently set and continually adjusted to match the capacity of the international fleet, rather than to satisfy well-defined economic or biological principles, but now with extended jurisdiction, quotas tend to be set so that they in fact constrain fishing effort. Management regimes that involve constraining quotas are often associated with considerable contention. Further, recent experience has shown that constraining quotas generate increased falsification of fishing statistics and costly enforcement programs. Indeed, there is a serious question as to whether the traditional fishery management paradigm—collect fishing statistics, set quotas, enforce quotas—can be cost-effective under a regime where quotas actually constrain fishing and are strictly enforced.

Extended jurisdiction has obviously extended or created a number of new boundaries between the coastal States and in some instances between the coastal State fishery zone and the High Seas. It was first thought that since distant water fishing was eliminated there would be no need for international commissions as there would be no international fishing problem. But this is obviously not the case. The establishment of one or more political boundaries (see Gulland, 1980, for a discussion of the "shared-stock problem") which intersect the range of the fishery stocks creates the situation where boundaries partition authority and thus provide opportunities for independent management regimes which may not work harmoniously toward common or compatible goals. Thus in a sense the problem of managing mackerel in the Northwest Atlantic, for example, now has common jurisdictional elements with the problem of managing highly migratory tunas.

Hence, there still needs to be institutional mechanisms to consider how fish are to be allocated among the coastal States, how technical information on the status of the stocks, which is critical to conservation strategies, is to be exchanged, and how mechanisms can be developed to most effectively undertake joint research on the stocks. Extended jurisdiction placing such obvious control on distant water fishing has, along with increased awareness of the general economic and social development problem, placed some focus on the needs of the coastal artisanal fishermen. This is important because estimates indicate that artisanal or traditional fisheries take a significant portion of the global catch of food fish. Because uncontrolled distant water fishing affected the catches of the artisanal fishery now that the distant water and domestic offshore industrial fishery is "manageable," questions arise as to the best method for utilizing this "manageability" to assist the artisanal fisherman. Finally, while before extended jurisdiction, there was considerable discussion of evaluating the performance of management, such evaluation was of more or less academic interests since if there were problems they could seldom be dealt with owing to the open-access nature of the resource. Now, however, when control is possible, evaluation is critical, because without evaluation control cannot be configured to yield optimal results. As is well known, however, evaluation needs to be based on the economic benefits that accrue from the catch and not just the physical quantity of fish caught although, as it is further well known, the kinds of data (e.g., cost of fishing) that are necessary to routinely evaluate the economic performance of management are exceedingly difficult to acquire.

Despite the substantial change in the global fishery management regime, there has been surprisingly little change in fishery management systems to make them more consonant with the new, apparently more controllable, closed-access regime. In the simplest analysis, society has two choices—it can take a *laissez-faire* approach or it can actively attempt to systematically formulate new requirements of fishery management under the new regime and, by so doing, influence the implementation of new and more efficient management systems. If society takes the first choice and attempts to muddle through with inchoate and ill-defined fishery management architecture, it stands the chance of foregoing benefits that would be attained from better management systems, while extant, less efficient management practices will become ingrained and difficult to terminate.

Because the global fishery management regime is still in transition, it is important that society does what it can now to establish a global institutional capacity to maximize returns from the fishery resources. In examining the question of institutional capability, we might begin by asking how could such an apparently simple matter as extracting full benefits from fisheries under a regime with authority for closing access be so difficult to deal with. How can problems articulated 10 years ago (see e.g., Rothschild, 1972) still exist and how could a regime have become established that apparently resolves the access problem but provides no new management architecture to realize the benefits of closed access? Where are the failures? Or perhaps, have we set our goals too high? Are there inadequacies in our approach to achieving fishery management goals?

Casting blame with respect to inadequacy of fishery policy and implementation performance may be unwarranted because the problem may not relate directly to fishery policy and implementation, but may be reflective of the general difficulties that seem to be attendant to the solution of complex societal problems. We can observe that the fishery policy problem is typical of a class of societal problems which seldom yield to easy or satisfactory solution. This class of problems may be typified in a number of ways—the problems are large and complex, difficult to circumscribe and formulate, not easy to quantify, abound in uncertainties, and are often characterized by considerable conflict among interest groups, each with different objectives and political contentions. Because of these difficulties, it is not surprising that in the address of the problems, despite the application of considerable human and budgetary resources to their solution, there are frequent failures to meet explicit and implicit policy expectations. In fact, the situation where public programs yield less than satisfactory results in a number of areas of social concern (e.g., world peace, hunger, balanced economic and social growth, energy management, resource conservation, health care and maintenance, and urban development) is so typical that it is now known as "policy failure" (see, for example, Ingram and Mann, 1980).

The prevalence of the policy-failure syndrome suggests that its full diagnosis and attempts for solution require a different approach than that which is commonly used in fisheries. This different approach would acknowledge that the less than satisfactory performance is the best that can be done given the problem setting (see Gulland, 1982) and that the only way to improve performance is to change the problem setting. Concern for the problem as it is presently formulated would thus be suppressed and factors which affect the structure of the problem and which are typically ignored would receive more direct attention.

Techniques for solution of complex societal problems were derived in the mid-1960s and are still quite relevant (see, e.g., Quade and Boucher, 1968). When examining the array of diagnostics for policy failure, one usually does not need to proceed further than the so-called pitfalls of systems analysis. These were identified by Quade and Boucher (1968) as (a) underemphasis on problem formulation, (b) inflexibility in the face of evidence, (c) adherence to cherished beliefs, (d) parochialism, (e) communication failure, (f) overconcentration on the model, (g) excessive attention to detail, (h) neglect of the question, (i) incorrect use of the model, (j) disregard of the limitations, (k) concentration on statistical uncertainty, (l) inattention to uncertainties, (m) use of side issues as criteria, (n) substitution of a model for the decision maker, (o) neglect of the subjective elements, and (p) failure to reappraise the work. Of these pitfalls perhaps the most important in regard to the present discussion is underemphasis on problem formulation. Problem formulation is the most complex and difficult task in policy development. The difficulty arises from the fact that in modern analysis the formulation of the problem is not taken as given but is the subject of considerable analysis itself, and as a result it is subject to frequent modification. Thus, in formulating a problem in a methodology where the problem statement is a variable, it is important to look beyond the bounds of the problem as it is conventionally defined. This art often involves melding information on the problem situation with the problem's contextual basis, its assumptions, its hypotheses, and its implied objectives (cf. Quade and Boucher, 1968). What is really important is not so much the study of the problem itself, but of the problem setting. This was well understood by John Herman Randall, Jr. (1926), who observed perhaps in a somewhat broader framework that:

> If men's minds are a mosaic or a palimpsest of belief upon belief, it is of the highest importance that they understand the life-history of those beliefs, why they are there, and whether they are justified in being there or should be discarded. What have been the great waves of thought and aspiration that have left these successive deposits? What did they mean when they were at the flood, what of value for today have they

> left, what must men seek out anew for themselves in the never-ending task of rebuilding civilization? When one has reached an understanding of what materials are furnished by the world around about him, and what resources he can hope for inside himself, it still remains for him to appraise the past as it is left to operate in the present, to understand it, to appropriate it, and to become its master.

As we assemble the materials for formulating fisheries policies in this decade we naturally tend to focus on the ideal or the desirable rather than on what is merely feasible. However, there may be a danger in this if there is no feasible pathway to attain the desirable. Majone (1975) had a good understanding of this problem. He observed that "The notion that, given sufficient ingenuity and resources, any social problem can be solved, in a manner consistent with announced policy objectives and with accepted institutional framework, has been a major source of failures and frustrations," and "In any field of human activity, nothing separates more clearly the amateur from the professional than the knowledge of what, under given circumstances, cannot be done in principle," and that "... the most important task of policy analysis, consists of submitting plans and objectives to the most stringent tests of feasibility."

Thus, we see the fisheries policy problem as being typical of many complex societal problems. We acknowledge that programs intended to address these problems often do not perform satisfactorily, but that this often is the result of the problem being defined too narrowly and consequently of too few intellectual and tangible resources being applied to understanding and attempting to modify the problem setting. If the setting cannot be changed, then there is little likelihood that we will significantly increase policy performance. It is important, however to consider that changing the problem setting is no mean task, so questions of feasibility must be paramount in the analysis.

In subsequent sections of this chapter I examine the materials that constitute the basis for fisheries policy development under extended jurisdiction; I then utilize these to develop a perspective on fisheries management in the 1980s.

Materials for Fisheries Policy Development

Up to this point I have emphasized that the development of fishery policy requires focus upon both the formulation of the fisheries management problem and the setting within which it rests. In order to do this we need to consider the notions and concepts fundamental to fishery management. In addition to considering these notions and concepts as they exist, it is important to also consider how they may change, or how they may be changed to improve fisheries management, for if we intend to effect change we need to understand what is mutable or controllable and what is not.

The concept of systems controllability was formalized by Churchman (1968). He observed that each system may be divided into a systems environment and systems resources. The systems environment are those properties of the system which are thought to be fixed and unchangeable; they are essentially taken as given. The systems resources are those properties of the system which are thought to be modifiable or controllable. Naturally, the systems designer and the policy maker have a common goal of transforming the systems environment into systems resources so that the total system is more manageable.

At the outset, the transformation of components of the environment into components of the systems resource sounds like a task of extreme difficulty. However, when one remembers that systems are, after all, simply logical configurations created and imagined to suit a particular purpose, then modification is bounded only by the imagination and creativity of the analyst and policy maker.

In this section I consider the following notions and concepts that will be important to the design of extended jurisdiction fishery management systems (see Rothschild, 1971; FAO, 1980, for discussion of fishery management systems): (1) the involved individuals in the fishery management setting and what their motivations are, (2) the structure of access to fisheries under the extended jurisdiction regime, (3) the technical information which is required under extended jurisdiction, (4) the role of government, (5) the roles of international and regional organizations, (6) fisheries as a basis for economic and social development, and (7) performance of the traditional fisheries management paradigm. As we examine these we shall search for ways to improve the management system.

Involved Individuals and Their Motivations

The design of a fishery management system requires that it include as components the appropriate individuals or groups of individuals and take into account the positions that they hold and cherish and how these motivate their actions and decisions.

The apparent traditional definition of the fishery system has tended to be too narrow, taking into account only the fishermen-fish interaction. While the fishermen and the fish are critical components of the fishery management system, they do not necessarily control it. In actuality, the individuals who invest or otherwise stimulate the introduction of capital (Anderson, 1980, discusses the economic implications of capital as an "input" into the fishery) into the fishery activity exert the most critical control of the fishery management system. The magnitude of capital investment is frequently an important determinant of whether a fishery produces benefits or a drain on society. Too much capital is economically wasteful and can promote in overfishing. In addition, there are many instances where investment has not accounted for stock fluctuations, so while the amount of capital utilized in a fishery may be contemporaneously optimal when the stock is at a peak in abundance, it is likely to be suboptimal and wasteful over the life of the capital equipment.

Capital investment in a fishery may be generated by fishermen, by processors, by boat builders, by development and investment banks, by individual investors who have no direct interest in the fishery, and by government programs. It is easy to recognize that each entity has a different degree of risk or discount rate, and generally operates on a different information base as to the effect of the investment on the industry as a whole.

For example, the fisherman wants a decent return on investment and undertakes considerable risk since it is he that must contend not only with the sea but with the vagaries of the fish stocks and the markets while at the same time he has the fewest options to engage in alternate activities. The processor encourages investment in fishing vessels to maintain a flow of raw material through his processing facilities and when he is financially involved with certain vessels, he can often guarantee that their production will be delivered to his processing plant. Boat builders, while not directly investing in fishing, influence governments to generate boat-building subsidies while they minimize impressions of economic or biological overfishing to create a continuing demand for fishing boats, even when additional boats would generate overcapitalization and other societal costs. Development and investment banks are sometimes misled by inaccurate analyses of the effects of additional capital in the fishery and so may stimulate overcapitalization. Governments may be misled and by not attributing capability to or by ignoring the conclusions of stock-assessment scientists, continue to stimulate investment into overcapitalized fisheries in the guise of fishery development programs. Finally, individuals who are far removed from fishing who are encouraged to invest in fisheries by broad-scale tax programs or special subsidy programs may contribute to overcapitalization at little risk and sometimes great short-term profit.

In addition to those who directly affect fisheries investment, there are groups of individuals who by virtue of their perspective and outlook have a strong influence on the way in which we think about fisheries management and hence on the structure of fisheries management institutions. It often seems that each group has its own and often disparate perceptions of the real world (Nabokov, 1980, pp. 252-253, has an interesting elucidation of "individual realities"). For example, the traditional fisheries manager has restricted his view to fishermen-fish interactions. The traditional fisheries manager concentrates on setting fishing effort to maximize average sustained yield, setting size-specific fishery mortality to maximize average yield per recruit, and maintaining a stock size that will maximize recruitment. As another example, economists have tended to restrict their attention to the conditions under which economic benefit to the fishermen is maximized (see, however, Anderson, 1980). A third example would be conservationists who have tended to focus attention and considerable societal resources on special interest problems rather than on the general and more important problem of making "wise use" of all fisheries resources.

In addition to investment motivation and intellectual motivation, there are political motivations and these are most evident in the public sector. The public sector has always exerted an important influence, perhaps because of the open-access nature of the resource, on the benefits that society obtains from its fishery resources. Those who have had experience in government will recognize that the public sector is often driven by noneconomic motivation. The motivation of the bureaucrats in the public sector is generally, in the final analysis, political. As a practical matter, these motivations tend to focus on problems which in the long run are not very important, but take on an urgency and immediacy in the world of bureaucracy and detract attention from policy and strategy development.

The relevance of these observations is that the systems environment includes a diverse array of individuals and entities. Each of these has a different point of view, outlook, and motivation. Because of this diversity one wonders how one should assess the frequently proffered advice, "that before one can manage fisheries one must establish *a* management objective." Further, it is not surprising, because of the mix of motivations, that when objectives are specified for a particular fishery, there are often several and the attainment of some stated objectives would prevent the attainment of other stated objectives. It is becoming clear that there is no single objective that should or can be attained in existing fishery management settings. It appears that our notions on objectives need to be transformed. The notion that the "objective" in any fishery is really a mix of conventional objectives, where the character of the mix changes as a function of almost continually changing biological and economic conditions.

The transformation of these systems properties into systems resources needs to be attained through taking into account the broader spectrum of individuals or entities who participate in the management system as well as their motivations. In addition, greater systems control may be attained by developing management systems that explicitly recognize that simple management objectives cannot be defined for most fisheries.

The Structure of Access

As pointed out in the Introduction, most observers of fisheries management believed that the open-access regime was the root cause of fishery management problems. Many also believed that some form of limited entry (an evaluation of limited-entry performance may be found in Rettig and Ginter, 1978, particular in Section VI) would remedy the problem. We now realize that it is not as simple as all this. The construction of a fisheries management policy requires an understanding of the way in which fishing access functions to generate benefits or disbenefits in a fishery.

Functionally, the problem of access is not so much related to whether an individual or group can fish a stock, but rather how many homogenous economic entities fish the stock. Access is closed if a stock can only be fished by a single entity; by definition, no one else can fish the stock. The functioning of open access can best be demonstrated by an example. Suppose there is a single entity fishing the stock. This entity has an optimal fleet size of 10 fishing boats. The entity, operating optimally, will not increase or decrease its fleet size. Suppose further that there is a fixed quota of fish. Each boat takes one-tenth of the quota. However, now suppose another entity is allowed to fish the stock. It uses one boat. Its boat takes one-eleventh of the catch and the share of the original entity drops to ten-elevenths of the total. Suppose that the one-eleventh share is profitable for the second entity, so it adds another boat. It nearly doubles its catch to two-twelfths of the total while that of the first entity drops to ten-twelfths of the total. The second entity adds to its fleet so long as it is profitable at greater and greater expense to the first entity.

If it is desirable to create conditions where a second entity can replace a first entity at any time or place, then the above system is satisfactory. However, only under the most limited circumstances will the above system be totally desirable for while it might be beneficial for the most cost-effective entity, the second entity, to catch the fish, the costs of capital in the first entity need to be accounted for. In addition, the hidden costs, associated with how investors behave when there is a possibility of a second entity entering the fishery need to be taken into account. It is just this situation that generates the wrong incentives in the fishing industry because there is an incentive to economically and biologically overfish the stock as long as there is the possibility of a second entity (see Clark, 1981).

Thus, as long as there are two more entities fishing the stock in an uncontrolled way, there there will be an open-access problem. However, even under circumstances of alleged control, undesirable incentives can be induced. This is because control, effecting allocation of resource shares among entities, is not a simple concept and it is therefore difficult to attain. One of the principal reasons for difficulty in developing a rationale for allocation is because the proper allocation algorithm should not simply divide the tons of fish among entities. This commonly used erroneous criterion (see, for example, the Draft UNCLOS Treaty) is at fault because the value of the fish are only partly represented by the physical quantity of the catch. A considerable component of the value of a fish stock resides in the density of the stock on the fishing grounds. All other things equal, a stock at a high density will be more profitable to catch and hence more valuable than the same stock at a lower density.

The density phenomenon is frequently ignored, but it arises in almost any allocation problem. Suppose there is a stock from which 50,000 tons of fish can be taken on a sustained basis. Entity A can take 30,000 tons and the management authority thus awards 20,000 tons to entity B as "surplus," with the implication that the award to B has no effect on A. However, the same 30,000 tons would be much more valuable to entity A without entity B than the same 30,000 tons with entity B, because the stock would be at a much lower density if 50,000 tons were caught than if 30,000 tons were caught. This proves that in situations where there is more than one entity fishing the stock there is no "surplus."

In addition to the difficult problem of control, in a situation which requires trading off the absolute quantity of the catch with the density of the fish, and hence the cost of producing a fixed quantity of fish, in an environment with limited data, there are a number of other technical difficulties that involve whether catch or effort should be controlled and the degree to which entities should be limited to particular space or time strata. For example, if there is a single entity, then the nature of control and access is relatively obvious. If there is more than one entity, and there is no control, we have the pure open-access system and all its faults. On the other hand, if there is more than one entity and control, the control can only be perceived as being imperfect, so even the so-called closed-access, multiple-entity system possesses open-access properties.

Thus, the declaration of extended jurisdiction has in one sense obviated the open-access problem. In another sense, however, it has created another sort of logically equivalent access problem. Before extended jurisdiction, anyone could fish a stock at any time. Extended jurisdiction eliminated that privilege but, on the other hand, it erected a number of interstate boundaries which intersect the range of many stocks. Since neither State has total control over the stock, another form of open access has been generated. When a single stock is intersected by an interstate boundary, fishermen from one State fish on one side of the boundary and fishermen from the adjacent State fish on their side of the boundary; thus either State can fish the stock in any way it wants just so long as it stays on its side of the boundary. While the nature of access has been reconfigured, the problem still remains. Thus, any feature which generates the existence of more than one entity fishing a stock creates an access problem. Particularly important, here, are the sovereignty of the various States and subdivisions of States. The existence of sovereignties implies boundaries, which, all other things being equal, will generally require that the resource be managed as at least two separate units. The benefits that would be obtained by managing the stock as a single unit will generally be greater than those from managing it as several units.

As an example, consider a stock which is distributed in a manner such that the smaller fish in the stock tend to be on one side of the boundary and the larger fish in the stock tend to be on the other side of the boundary. An optimal yield, in this instance, would involve catching mostly large fish. In other words, small fish are caught only at a cost to the total catch. What incentives are there for the small-fish state to give up its catch so that the over-all catch becomes larger but is taken essentially by the large-fish State? If there were no boundary, then this problem might not arise.

The relevance of these observations to systems design is that the extensions of jurisdiction have not entirely eliminated the open-access problem; they have created a new access problem which is generally not yet well understood. The maintenance of interstate boundaries has turned out to be oriented toward one of the most sensitive issues that arises among the global States, and this is the issue of sovereignty. Transforming these properties of the system into systems resources cannot proceed unless they are thoroughly analyzed and methods for dealing with sovereignty developed. One possibility would be for one state to compensate the other in instances where their actions could deleteriously affect the benefits of the other (the concept of transfer payments was mentioned by Gulland, 1982). Such compensation would maintain sovereignty, reduce over-all costs, and result in greater over-all benefits for both States.

Technical Information

Management performance depends upon the information available to the manager. The development of management information proceeds along two tracks. Along the first, standard techniques are used to analyze data and provide the analyses to the manager. In the second, new techniques are developed under the presumption that these new techniques will improve the cost effectiveness of management.

With respect to the utilization of standard techniques, it is not uncommon to find that these are frequently not fully utilized. Quotas and size limits are often set without reference to extant techniques. Essential data are also often not collected, and if they are, information systems for their assembly and dissemination do not exist. It is obvious that there are constraints in bringing the extant global fisheries management machinery to the state of the art level.

The development of new techniques has not proceeded rapidly. This is unfortunate because the increased imputed value of fish resources under extended jurisdiction suggests that not only the more enthusiastic application of standard techniques, but the development of new techniques would greatly enhance benefits from management.

Under the new extended jurisdiction regime various information issues become even more critical. These include (a) being able to forecast the variability of fish stocks, simply to improve management, (b) to determine whether depleted stocks have an inherent propensity to increase, (c) to determine the effects of pollution (including modification of the coastal zone), and (d) to understand interactions among the species in multiple-species fisheries (FAO, 1978, discusses the multiple-species problem).

All of these issues may be dependent upon the same problem, that of understanding mechanisms that cause fish populations to fluctuate (for a recent review of the extensive temporal fluctuations that commonly occur in fish populations, see Cushing, 1982). Most workers feel that variations in recruitment, that is, variations in year-class strength or the number of fish that enter the fishery each year, are caused by physical and biological phenomena that operate when the young fish are very small, just after they are hatched. If these recruitment events are, as we suspect, the critical forces that drive the fluctuations in fish populations then an understanding of these events is critical to solving the four problems mentioned above.

First, it is clear that an ability to forecast recruitment fluctuations will enable forecasts of fish stock variability. Second, some depleted stocks increase in magnitude when fishery pressure is eased but other depleted stocks tend to remain at relatively low levels. The propensity to increase is thought to be associated with recruitment phenomena. It is important to understand these phenomena because if the stock had a propensity to increase then an appropriate management strategy would be to decrease fishing to increase the abundance of the stock and to bring it to a more profitable level, but if the stock had no propensity to increase, then a failure to adjust the industry as a whole to the lower stock level would have costly repercussions. Third, as indicated, fish stocks continually fluctuate in abundance and it is important to know to what extent these fluctuations result from various forms of pollution and habitat modification. In order to separate anthropogenic causes from natural causes, it is important to understand the mechanisms that produce natural fluctuations in fish populations, the recruitment mechanisms. Such understanding will enable improved decisions on pollution control. Fourth, if recruitment drives the fluctuations of each species, then recruitment must account for most variability in the multiple-species setting.

Unfortunately, our present understanding of the causal factors that affect recruitment is quite limited. This lack of understanding does not ncessarily derive from a lack of funding, as there are locations in the global ocean where these sorts of problems have been studied for several decades (see, as an example, Hempel, 1978). Despite intensive study in these areas, there is little predictive insight into the mechanisms that cause fish stocks to vary. It is clear that the multitude of observations that have been made in these intensively studied areas have not lent much insight into the processes and, hence, making the same kinds of observations, in the same way, for yet another decade is not likely to be helpful.

Most insight into the recruitment problem is likely to be obtained by a systematic examination of functional interrelationships of the critical biological events (e.g., egg production, abundance and quality of food for larval fish, predation on larvae, genetic factors) and the simplest physical features that effect larval survival (heat, light, and motion). Such studies on scales (e.g., fine and microscale) relevant to larval survival clearly need further emphasis (Rothschild, 1981).

The relevance of these observations to systems design is that much can be done to bring technical information to the state of the art. In addition to applying information already available, substantial gaps in our knowledge of recruitment processes need to be addressed. A systematic attack on these questions is no simple affair, since such research will likely require increased coordination among the global research organizations.

The Role of Government

Governments, among their various functions, play an important role in affecting the economic efficiency with which natural resources are transformed into wealth as well as affecting how and to whom the wealth is distributed. The control that governments exert regarding the generation and distribution of wealth varies from government to government. At one extreme, fishing can be a State industry and hence the government has, in a sense, complete control. At the other extreme, government may exert little control. Typical situations exist inbetween these extremes where, for example, fishing is a private-sector enterprise with substantial public-sector control. Public-sector controls operate often directly and indirectly with direct control being exerted in the form of regulations on fishing, for example, and indirect control arising through some general tax policy, for example, that is not intended to affect fishing, but nevertheless does, because fishing is part of the general enterprise to which the tax policy refers.

The key issue is whether government actions, or inactions as the case may be, either enhance or diminish societal benefits from fishing. In the case where there is total government control the analysis of government performance can be relatively simple, but in cases where the public sector influences the private sector, the analysis can be quite complex. The main complexities that arise under the total government control regime are that the control usually resides in a variety of organizational and administrative hierarchies which are often in conflict and that the costs of government programs, some of which may only partially relate to fisheries, are difficult to calculate. Considerably more complexity is involved in analysis of the regime where the public sector controls or manages the private sector, because in addition to complexities associated with analyzing the role of the government, there are significant complexities involved in the varying extent of government control and in the fact that while government control is often intended to be in the best interest of the public, its configuration and existence may be such that it actually reduces societal benefits (excellent discussions of disincentives generated by government may be found in Friedman, 1973; Schultze, 1973; see also Rothschild, 1979).

One particularly important aspect of the role of government is the way in which it structures its programs to provide incentives to the private sector to adopt public-sector goals. Fishery management has long assumed that the industry has public-sector goals, but this is clearly not the case. The only way to develop public-sector goals in an industry is to develop an economic incentive-disincentive system to insure that public goals are met. The incentive-disincentive system was first called to my attention by J. Dykstra, former President of the Point Judith Fishermen's Cooperative. If there were, for example, sufficient public as opposed to private incentives in fishing industries, there would be no need for enforcement or for special data collection programs.

Finally, it is important that government organize itself in such a way so that it can be efficient in delivering the services which it intends to deliver. This includes, for example, generating an organizational format that maximizes capability and legitimacy and allocates the appropriate amount of human and financial resources to get a job done (Allison and Szanton, 1976).

The relevance of this cursory discussion on the role of government in systems design is to an area of policy development that is seldom given explicit attention. Usually the role of government is seen as whatever government does and because of this there is little normative guidance. The idea of development of industry incentives through providing industry with management authority will continue to surface as the unworkability of traditional schemes becomes more and more clear. As the flow of regulatory activity moves from government into the hands of industry, the function of government changes from one of actually regulating to one of providing information so that internal regulation can be efficient.

International and Regional Organizations

Under the open-access regime, regional and international organizations provided a forum for distant water fishing states and coastal States to consider the effect of fishing on the stocks and to make management recommendations. The management recommendations generally served to effect a single, over-all annual quota for each stock and perhaps establish a mesh size or a minimum size below which fish could not be caught or retained. The quotas set by the organizations were generally ineffective since they did not constrain the amount of fishing, and they served to actually promote economic waste in the international fleet as a whole. Further, the ineffectiveness of the quota system made the oragnizations that seemingly promulgated the quota approach to management seem unresponsive to management needs.

The malaise was not particularly evident in the early 1960s, when stocks were sometimes not fished near their capacity, but as fishing intensity increased the difficulties became more and more apparent. Quotas that actually constrained fishing had to be set, but this only intensified the vicious spiral wherein the very existence of a quota caused each state to add more and more boats to the fishery so that they could maximize their share of the catch, but at considerable economic waste. In addition, quotas that actually constrained fishing intensified data collection and enforcement problems, particularly at the interntional level.

It is important to recognize that the quota system, and the economically inefficient "scramble" for fish which it generated, was not really the fault of the organizations themselves; the problem was in the way that the organizations were structured. The structure was basically that each of the organization's member states had essentially an equal vote. With an equal vote, each state could insist on having the quantity of fish necessary to support its industry with a component of the over-all quota. However, this quantity generally kept increasing to the edge or beyond the edge of a biologically supported maximum catch because the quota system stimulated introduction of excess capital into the fishery. This was, of course, at great cost to the fishing states.

So it is not surprising that under the extended jurisdiction regime the coastal States, not enthusiastic about the performance of international and regional organizations, and recognizing their own authority to eliminate foreign fishing from their fishery zones, diminished or eliminated their participation in a number of international and regional organizations. This is unfortunate because extended jurisdiction has in many ways increased, rather than diminished, the need for international and regional organizations and cooperation. The needs stem from the fact that, as has been pointed out under the section on access, there is now a different form of access problem than before extended jurisdiction, in the sense that single stocks or stock complexes exist under the authority of more than one coastal State and management therefore needs to pay explicit attention to new forms of cooperation.

It is clear that management is going to center itself, eventually, more and more on questions of allocation. Under the open-access regime, planned and binding allocations were not prominent in fishery management regimes. Under some circumstances, total catch quotas were set and the various entities involved in the fisheries scrambled for their share. In other circumstances the total catch quota was more or less partitioned on the basis of what each entity involved in the fishery thought they could catch. Now, however, under extended jurisdiction, international organizations will need to look more carefully at allocations and while possibly not having the authority to allocate, they will most certainly be asked to evaluate various allocational alternatives.

Another important aspect relating to international organization involves the possibility of obtaining economies of scale in research, management, and training. Each of these activities is expensive and there is no need for each coastal State to duplicate activities on the same stock and in addition there needs to be some mechanisms to undertake the kind of research that will improve management. Research oriented toward improving management techniques is particularly expensive and likewise there seems to be little sense in each coastal State duplicating the research of every other coastal State or bearing on its own account the cost of such research. The same situation occurs in the context of training and again better training could be afforded on a regional basis rather than on a national basis. A careful analysis of training needs is given by Brewer (1980).

Another role for regional cooperation is in commerce and trade of fisheries products. Again, new fisheries institutions might form a basis for a more systematic development of cooperative economic ventures among the coastal States and, perhaps, distant water fishing nations to make the best uses of their resources in terms of the cheapest capital, labor, or technology.

The relevance of these observations to systems design is that new or reconfigured international and regional organizations are needed for more effective fishery management. The new organizations need to provide management advice on stocks that occur in the fisheries zones. Beyond its traditional role, they need to operate to reduce the costs of routine management analysis. In addition, they should cooperate to identify and perform the kinds of research that will improve management techniques, research that very well might be more expensive than any single state could undertake. As management directs itself toward allocational questions, the economic behavior of the fleets becomes a matter of special concern and so, where in the past most research was focused on biology, future activities may need to involve both biological and economic analyses. Needless to say, too, the new organizations will be constituted with higher levels of legitimacy than those of the earlier days. This is because special attention will be focused upon the degree to which international institutions contribute to national capabilities and goals and indeed it will be essential for these organizations to demonstrate that their existence provides greater benefits to the coastal States than the coastal States could obtain independently.

Fisheries as a Basis for Economic and Social Development

In innumerable coastal villages and hamlets, fisheries can be locally extremely important; they provide an important source of food and livelihood for artisanal fishermen. Globally, the landings made by small-scale traditional fishermen account for a significant proportion of the world catch of food fish. On a national or regional scale, the artisanal fisheries account for large quantities of fish and are an important basis for economic and social growth.

Policies relating to the management of the small-scale artisanal fisheries are difficult to develop. The basic issue often involves allocation of stocks between artisanal fishermen and industrialized fishermen. Oftentimes a particular stock of fish will be accessible to both artisanal fishermen and more industrialized offshore fishermen. Depending upon the relative magnitude of the catches, the industrial fishery could seriously affect the artisanal fishery. On paper a rapid move toward industrialization might sound as if it were a good policy; but the artisanal fishermen could generally suffer because they probably would not be able to enter the industrialized fishery and thus would have reduced opportunities. In the context of extended jurisdiction, the exclusion of distant water fishermen from the waters of the coastal State could serve to benefit the artisanal fishermen in the sense that the abundance of fish available to artisanal fishermen should increase in those instances where the distant water fishermen and the artisanal fishermen fished a common stock. In some instances, the coastal State will not have the technological capability to harvest the stocks that were previously harvested by distant water fishermen and in these instances consideration must be given as to whether the stocks should go unfished or whether it is in the long-run benefit of the coastal State to develop programs with other States or industrial entities to harvest the resources so that the coastal State could share in the economic benefits which would derive from the fishery or processing operation.

In considering these issues, it is well to remember that they do not exist in a fishery setting per se; instead, they exist in a setting of concern for agrarian reform and rural development. The flavor for such a setting is framed by the Declaration of Principles expressed by the World Conference on Agrarian Reform and Rural Development (FAO, 1979b). The Conference called for a "Programme of Action" founded on these guidelines and principles:

(i) that the fundamental purpose of development is individual and social betterment, development of endogenous capabilities and improvement of the living standards of all people, in particular the rural poor,
(ii) that the right of every State to exercise full and permanent sovereignty over its natural resources and economic activities and to adopt the necessary measures for the planning and management of its resources is of vital importance to rural development,
(iii) that the use of foreign investments for agricultural development of developing countries, in particular that of transitional corporations, must be in accordance with national needs and priorities,
(iv) that national progress based on growth with equity and participation requires a redistribution of economic and political power, fuller integration of rural areas into national development efforts, with expanded opportunities of employment and income for rural people, and development of farmers' associations, cooperatives and other forms of voluntary autonomous democratic organizations of primary producers and rural workers,
(v) that appropriate population policies and programmes can contribute to long-term social and economic progress,
(vi) that maximum efforts should be made to mobilize and use productively domestic resources for rural development,
(vii) that governments should introduce positive bias in favour of rural development and provide incentives for increased investment and production in rural areas,
(viii) that equitable distribution and efficient use of land, water and other productive resources, with due regard for ecological balance

and environmental protection, are indispensable for rural development, for the mobilization of human resources and for increased production for the alleviation of poverty,

(ix) that diversification of rural economic activities, including integrated crop-livestock development, fisheries and aquaculture and integrated forestry development, is essential for broad-based rural development,

(x) that location of industries in the rural areas, in both the public and private sectors and particularly agro-industries, provides necessary and mutually reinforcing links between agriculture and industrial development,

(xi) that policies and programmes affecting agrarian and rural systems should be formulated and implemented with the full understanding and participation of all rural people, including youth, and of their own organizations at all levels, and that development efforts should be responsive to the varying needs of different groups of rural poor,

(xii) that understanding and awareness of the problems and opportunities of rural development among people at all levels and that improving the interaction between development personnel and the masses through an efficient communication system are prerequisites for the success of rural development strategy,

(xiii) that constant vigilance should be kept to ensure that benefits of agrarian reform and rural development are not offset by the reassertion of past patterns of concentration of resources in private hands or by the emergence of new forms of inequity,

(xiv) that women should participate and contribute on an equal basis with men in the social, economic and political processes of rural development and share fully in improved conditions of life in rural areas,

(xv) that international cooperation should be strengthened and a new sense of urgency introduced to augment the flow of financial and technical resources for rural development,

(xvi) that all governments should undertake new and more intensive efforts to ensure world food security and overcome inequities and instability in the trade of agricultural commodities of particular importance to developing countries, and

(xvii) that developing countries, with the support of international development organizations, should strengthen their technical cooperation in rural development and foster policies of collective self-reliance.

We can easily see that the "Guidelines and Principles" articulated form a basis for structuring policies which relate to artisanal fishing. Fleshing out a "Programme of Action" with respect to the principles will require the address of the following technical issues (see also Smith, 1979):

1. How can the wellbeing of the small-scale artisanal fishermen be enhanced?
2. What is the appropriate balance between small-scale artisanal fishing and more industrialized fishing operations?

3. What is the optimum amount of fishing effort that should be expended in both the artisanal and the industrial fishery and how should fishing effort be monitored?
4. What are the tradeoffs between allowing distant water fishermen to fish the stocks or engaging in joint ventures and restricting fishing entirely to coastal state nationals. Under the latter circumstance, what should be the appropriate levels of compensation to the coastal state?
5. What steps can be taken to assist the coastal state to develop self-sufficiency in management, technical expertise, and fishing?
6. What are the mechanisms for developing a data collection and fishery management system to insure maximal return to the coastal state?
7. In what areas can regional cooperation be cost effective for the individual coastal states?

In terms of systems design, it is clear that where fisheries exist as a raw material for economic and social development that a number of issues arise which are not ordinarily within the realm of fisheries management, as it has been traditionally conceived. These development issues need to be moved into the "resource" of the traditional fisheries problem and in this way the standard fisheries work can contribute most effectively to resolution of the critical economic and social development issues.

Performance of the Traditional Fisheries Management Paradigm

At first appraisal, the traditional fisheries management paradigm appears to be relatively simple. Management objectives are set; fishery statistics are assembled; the fishery statistics are used to compute an "optimal" level of size-specific fishing mortality or catch; regulations are set to insure that the optimal catch is not exceeded (note, this is distinct from actually targeting on optimal catch); and the adherence of the fleet to the regulations is monitored and enforced. The performance of the traditional fisheries management paradigm is mixed and seems to depend upon whether management is passive or active.

Passive management has little direct effect upon the fishermen or industry in the sense that regulations are either set so they do not change fishermen or industry actions (e.g., a quota set at 30,000 tons a year would have little effect on a fleet with an annual capacity of 29,000 tons) or, if the regulations are set so that they would affect fishermen or industry actions, then they are either not adhered to or not enforced (indeed, some regulations cannot be enforced unless there is a fisheries officer on each boat).

Since passive management is *pro forma* and changes little, it must operate at relatively great net cost. In contrast, active management actually constrains the behavior of the fishermen in the sense that regulations actually modify the catch or the actions of fishermen; further, there are mechanisms either in the nature of the regulations or in enforcement that insure adherence. Because of the complexity of fisheries and multiple or shifting management objectives, it is sometimes difficult to assess the performance of active management; while results are often in the right direction, it is often not clear as to whether the results are the "best" or just good.

As implied earlier, most fishery management in the open-access regime tended to be passive rather than active. This was because passive management derived naturally from the open-access regime. However, it is not certain as to whether truly active management oriented toward and capable of extracting maximal benefits from the fishery, will be a reality under the extended jurisdiction regime as we now know it.

Whether active management works or not depends upon the utility of the traditional management paradigm in an active management setting. Is the traditional paradigm much too simple to control or even address the previously discussed complexities or real-world fisheries, such as highly variable stocks, many species in the catch, and management objectives which are multiple, often in conflict, and continually changing?

In fact, the traditional management paradigm does not work well even in a simplified setting. Consider the following rather typical scenario: suppose there is a single-species fishery that has been fished for 20 or 30 years. In the early days, the fishery yielded 50,000 tons a year, but in recent years fishing effort has increased and the catch has dropped to 20,000 tons. The management agency decided to "rebuild the stock," through active management, and in order to do this it sets an annual quota of 10,000 tons. The reduction in the catch results in a sharp increase in the exvessel price of fish and this causes the fishing industry to place considerable pressure on the management agency to increase the quota.

Strongly held views in the management agency sustain the 10,000-ton quota. To monitor the catch of each fisherman, the agency plans a log-book system. It also orders a careful study to determine appropriate levels of catch and fishing mortality. A task force is organized to execute the study. The task force finds that there are several thousand fishermen involved on a full- and part-time basis in the fishery. They find that the reporting was incomplete and that records that are available contain many obvious errors and archiving is so poor that a computer information system needs to be designed just to handle the fishing data. It will take substantial sums of money to develop the system and to maintain it.

The task force also becomes concerned because they fear that the decline in catch may not be real, but a function of increased incompleteness in reporting as the number of fishermen increased. They further learn that the contemporary catch may actually be even higher because fishermen underreport catches in the new log books in order to forestall closing the fishery by apparently delaying the attainment of the quota.

The task force recognizes that the faulty fishing statistics affect the quality of their assessment of the appropriate catch levels, but they are compelled to make recommendations and they suggest the 10,000-ton quota as being prudent and appropriate.

The log book and information system is put in place, but there are insufficient funds to keep it operating properly and since the fishery is taking only 10,000 tons it does not receive high priority in any event. Furthermore, the fishermen continue to underreport their catch and it is a problem of considerable technical difficulty just to determine when the 10,000-ton quota is reached and to halt fishing activities. Obviously, more enforcement is needed, but this too is quite expensive since it involves personnel and expensive capital equipment including fishing patrol boats.

The image that we develop even in this simple setting is that active management under the traditional paradigm builds disincentives among fishermen to comply with management regulations and can involve costly statistical and enforcement requirements. Further, when such management has been attempted even in the highly recommended limited-entry setting, the results have not generally been satisfactory.

While active management is expensive, it is not known whether it is cost effective, although many observers feel intuitively that it is not. These are, of course, judgments in the abstract since specific fisheries need to be examined and the benefits that accrue from management, e.g., increased catches, more valuable fish, need to be weighed against the costs of obtaining these benefits.

In terms of systems design, it does seem appropriate, however, to develop a management system in which fishermen have incentives to supply management information and abide by regulations.

A Perspective on Fisheries Management Policy

As a result of extended jurisdiction, the essential structure of the global fisheries regime has changed. The change has endowed the coastal States with exclusive authority to manage the stocks of fish in their coastal waters. The generally increased authority implies a capability for increased management control and this in turn implies that there is considerable potential for increasing the performance of fisheries management and the benefits which would accrue to society from harvesting the fishery resources.

Realization of this potential, however, will require considerable thought, policy formulation, and implementation to build the kinds of institutions that have the capability to extract the full benefits from the resources. At this point in time, our experience has generally been with what we have called passive management. We have pointed out that managers would have liked to engage in active management but this was essentially precluded by the structure of the open access system. In whatever limited experience we have had with active management, the results have not been particularly encouraging in the sense that we are realizing that more controlled and finely tuned management requires considerable data and these are expensive to obtain, store, and process; active management constrains the fishing industry in such a way that little incentive is induced in the industry to adhere to regulations, which are difficult and expensive to enforce anyhow. If it is at all possible to significantly improve benefits from fishing, we will need to develop a new architecture for fishery management. This architecture will need to be based on a policy structure that takes into account not only the changes associated with extended jurisdiction, but the requirements which extended jurisdiction impose on management, if increased benefits are to be obtained.

The policy structure will need to take into account the concepts identified in the previous section. For example, policies need to reflect that:

1. The essential economic performance of fisheries is related to forces that stimulate investment or disinvestment in fisheries and so control of performance may relate more to controlling these forces than to controlling the most distal problem of fishermen-fish interactions.
2. The nature of access has changed in two fundamental respects. First, the coastal States now have exclusive authority to control fishing in their fishing zone and second, authority over a number of stocks is not exclusive, since the range of some stocks crosses interstate boundaries and the High Seas. The new authority in the fishing zones places emphasis on determining the best way that the authority might be used to increase benefits, and the lack of exclusive control over stocks places emphasis on the development of new arrangements among the sovereign States regarding management.
3. The requirements for information, data, and new management concepts under an active management, extended jurisdiction regime are bound to be different from those for a passive management, open-access regime. Paradoxically the higher imputed value of fish in the extended jurisdiction regime should warrant a greater expenditure in this area, but it may be that the expenditures on developing new concepts and information have to increase exponentially rather than linearly and hence may not be warranted. In terms of concepts the problem of predicting recruitment is of intrinsic importance because it lies at the heart of developing year to year management strategies; rebuilding depleted stocks, evaluating anthropogenic effects on fish stocks, and understanding the interactions among species in multiple-species fisheries. Serious in-depth evaluations of the data, information, and adequacy of available concepts are needed and support needs to be generated in areas that are promising.
4. Whatever their form, governments generally play a critical role in the peformance of fishery management by their actions or inactions. There is a building opinion in a number of quarters that where there is public-sector control of the private sector, management of the private sector could be made more efficient by reconfiguring public-sector control to operate only at the bounds of what is acceptable to the public. This would provide the private sector with incentives to behave more in the interest of the public. Governments might then redirect their activities into developing information to make the fisheries system work more effectively.
5. International and regional organizations will ascend in importance. The lack of exclusive control over many stocks, the need to gain economies of scale in research, the need to exchange data, and the possibility of economic interaction suggest that these organizations can play an exceedingly important role in the extended jurisdiction regime. For maximum effectiveness, strategies for

organization need to be developed which take into account the successes and the failures of the organizations that operated under open-access as well as the special needs of the coastal States.

6. Fisheries constitute an important basis for economic and social development not only in the developing countries, but in economically disadvantaged sections of the coast in developed countries as well. The full utilization of fisheries for economic and social development will require viewing the problem as more of a problem in rural development and reform than as a fishery problem. In other words, the fishery policy will be subsidiary to the development policy.
7. The traditional management paradigm operating in an active management, extended jurisdiction setting may have serious difficulties. The sources of these difficulties need to be fleshed out, evaluated, and changed. It may be that change is not possible and hence the extant performance of fisheries management will need to be deemed satisfactory.

A consideration of these policy components leads us to a plan for developing a "desirable" fisheries management system. The first step is to circumscribe or define a fishery that will be managed as a unit. The fishery should be defined so that it is relatively homogeneous biologically. This means that the boundaries of the unit should be set as much as practicable so as not to exclude any large components of individual stocks. The second step is to agree that for the purpose of fishing, there are no political boundaries that intersect the fishing unit. Steps one and two would mean that the access problem as far as managing the stock(s) as a whole would be eliminated, and a single entity would have authority to manage the fish.

Step three would involve leasing the unit to a single industry entity. The entity would be chosen on the basis of its capability, qualifications, and potential performance. The principle of controlling only at societally acceptable bounds could be adopted. Hence, the industry entity would be accountable to government(s) to maintain production above a certain level (to avoid a monopolistic position) and to prevent the stock from declining below a certain level (to maintain conservation of the stock). The selection of an entity would not mean that it had a perpetual right to fish the unit; its performance would be continually reviewed and if found lacking it would be replaced by another entity. The implementation of step three would provide the entity with incentives to become economically efficient. In being efficient it would be well managed and utilize the proper amount of capital to harvest the stocks. In addition, it would perform as an ideal fishery manager—it would collect the appropriate statistics; it would carefully archive the statistics and generate its own stock assessments. There would be no need for enforcement in the usual sense because it would be in the self-interest of the entity to abide by its own rules, which are, after all, intended to maximize its profits.

It is obvious that under the idealized fishery system the role of governments must change. Under the traditional system the government attempted to regulate the fishery as best it could under conditions of open-access and little industry-public-sector incentive. Performance could not have been good. Under the idealized system, the role of government as a regulator per se is diminished; rather, it is one of monitoring the performance of the industry; it insures that the industry is working within acceptable bounds; and it sets standards of performance, both economic and regarding conservation. The industry is given wide latitude to operate within the economic and conservation bounds. In addition, in some instances, the operation of government will change with respect to information development and research. Under the new system the role of government will be to provide information to keep the market working efficiently and to conduct the kinds of long-term research that individual entities could not afford to undertake and which would, when accomplished, improve the fishery management process.

If the stocks in a management unit are unfished, then the establishment of a fishery entity should be relatively simple. If, on the other hand, the stocks are already fished, then the implementation of a fishery entity will be a much more difficult process. Again, if implementation is to be at all successful, the existing participants must be compensated for their capital and livelihood. Financing for compensating the existing participants might be generated by the sale of bonds using the potential rent from the fishery as collateral as well as public-sector funds. The existing participants are compensated either directly with cash payment or with shares of stock in the fishery entity. Initially, fishermen who participated in the fishery and wanted to continue to fish could. While they would not own capital they would own shares of stock and could continue to fish on their former vessels.

At some point in time, the management entity would make the business decision to either increase or decrease its capital stock. Strategy for increase would follow standard capital development procedures. A decision to decrease capital, however, is more difficult to deal with because this would involve dislocation of fishermen. However, the problem may not be as difficult as in the traditional system because the new system has been designed with incentives for fishermen to leave an overcapitalized system. That is to say, if the fishery is overcapitalized the management entity would decide on the optimal amount of capital and sell or otherwise salvage the inefficient units. Inefficient fishermen would have an incentive to leave the system because the value of their shares of stock would increase.

The system described above provides a different setting than that found in traditional fishery management. The system, at this point in its development, is too simplistic and it would be opposed by a number of interest groups. On the other hand, as indicated earlier, feasible systems should be oriented as much as possible toward desirable systems. Indeed, prototype systems are beginning to emerge which show characteristics of the "desirable" system. For example, Canada is experimenting with managing fisheries as economic units and the European Economic Community is suppressing individual State sovereignty as it strives to develop a common fishery policy.

When all is said and done, it is problematical how fast global fisheries policies in the 1980s can come to grips with the inadequacy of the traditional fishery management paradigm. It is evident that if we are to see a large improvement, we will need to see a change in the setting under which fisheries operate. This change in setting will need to be built on a philosophical foundation of just what fisheries management should control and what it should not control; of sovereignty; of the scientific and technical basis for management decisions; on the role of government in fisheries management; and of the role of fisheries in economic and social development. To be sure, questions of feasibility versus desirability are of critical importance, but right now we need to focus our attention on what is desirable, remembering that, most often, closing the gap between what is merely feasible and what is desirable requires only one ingredient—and that is leadership.

I would like to thank John A. Gulland, FAO, Rome, Jean-Paul Troadec, ISTPM, Nantes, and Susan L. Brunenmeister, University of Maryland, for their comments on the manuscript.

References

Allison, G., and P. Szanton. 1976. Organizing for the decade ahead. *In*: H. Owen and C. L. Schultz (eds.), Setting National Priorities. The Next Ten Years. The Brookings Institution, Washington, D.C., pp. 227-270.

Anderson, L. G. 1980. A comparison of limited entry fisheries management schemes. *In*: Report of the ACMRR Working Party on the Scientific Basis of Determining Management Measures. FAO *Fisheries Report* No. 236. FAO, Rome.

Brewer, G. D. 1980. An international institute of fishery management. *In*: Report of the ACMRR Working Party on the Scientific Basis of Determining Management Measures. FAO *Fisheries Report* No. 236. FAO, Rome.

Churchman, C. W. 1968. The Systems Approach. Dell, New York, 243 pp.

Clark, C. W. 1981. Bioeconomics of the ocean. BioScience 31(3):231-237.

Cushing, D. H. 1981. Temporal variability in production systems. *In*: A. R. Longhurst (ed.), Analysis on Marine Ecosystem. Academic Press, New York, pp. 443-471.

Driscoll, D. J., and N. McKellar. 1979. The changing regime of North Sea fisheries. *In*: C. M. Mason (ed.), The Effective Management of Resources. Nichols Publishing Co., New York, pp. 125-167.

FAO. 1978. Expert Consultation on Management of Multispecies Fisheries. Some scientific problems of multispecies fisheries. Report on the Expert Consultation on Management of Multispecies Fisheries. Rome, 20-23 September 1977. FAO *Fish. Tech. Pap.* 181, 42 pp.

FAO. 1979. Report of World Conference on Agrarian Reform and Rural Development. Rome, 12-20 July 1979. WCARRD/REP, July 1979.

FAO. 1981. Marine Resources Services, Fishery Resources and Environment Division, Fisheries Department. Review of the State of World Fishery Resources. FAO *Fish. Circ.* 710, Rev. 2, 51 pp.

FAO. ACMRR Working Party on the Scientific Basis of Determining Management Measures. 1979. Interim Report of the ACMRR Working Party on the Scientific Basis of Determining Management Measures. Rome, 6-13 December, 1979. FAO *Fish. Circ.* 718, 112 pp.

FAO. ACMRR Working Party on the Scientific Basis of Determining Management Measures. 1980. Report of the ACMRR Working Party on the Scientific Basis of Determining Management Measures. Hong Kong, 10-15 December 1979. FAO *Fish. Rept.* 236, 149 pp.

Friedman, M. 1973. Cutting government back to size–eight guidelines. *In*: R. H. Haveman and R. D. Hamrin (eds.), The Political Economy of Federal Policy. Harper and Row, New York, pp. 23-25.

Gulland, J. A. 1980. Some problems of the management of shared stocks. FAO *Fish. Tech. Pap.* 206, 22 pp.

Gulland, J. A. 1982. Long-term potential effects from management of the fish resources of the North Atlantic. J. Cons. Int. Explor. Mer. 40(1):8-16.

Hempel, G. 1978. North Sea fish stocks–Recent changes and their causes. (Editorial) Rapp. Proc. Vert. Reun. Cons. Int. Explor. Mer 172, 449 pp.

Ingram, H. M., and D. E. Mann (eds.). 1980. Why Policies Succeed or Fail. Sage Publications, Inc., Beverly Hills, 312 pp.

Majone, G. 1975. The feasibility of social policies. Policy Sci. 6:49-00.

McHugh, J. L. 1982. Jeffersonian Democracy and the Fisheries Revisited. *In*: B. J. Rothschild (ed.), Global Fisheries: Perspectives for the 1980's. Springer-Verlag New York.

Nabokov, V. 1980. *In*: F. Bowers (ed.), Lectures on Literature. Harcourt Brace Jovanovich, New York, 385 pp.

Pearse, P. H. 1981. The Commission on Pacific Fisheries Policy. Conflict and opportunity toward a new policy for Canada's Pacific fisheries. A preliminary report of the Commission on Pacific Fisheries Policy, Vancouver, B.C.

Quade, E. S., and W. I. Boucher (eds.). 1968. Systems Analysis and Policy Planning. Applications in Defense. Elsevier, New York.

Randall, J. H., Jr. 1926. The Making of the Modern Mind. Columbia Univ. Press, New York, 720 pp.

Rettig, R. B., and J. J. C. Ginter (eds.). 1978. Limited Entry. As a Fishery Management Tool. Univ. Washington Press, Seattle.

Rothschild, B. J. 1971. A Systems View of Fishery Management with Some Notes on the Tuna Fisheries. FAO *Fish. Tech. Pap.* 106, 33 pp., Rome, August 1971.

Rothschild, B. J. (ed.). 1972. World Fisheries Policy. Multidisciplinary Views. Univ. Washington Press, Seattle.

Rothschild, B. J. 1979. Federal reorganization for marine affairs (fisheries and aquaculture). *In*: A Special Report to the President and the Congress, Vol. II. Reorganizing the Federal Effort in Oceanic and Atmospheric Affairs. National Advisory Committee on Oceans and Atmospheres, March 1979, pp. 107-126.

Rothschild, B. J. 1981. More food from the sea. BioScience 31:216-221.

Saetersdal, G. 1980. A review of past management of some pelagic stocks and its effectiveness. Rapp. Proc. Verb. Reun. Cons. Int. Explor. Mer 177:505-512.

Schultze, C. L. 1973. Perverse incentives and the inefficiency of government. *In*: R. H. Haveman and R. D. Hamrin (eds.), The Political Economy of Federal Policy. Harper and Row, New York, pp. 15-22.

Smith, I. R. 1979. A research framework for traditional fisheries. ICLARM Studies and Reviews No. 2. International Center for Living Aquatic Resources Management, Manila, 45 pp.

Székely, A. 1982. Implementing the New Law of the Sea: The Mexican Experience. *In*: B. J. Rothschild (ed.), Global Fisheries: Perspectives for the 1980's. Springer-Verlag New York, Inc.

8
Managing Fisheries in an Imperfect World

John A. Gulland

Introduction

The typical system of present-day fishery management is generally thought of as consisting of several stages. First, data are collected from fish markets and research vessels; second, these are analyzed at biological research institutions where the effects of alternative management measures are predicted; third, on the basis of these predictions, and in the light of social and economic objectives, one particular management measure is chosen; and finally, appropriate legal measures are introduced and duly enforced.

If this were really what happened in practice, and if the measures chosen did in fact have the effect predicted by the biologists, and if these resulted in the social and economic objectives being achieved, then all concerned with management (not least the fishermen, and the community at large) would be very content. This is rarely the case. The measures may not in practice be implemented or properly enforced. If enforced, they may fail to have the effects on the stock predicted by the biologists. If they do, the social and economic impact may (especially in the long term) be quite inconsistent with the stated objectives.

John A. Gulland's early training was as a mathematician. He worked for a number of years at the Fisheries Laboratory, Lowestoft, England, where he specialized on the population dynamics of demersal fish. He has been concerned with providing scientific advice to various international commissions, such as the Northeast Atlantic Fisheries Commission, the International Commission for the Northwest Atlantic Fisheries, and the International Whaling Commission. Since 1966 he has been with the Department of Fisheries, Food and Agriculture Organization of the United Nations, Rome, concerned with fisheries management and planning of fisheries development programs throughout the world.

Finally it can happen that the stated objectives are no longer what are really required from a well-managed fishery.

This is a lengthy listing of possible areas of failure and it is not surprising that most management systems have failed to operate perfectly. Admittedly many management systems, while operating less than perfectly, have been successful in making things better than they would have been if there had been no management—"better" meaning here improvement in at least one of the important characteristics of a fishery, such as the size of the total catch, the abundance and stability of the population, or the catches or net income of the individual fisherman. I am not suggesting here that the imperfections that exist in all parts of the system are, for this reason alone, matters for criticism, or that their existence should be an excuse for postponing action until matters are improved. Nevertheless, the benefits that have been obtained have been, in nearly all cases, much less than would have been possible.

Certainly the best should not be enemy of the good. The recognition that the management actions considered are likely to be less than perfect should not induce an intellectual paralysis so that no action is taken at all. The point that is being made in this chapter is that the management system (from the collection of basic data to implementation and enforcement of regulations) is likely to be different, and more effective than the present system, if there were to be greater recognition that each of the elements in the system is imperfect and is likely, despite all improvements, to remain to some degree less than perfect. For example, the relations between the fishery administrator or manager and his scientific advisors would be different if there were better recognition that the predictions of the scientist have the reliability typical of the weather forecaster, or the economist predicting next year's rate of inflation. At present there is too often the feeling that in a properly run fishery the scientific advice should have the reliability of predictions of the time of sun rise or of eclipses. While this optimistic but wholly unrealistic feeling lasts the administrator feels let down when for the third year in succession predictions of total catch are substantially different from events, even if (or especially if) the differences are explained, after the event, by the existence of an unusual year-class, or unexpected increases in the efficiency of the vessels in the fleet. Equally the scientist feels let down when he is not provided with the research facilities, or sufficiently detailed statistical data, that might have enabled him to make the "perfect" prediction.

This chapter therefore explores how a more general acceptance of imperfections may affect the various stages of fishery management. This will be done largely from the point of view of the biologist. The main attention will therefore be paid to the implications of accepting that the biologist will never have complete information (concerning either the activities of the commercial fishery or the events in the natural environment), or entirely satisfactory methods of analyzing what information he does have. The main impact of this will be on how data are collected and analyzed, but it must also affect how the managers use the scientific advice. Equally the uncertainties in the later stages of management must affect how the biologist goes about his work. Imprecision in objectives (or at least in how these are applied in practice), uncertainties in the economic analyses used in determining the chosen policy, and the inability to achieve exact implementation of most measures will have their effect on the nature of advice required from the scientist. In particular I will address two groups of questions: how should the biologist best arrange his studies, bearing in mind that he will never have as much information, or as complete methods of analysis as he would like, and how should biological advice to the administrator be formulated and used, bearing in mind that it cannot be wholly comprehensive and accurate, nor will it be used in a wholly perfect manner.

To ask for recognition that fishery management operates in a less than perfect world is, of course, not a new demand. Several authors have pointed this out in relation to all or part of the management system. In particular Larkin (1972) in his chapter in the antecedent to this book made an eloquent plea for the fishery manager to be quite open in accepting that he has less than perfect knowledge of the system. Larkin asked the manager to state clearly the hypothesis on which the proposals for the coming season would be based, and the ways in which the events during the season would test the hypotheses. He expected—without any obvious regret—that such a procedure might lead to a high replacement rate of managers. Doubleday (1976), May *et al.* (1978), and others have examined problems raised, for example, by setting a constant annual catch quota, when the population is treated as constant (except for the effects of fishing), but in fact is subject to natural variation. However, less attention has been given the effects on policy if the possibility of variation is accepted but the actual variations cannot be observed accurately.

The purpose of this chapter is not therefore to break new ground but rather by emphasizing the problems to reduce the number of occasions when, on the one hand, scientific advice (e.g., on the possible interactions between fisheries on associated species) is not offered because information is not complete or the analysis is less than the standard expected of a paper for a scientific journal, and on the other, known or suspected inadequacies in the science are used by the manager as excuses for failing to take action.

In discussing imperfections in the various elements of the management system it is useful to distinguish between actual error and variations. Possible sources of error include inadequacies in the basic information, e.g., incomplete statistics of the total catch; poor understanding of the system, e.g., a poor model of the population dynamics of the species being studied; and a failure to look at the whole system, e.g., looking only at the fishery on one species when that species is being affected by a fishery on another species on which it feeds. Sources of variation include natural variation (good or bad year-classes or longer term changes in abundance); changes in the human system, e.g., jumps in fuel price; and changes in objectives, e.g., the recent emphasis on considering whales and seals as objects to be protected rather than sources of oil, meat, or fur.

The Nature of the Imperfection

Before attempting a prescription for improving the way we manage fisheries in an imperfect world it is necessary to look a little more closely at what these imperfections are. In doing this it is convenient to follow, with minor modifications, the breakdown of the steps involved in management as described in the report of the Advisory Committee on Marine Resource Research (ACMRR) Working Party (FAO, 1980). For each stage the kinds of variation will be outlined, then their likely effects on the "success" of management examined, and some possibilities of improving the situation—which need not involve attempting to remove the imperfections—briefly set out.

Objectives

Not long ago little attention was paid to defining the objectives of management. It was usually assumed that catching more fish was all that was needed, and if a defined objective was needed, then maximum sustained yield (MSY) was good enough. More recently the weaknesses of MSY have been pointed out, at first mainly by economists, who stressed the importance of looking at the net economic yield and the question of costs (e.g., Scott, 1955), and later by biologists concerned with broad interests of conservation (e.g., Holt and Talbot, 1978). Much of this is in fact not new. Biologists as long ago as Graham (1939), who were perhaps in the simpler world of those days closer in touch with the needs of the fishermen than today's specialists, were well aware that there was more to a successful fishery than the greatest possible total catch. What is new is the general recognition that the ultimate objectives must be identified, if only implicitly. This has removed the earlier imperfection of having no objective or pursuing a clearly erroneous objective, but problems still remain.

The most obvious problem is in finding a definition of the objectives of fishery management. In the perfect world it would be possible, as the first stage in any management program, to set out clearly defined objectives, e.g., in the articles of an international fishery convention, or in the basic text of national fisheries legislation. To state objectives in broad terms, e.g., "that measures should be taken to promote the conservation and rational utilization of the resource," is relatively easy but still leaves the question open of what particular objective or balance of objectives—high total catch, high employment, high income for fishermen, etc.—should be taken as guides in connection with a given management decision, e.g., next year's catch quota. The danger then is that expediency will dictate that short-term objectives and those of immediate political attraction will be dominant. This problem has been attacked by attempting to set out objectives and basic policy in sufficient detail that they determine precisely what actions should be taken. An example is the New Management Policy (NMP) adopted by the International Whaling Commission (IWC) in 1975, when it was found that the existing convention, backed up by the broad concept of MSY, left considerable argument as to what should be done. Possibly the system would work if the scientific advice were perfect and could determine precisely and without argument the state of the stock, the value of MSY, etc. Then the NMP would determine what measures should be taken. However, it is not easy to assess the state of whale stocks, so there is considerable argument at every IWC meeting which, though expressed in terms of scientific interpretation of data, is basically over objectives (protection vs. harvesting).

A problem that is likely to be more important in the long term is that in the imperfect world it is highly unlikely that any detailed definition of objectives and policies will have lasting validity. As time passes new objectives or a new balance between objectives may arise; it may become clear that the original definition was based on an incomplete understanding of the interlocking systems involved in resource management, e.g., in the biologic system of the interaction between different species supporting different species; or the definitions may turn out to be more rigid and restrictive than originally intended.

The System

Fisheries operate as part of a complex system, or rather the act of catching and selling fish is part of a number of overlapping and interlocking systems, which extend in one direction to all the biological, and also physical and chemical, events in the ocean, and in the other to the economic and social problems of isolated fishing communities or to the political influences which determine who should be given licenses to operate in a profitable but potentially overcapitalized fishery. In the past many of the failures to manage fisheries were due to dealing with too restricted a system. Though many fishery biologists from Michael Graham (1939) onwards have been active in pointing out the wider social implications, there often has been the tendency to regard management as almost entirely a matter of the population dynamics of some particularly preferred species (salmon, plaice, etc.). This had two results. First, the advice tended to be impracticable and unrealistic–and to a large extent ignored–because it did not address the real obstacles to better management, or the real objectives of those responsible for taking management decisions. Second, the advice could be actually erroneous, even in biologic terms, because of the impact of other events in the sea–fisheries on other species or changes in the natural conditions. The Peruvian anchoveta fishery provides an example of both failures. The effectiveness of the biological advice, as prepared by the Instituto del Mar del Peru (IMARPE), and also presented in the reports of advisory panels (IMARPE 1970a, 1972, 1973) in influencing basic government policy was greatly increased when the scope of advice was widened through a panel that included economists as well as biologists (IMARPE 1970b), which clearly identified the overcapacity of both vessels and fish meal plants as the problem that had to be solved. The collapse of the Peruvian anchoveta fishery showed that the simple single-species production model, e.g., that of Schaefer (1954), failed to predict what might happen, though the scenario suggested by Paulik (1971) and by the 1972 panel report came very close to the actual events in 1972-1973. Though it is still not possible to separate the roles of environment, especially "El Niño," and of heavy fishing in triggering the collapse, it is clear that the scope of traditional single-species models needs to be widened to consider questions of stability, the impact of environmental fluctuations, and the influences of other species. The events in

Peru, and still more in Chile, since 1972 show also that this widening of scope to include several species applies also to the economic side. The decline of the anchoveta has coincided with (even if it is not yet proved to be the cause of) increases in sardine and other species. Catches in Chile changed from 960,000 tons of anchovy and 175,000 tons of sardine in 1971 to over 1.6 million tons of sardine and only 50,000 tons of anchoveta in 1980. Since anchovy can, with very minor exceptions, be used only as a low-priced raw material for reduction to meal and oil, and the sardine can be used, at a much higher price, for canning and other forms for direct human consumption, the decline in anchovy has not been an unmixed disaster.

There is now enough appreciation of this to be reasonably hopeful that many fishery managers will avoid the mistake of looking at too restricted a system. The danger now may be that attempting to look at the whole system in a comprehensive way can either induce an intellectual paralysis or lead to undue trust in some specific model that attempts, for example, to link together most of the elements in the natural ecosystem, or to link biological models with most of the economic and social processes. A number of models, notably those of Andersen and Ursin (1977) and Laevastu and his colleages (e.g., Laevastu and Larkins 1981) exist that describe the interactions between species but are not yet sufficiently well developed, nor have they got sufficient general acceptance, to be used as immediate guides for managing multispecies fisheries. As a result many decisions in such fisheries, e.g., in the North Sea, are based on the better developed single-species models, with the implicit assumption that interspecific effects are not important. The results may be far from optimal (e.g., Gulland 1981). Economists have been successful in bringing together biological and economic factors (e.g., Scott 1955; Clark 1976) which show clearly how to determine where the perfect fishery would be, as seen by the economist. However, to make the models tractable many aspects, including the real biological complexities and the fact that many peoples' ideal world can be very different from the economic optimum, are ignored. Thus, while useful in suggesting what might be done and in pointing out the likely consequences, which might otherwise be unexpected, these models also are not particularly useful as immediate guides to what should be done.

I conclude from this that it is not going to be possible to deal with the complete system in a single comprehensive analysis. Any attempt to do so will run into intractable problems of model building and data collection. Advice on management will have to continue to be determined, particularly in a quantitative sense, by analyses of small parts of the whole system, e.g., the population dynamics of a single species. What is important is that the advice includes qualitative consideration of how other elements of the system might modify the results of the narrowly based analysis, e.g., the degree to which increasing the abundance of cod in the North Sea, while desirable for the cod fisheries, might have undesirable effects on fisheries on species that cod eat.

Data

A constant feature of discussions on fisheries and fisheries management are complaints about the inadequacy of the basic data, particularly catch and effort statistics. Certainly in nearly all fisheries in the world today it would be easy to improve the catch statistics, and this would result in an improvement in the understanding of the dynamics of the stock and in the reliability of the advice given. Improvements along these lines should be of high priority in the next few years. However, there is a limit to how far this improvement should go. While the total catch is not known better than, say, to within 30%, with virtually no details of catch or size composition or of fishing effort, improvements which reduce the probable error in total catch to no more than, say, 5% and to provide species, size, and effort data would improve the situation. They would enable at least the simpler models to be applied with confidence and could, with careful design, be done at a cost much less than the benefits achieved.

Once these improvements are achieved, it is natural, as has been the case in the few fisheries where the basic data are not wholly inadequate, for the scientists to see where further improvements could be achieved. It would be nice if the remaining sources of bias or variance in the catch statistics were removed. Details of catches and corresponding fishing effort broken down according to increasing fine time and area divisions would undoubtedly provide better indices of fish abundance. Increased sampling of the sizes and ages of the fish landings would enable more detailed methods of analysis to be applied with greater confidence. Apart from data from the fisheries themselves more collections of many kinds of research data—acoustic surveys of pelagic stocks in more detail and at shorter intervals, trawl surveys of small prerecruit fish, etc.—would undoubtedly be useful. There must, however, be a limit in the amount of data collected. In the extreme case of whaling has it been possible to collect and compile information on the species, size, and sex of every animal killed, at least in the major fisheries. Even with these data it has still been very difficult to be sure what has been happening to the whale stocks, except that most of them have been decreasing.

Current changes in the supply of data are not all improvements. The introduction of regulations has made it attractive to misreport data of different kinds, according to the nature of the regulations. Sometimes the misreporting is quite subtle. Off California the anchovy fishery has been controlled by catch quotas for a number of distinct areas. In this fishery it is believed that fishermen initially report most of their catches as coming from the more distant areas, so as to keep the fishery in the nearer areas open for as long as possible. In the North Sea the application of catch quotas based on estimates of total allowable catch have meant that the reported catches of the most valuable species (sole) may be not much more than half the actual catches. These are problems of enforcement, but the growing number of such cases, in the extreme introducing serious doubts over the scientific analysis on which the regulations are based, suggest that more attention needs to be paid to the impact of regulations on the supply of data and the possible adjustment of the regulations to take this into account.

Analysis

Some of the shortcomings of current methods of analysis have already been noted. Those models that are simple enough to use ignore all but a small part of the whole fishery system, while those that provide a description of most of the system (e.g., multispecies biological models or bioeconomic models) tend to require more data than can be made available in order to give usable results.

Much of the emphasis in the theoretical side of current fishery research is toward the development of improved models that better describe the fishery system (or at least a greater part of that system) and that are less demanding on data. Similarly, much of the practical side of research is concerned either with improving the supply of basic data, so that existing models can be more widely and more realistically applied, or with making the observations that will test the newer models and theories as they are developed. Progress is being made, and undoubtedly it will continue to be made. For example, there are prospects that the belief held by many oceanographers and fishery biologists—that knowledge of the physical environment in which fish stocks live, and of the variability of that environment, will be of direct value to the fishery manager—will soon be more than an article of faith and will be based on actual operational experience. Some of the papers in a recent volume edited by Glantz and Thompson (1981) point out the possible opportunities for this in relation to the impact of El Niño on the Peruvian anchoveta fishery. Similarly there is a better understanding of the general nature of the interaction between fisheries on different species (see, for example, May *et al.*, 1979; FAO 1978).

There is, however, no sign that the gap between what current models can do and the current demands by fishery managers will shrink. Any improvement in the methods of analysis has been more than matched by demands from the managers for increased accuracy, and for the analyses to take account of a wider range of factors.

This continuing gap is not a disaster. As pointed out by Walters (1981), there is "a prevailing myth in resources management . . . that one should construct a model that best reflects available data, and then operate as though this model were correct." The aspect of this myth I want to stress here is the pressure, on both scientific adviser and administrator, to adopt some model and method of analysis as the standard, and therefore correct, procedure. Thereafter this procedure is used without much more question, although the model may become less and less representative of reality and less and less appropriate for giving the type of advice that the manager needs. The sequence of events—and the problems of the scientific community, individually or in groups, in having to do too much in too short a time—are well illustrated by the use of cohort, or virtual population, analysis in the northeast Atlantic. This method was developed independently in several places (Gulland, 1965; Jones, 1964; Murphy, 1965). In the context of the advice being provided by International Council for the Exploration of the Sea (ICES), the application was developed to deal with a problem faced by the ICES Working Group studying the Arcto-Norwegian cod. With the mortalities then believed to be occurring in the juvenile and feeding fisheries (mostly on cod from 3 up to around 8 years old), there could not be enough survivors at 9 years and upwards to provide the recruits to the spawning stock and the spawning and prespawning fishery around the Lofoten Islands. Virtual population analysis resolved this inconsistency and enabled the working group to show that part of the apparent mortality in the juvenile fisheries was a decline in fishing mortality (or more precisely the age-specific catchability coefficient q_i) among the larger fish.

Cohort analysis is particularly suited to dealing with long series of data giving the total number of each age caught in fisheries on medium- to long-lived fish—the situation in most fisheries of the northeast Atlantic. It was also found scientifically attractive because it did not require effort data and was developed at a time when the shortcomings in catch per unit effort as a measure of abundance, especially in herring and similar fisheries, were becoming widely recognized. It also had some more subtle things in its favor as a standard technique; the basic procedures used were readily understood, but the calculations took some time (especially when repeated revisions were made to the basic catch at age data). The calculations can then neatly involve most of the time available during the meeting of 5 days or so that are typical of the several dozen international working groups on individual stocks that do most of the preparation of scientific advice in respect of the northeast Atlantic fisheries. Finally the method is well suited—on condition that reasonable assumptions can be made about the value of the fishing mortality in the most recent year or two—to providing estimates of the magnitude of the total allowable catch (TAC) appropriate to whatever is taken to be the long-range management strategy, e.g., set fishing mortality at the level that will give the maximum yield per recruit, F_{MSY}.

The work of providing scientific advice in the northeast Atlantic has therefore been very largely concerned with carrying out cohort analysis and calculating TACs. Management has equally largely been concerned with implementing these TACs, especially deciding how the total should be allocated between and within countries. There has been little questioning—largely because the pressures of the successive stages in the management process have not allowed for questioning—of whether the analysis was giving the correct answer, of whether the analysis was that best suited to giving the most useful advice, or whether the management policies were best suited to achieving the real long-term objectives of the manager.

The results have been what might have been predicted. The working groups assessing individual stocks have stepped on the occasional scientific banana skin, producing conclusions which later turned out to be substantially in error, and the managers have failed to develop policies which would tackle directly the major problems in the fisheries. Ironically one of the more striking failures occurred in relation to the Arcto-Norwegian cod stock, for which the cohort analysis technique had been first developed. In the late 1970s a sudden increase in fishing effort was not detected in the cohort analysis, so that the fishing mortality in the most recent years (not dealt with well by the analysis) was underestimated, and hence the stock overestimated. This in turn allowed catches to be too high, and the divergence between estimates (of population or of fishing mortality) to increase to the point that when other data (from surveys and commercial catch and effort) were examined the estimates of allowable catch had to be revised downwards very greatly.

I am not concerned here with the suitability of cohort analysis (which is as good as most other approaches, if not better), or with whether one or other working party should have used better methods (it is always easy to be wise after the event), but with the mechanics of preparing advice and the fact that the system definitely encourages the semiroutine application of accepted methods of analysis and discourages careful examination of these methods. This situation is not confined to ICES and the northeast Atlantic, but is a feature of most international commissions, as well as many national systems. One exception in recent years has been the International Whaling Commission, where methods of analysis have been under regular challenge and new techniques frequently proposed. There is an honorable reason for this. Disagreement among scientists has often been the first line of defense among administrators not wishing to take difficult management decisions. In response the scientists who want to see action taken have put emphasis on consensus. Again the IWC is the exception. For whales the burden of proof has swung the other way, so that to some extent the ability of the stock to withstand harvesting has to be established before catching can continue. In the IWC therefore it is, again only to some extent, in the interests of those seeking more vigorous conservation to question current methods of analysis and to avoid the growth of a consensus.

The influence of an unchallenged standard method of analysis on the nature of the advice and on the general pattern of management is probably more serious, though less easy to demonstrate. The drastic changes in the cod quota showed everyone that more than routine cohort analysis was needed to calculate reliable TACs for the Arcto-Norwegian cod. It is less easy to show to what extent the current state of management in the North Sea—which is virtually unanimously agreed to be highly unsatisfactory—is because the scientists are accustomed to processing data to produce recommended TACs for the coming season for each stock, and therefore the manager's attention is focused on these TACs. Nevertheless it is reasonable to suppose that different systems of analysis—whose principal outputs were, for example, estimates of the desirable level of fishing effort in terms of number of vessels—would have given a different thrust to how the management decisions were taken.

The damage done in the situations described here has come less from the method of analysis used being worse than any other than from the fact that it has continued being used for too long. This exposed the weaknesses of the method: that certain types of information on the resources were not effectively used, and that attention of the managers became focused in particular methods of management to the exclusion of other, and possibly more suitable, methods. The moral is that methods of analysis should be changed regularly, or at least alternative methods should be used in addition to traditional methods. The search for alternatives should not wait until those currently used have been clearly demonstrated as inadequate but should be made at regular intervals. Also use of an alternative method should not wait until it has been demonstrated as superior. In fact many alternatives will not be any better, as a regular method analysis, than that being currently employed, but nevertheless their use will still be valuable.

Decision Making

The point at which decisions—to increase the mesh size to 70 mm, to limit the total catch next season to 120,000 tons, etc.—are taken is clearly one of the key points in the whole management system. The fishery biologist is often trained to believe that these decisions are taken by administrators who coolly weigh the information given to them on the state of the fishery, and the consequences of different action, take account of the declared national objectives for the fishery, and in a logical fashion decide on the actions to be taken. Perhaps somewhere this is the case. More often decisions have to be taken by harassed officials under heavy and conflicting pressures, often only indirectly concerned with fishery issues—for example, from a politician about to face a difficult election in a fishing town.

It must be accepted that the decisions taken under these circumstances will not be the theoretical best. Equally it must be accepted that these circumstances are not likely to change—and certainly will not change merely because changes would make life easier and less frustrating for those providing scientific advice on management. With differences according to national customs and political systems, the political and similar pressures are likely to remain much the same as they are. Those preparing scientific advice must accept this situation and consider how their advice may be altered to take it into account. Probably not very much change is needed, at least from well-presented advice. A recommendation for action, presented without clear explanation is likely to be ignored if it does not appear to match immediate political needs, however beneficial it might be in the long run. However, if advice is expressed in a clear and simple manner that can get the attention of the politicians and decision makers, the possibilities of adopting good long-term policies are greatly increased.

Implementation and Enforcement

Fishermen are probably no greater law breakers than any other group of people. However, fishing does encourage the independent view and reluctance to accept, without proper explanation, rules and regulations, especially if they come from bureaucrats in a distant capital. Further it is not easy for a government official to check on what the individual fisherman is doing, perhaps in a small boat in poor weather some way from land. Only in a perfect world therefore is it reasonable to assume that rules and regulations to manage fishing would, once adopted, be necessarily carried out correctly. In the real, but imperfect, world some types of regulation are clearly extremely difficult to enforce. To take an example from an area where enforcement is probably more effective than most other places, the actual catch of sole in the North Sea is much greater—probably twice as great—as that officially recorded—the latter matching the official quota.

Increased attention is therefore being now given to problems of enforcing regulations. In the first instance particular emphasis has been given to controlling the activities of foreign fishermen, but it is generally recognized that a country's own fishermen are, if uncontrolled, equally capable of damaging resources, and the livelihood, as any other fishermen. Though monitoring, control, and surveillance (FAO 1981) are largely technical matters for coastguards, fishery officers, and similar authorities, they do interact with other aspects of management. The enforcement officers need to decide, given certain fishery regulations, what are the most effective ways of ensuring that the fishermen comply with them. It is soon obvious that to attain full compliance, e.g., to patrol a closed area to be sure that no fishing vessel enters it, can be an extremely costly matter. Attention then has turned to questions of finding the most cost-effective method; and, accepting that full enforcement is not possible, determining what rate of infringement is acceptable; and finding what pattern of enforcement can ensure that this rate is not exceeded at the least cost.

At this stage considerations of enforcement procedures become of wider interest. Different management measures can have the same broad objective (to protect small fish or to reduce the total fishing effort), but the costs of enforcing them to the equivalent levels of compliance can vary greatly. Once it is recognized that compliance with regulations will not be perfect, and that the costs of enforcement can be significant relative to the benefits to be obtained, then consideration of enforcement must become an important element in choosing what measures should be adopted.

This may mean reexamining some of the accepted truths of fishery management, including the belief that the best way of controlling the amount of fishing (i.e., keeping the fishing mortality at some desired level) is by setting a figure for the total allowable catch. The most careful examination of the choice between catch quotas and controls on effort was probably done in the mid-1960s, at a time when International Commission for Northwest Atlantic Fisheries (ICNAF) and other international commissions were considering extending their measures from controls on the quality of fishing (mesh size and minimum landing size of fish) to controls on the amount of fishing. In the international fisheries of that time the problems of setting controls that would take account of the varied fishing power of large Russian factory trawlers and small inshore Newfoundland fishermen, and also of changes in efficiency, or fishing power, of each gear from year to year, were horrendous. Catches which could be directly compared were the most promising method of control. They were generally adopted, even though they have disadvantages, e.g., the need to adjust catch quotas each year according to variations in year-class strength and other changes in the stock. Little account was taken at that time of difficulties of enforcement, it being tacitly assumed, in polite international circles, that regulations once agreed upon were thereafter obeyed.

This assumption is no longer widely credited. Much more common is a suspicion by fishermen that all other fishermen are failing to comply with regulations. Thus suspicion is shared by most other people connected with fisheries. Casual conversation with senior fishery officials in Europe show that they fall into two classes—those who believe that foreign fishermen are exceeding catch quotas, and those who, off the record, in addition accept that perhaps their own fishermen are bending, if not actually breaking, management regulations. In these circumstances effort controls have a considerable advantage over catch quotas. Even with quite good controls at each landing port, it can be difficult to be sure whether a country's total catch falls within an agreed limit. It is even more difficult to determine whether an individual vessels is complying with the regulations, except in the case of any vessels that are clearly continuing to fish after the quota is reached and the fishery is closed. On the other hand, if the control on effort is expressed in terms of allowing a certain number of vessels to operate and all those interested are issued with the list of licensed vessels, with the port registration numbers, radio call-signs, or other identification marks, it is easy to check whether the operation of a vessel observed to be fishing is in fact within the agreed limits. Of course this is a simple case, and in practice effort controls (and indeed catch controls) are subject to a number of complications–the need to adjust the total effort which is allowed each year for increases in fishing power, the question of fishing directed to a mixture of stocks, etc. Nevertheless one of the aspects of management that clearly needs reexamination in the light of imperfect and costly enforcement is the current emphasis (particularly in the text of the new draft convention on the Law of the Sea) on setting limits on total catch as a method of controlling the total amount of fishing.

Discussion

The preceding sections have discussed the imperfection that often, indeed in some cases almost inevitably, occur in the various stages of fishery management. Suggestions are made for modifying actions in the light of these imperfections. In doing this emphasis has been given to recognizing these imperfections, and learning to live with them, rather than attempting to eliminate them. What is true of the individual stages is true *a fortiori* of the management process as a whole. No fishery ever has been, or ever will be, perfectly managed. The fishery manager should therefore not attempt to determine and implement the ideal management, but determine actions which are practicable and which will make the fishery better (in whatever senses are thought to be important) than it would be if he took no action.

An acceptance of imperfection in the process as a whole–rather than a belief that if only one or other imperfect element, such as the biologists' analysis, could be perfected all would be well–must color the basic approach to management. The manager will need to have much closer, and sympathetic, links with the fishermen and the fishing industry, as well as with his advisors and others that can influence the way the fisheries develop. If the regulations set by the manager are the correct ones and do not need to be corrected and amended at intervals, the manager can afford to distance himself from those he is seeking to benefit and control. If, however, he has to keep changing the regulations then it is important that those affected understand the reasons for the change and the problems and uncertainties faced by the manager and his advisor and can thus be persuaded that the actions being proposed are more likely to be beneficial than others. Acknowledgment of the uncertainties and imperfections of management, particularly in the narrow sense of applying regulations, therefore lends support to the pleas being made (e.g., by the ACMRR Working Party, FAO 1980) to consider the wider framework. Fishery management should look at all the factors determining the way fisheries operate. If the regulator will have difficulty in controlling a fishery on a heavily fished stock, it is important that the growth of the fishery proceed in such a way as to reduce his difficulties. For example, if grants or subsidies are used to encourage the development of a new fishery, they should be phased out early, soon after the fishery has shown itself to be economically viable, before their main effects are to encourage overcapacity in a fully developed fishery and to exacerbate the problems of the regulator.

The chief change in approach that seems called for, however, comes from a wider recognition that whatever is being done at the moment will have to be changed. The methods of biological analysis, or economic modeling, or the form of the regulation may be the best that can be thought of in the light of current knowledge—though this itself is a bold assumption—but new data will turn up, new theories will be generated, or the balance of national objectives will change, and other models will have to be developed and new regulations introduced. One lesson is that current procedures have to be regularly questioned and alternatives considered. For example, the fact that cohort analysis has for the last decade or so been widely used by international biological working groups in the northeast Atlantic, almost to the total exclusion of other methods of analysis, would suggest that very high priority should be given to other approaches, at least in parallel. A related convention widely used in the northeast Atlantic, but also elsewhere, is that the key step in managing any fishery is the consideration of the total allowable catch for the coming season. Again it would appear that other approaches, which on the one hand give the biologists more incentive to look at the long-term effects on the stock, and on the other address more directly the problems of overcapacity, would be valuable.

Another lesson, being expounded very vigorously by Holling and his colleagues at the University of British Columbia and at the International Institute for Applied Systems Analysis (Holling, 1978; Walters and Hilborn, 1976), is that management should become more adaptive. At the simplest this implies that one aim of management should be that the fishery and the management system are in a position to deal with unexpected events. Of course by definition it is difficult to plan in advance for unexpected events, though flexibility is always valuable, and often the general pattern of some future events can be predicted, even if the specifics cannot be. For example there is enough experience of the large stocks of clupeoid fish in upwelling areas and elsewhere to make it fairly likely that a period of high catches of one species will not continue for long. The causes may vary—heavy fishing, a shift in environmental conditions, or a combination of both—but such stocks have often collapsed and often this collapse has coincided with, or been followed shortly afterwards by, a rise in a related species (sardine/anchovy in California, anchoveta/sardine in Peru). Apart from predicting, or better, promoting action that will prevent such collapses (and more than partial success in either of these is some way away), the manager should see that the fishery is arranged so that the damage of a possible collapse is not too great and that it can take advantage of the rise of any alternative species. In practice most regulations, insofar as they have any effect on the flexibility of the fishery, have the opposite effect. Regulation distorts the relative attractiveness of different fishing methods and in the long term tends to produce a fleet that is well matched only to the peculiar conditions set by the regulations. So far as possible the form of the regulations should be chosen to minimize this tendency.

The approach to adaptive management stressed by Holling and others goes beyond this and requires a greater degree of flexibility in the management process itself. They also stress the need to use management to reduce the amount of uncertainty, and to set the regulations for one season so as to produce (within the constraints of the other objectives of management) the greatest amount of new information for setting the regulations in future seasons. Without some such approach it is highly likely that regulations are repeated from season to season, with the fishery remaining unchanged in a state which appears to be satisfactory but is a long way from the optimum.

In conclusion, I do not believe that the imperfections in the ways fisheries are managed can be completely removed. Objectives will continue to be poorly identified; managers will continue to look at only part of the whole problem; their advisors will have poor data and use inappropriate methods of analysis; regulations will be enforced only partially. I do believe that if the existence of these imperfections is properly recognized some of them can be reduced, and even if this cannot be done, their impact on the overall success of management can be made much smaller. Even if fisheries can never be managed perfectly, they can be managed much better than at present.

References

Andersen, K. P., and E. Ursin. 1977. A multispecies extension to the Beverton and Holt theory of fishing, with accounts of phosphorus circulation and primary production. Meddr. Dan. m. Fiskeri-og-Havunders N.S. 7:319-436.

Clark, C. W. 1976. Mathematical Bioeconomics: The Optimal Management of Renewable Resources. John Wiley, New York, 352 pp.

Doubleday, W. G. 1976. Environmental fluctuations and fisheries management. Collected Papers. Int. Common. Northw. Atl. Fish. 1:141-150.

FAO. 1978. Some Scientific Problems of Multispecies Fisheries. Report of the Expert Consultation on Management of Multispecies Fisheries, Rome, Italy, 20-23 September 1977. FAO Fish. Tech. Pap. 181, p. 42.

FAO. 1980. ACMRR Working Party on the Scientific Basis of Determining Management Measures. Report of the ACMRR Working Party on the scientific basis of determining management measures. Hong Kong, 10-15 December 1979. FAO Fish. Rep. 236, p. 149.

FAO. 1981. Report of an expert consultation on monitoring, control and surveillance systems for fishery management. (Mimeo.)

Glantz, M. H., and J. D. Thompson (eds.). 1981. Resource Management and Environmental Uncertainty: Lessons from Upwelling Fisheries. New York, Wiley Interscience. 491 pp.

Graham, M. 1939. The sigmoid curve and the overfishing problem. Rapp. Proc.-Verb. Cons. Int. Explor. Mer 110(2):15-20.

Gulland, J. A. 1965. Estimation of mortality rates. Annex to Arctic Fisheries Working Group Report. Int. Cons. Expl. Sea, Ann. Meeting, 1965. (Mimeo.)

Gulland, J. A. 1981. Long-term potential effects from management of the fish resources of the North Atlantic. J. Cons. Int. Expl. Mer. 40(1):8-16.

Holling, C. S. (ed). 1978. Adaptive environmental assessment and management. Wiley, Int. Inst. Applied Systems Analysis, Chichester, U.K., 377 pp.

Holt, S. J., and L. M. Talbot. 1978. New principles for the conservation of wild living resources. Wildlife Monogr. 59:1-33.

IMARPE. 1970a. Report of the panel of experts on population dynamics of Peruvian anchoveta. Bol. Inst. del Mar del Peru 2(6). 324-371.

IMARPE. 1970b. Report of panel of experts on the economic effects of alternative regulatory measures in the Peruvian anchoveta fishery. Inf. Inst. Mar Peru, Callao 34, p. 83.

IMARPE. 1972. Report of the second session of the panel of experts on the population dynamics of Peruvian anchoveta. Bol. Inst. del Mar del Peru 2(7) 377-457.

IMARPE. 1973. Report of the third session of the panel of experts on the population dynamics of Peruvian anchoveta. Bol. Inst. del Mar del Peru 2(9). 525-599.

Jones, R. 1964. Estimating population size from commercial statistics when fishing mortality varies with age. Rapp. Proc.-Verb. Int. Cons. Expl. Mer, 155: 210-214.

Laevastu, T., and H. A. Larkins. 1981. Marine Fisheries Ecosystem: Its Quantitative Evaluation and Management. Fishing News Books Ltd., London, U.K.

Larkin, P. A. 1972. A confidential memorandum on fisheries science. *In*: B. J. Rothschild (Ed.), World Fisheries Policy. Univ. of Washington Press, Seattle.

May, R. M., J. R. Beddington, J. W. Howard, and J. G. Shepherd. 1978. Exploiting natural populations in an uncertain world. Math. Biosci. 42:219-252.

May, R. M., J. R. Beddington, C. W. Clark, S. J. Holt, and R. M. Laws. 1979. Management of multispecies fisheries. Science 205(4403):267-277.

Murphy, G. I. 1965. A solution of the catch equation. J. Fish. Res. Bd. Can. 22(1):191-202.

Paulik, G. J. 1971. Anchovies, birds and fishermen in the Peru Current. *In*: W. W. Murdoch (Ed.), Environment: Resources, Pollution and Society. Sinauer, Stanford, Conn.

Schaefer, M. B. 1954. Some aspects of the dynamics of populations important to the management of marine fisheries. Bull. Inter-American Tropical Tuna Comm. 1:25-56.

Scott, A. D. 1955. The fishery: the objectives of sole ownership. J. Polit. Econ. 63:116-124.

Walters, C. J. 1981. Optimum escapements in the face of alternative recruitment hypotheses. Can. J. Fish. Aquat. Sci. 38:678-689.

Walters, C. J., and R. Hilborn. 1976. Adaptive control of fishery systems. J. Fish. Res. Bd. Can. 33:145-159.

9

The Management Challenges of World Fisheries

Garry D. Brewer

The Setting

By now the realization that an additional third of the globe has passed to national control, in the wake of the Third United Nations Conference on the Law of the Sea (UNCLOS), has sunk in. Implications of this unprecedented "enclosure movement," as it affects the world's fisheries, have been recognized since at least 1978:

> Our concern is that the scientific and technological communities will be insufficiently prepared to deal with the ramifications of the world enclosure movement. . . . Fishery management systems will have to face regimes amounting to national property rights in various stocks of fish, as opposed to the open access of the past. . . . It is therefore necessary for national and international marine policies to be formulated and implemented in a more coherent and coordinated fashion than they have been in the past (Ross and Miles, 1978).

However, positive steps to formulate and implement the "coherent and coordinated" policies deemed essential have been few. With slight exception, formula-

---

Garry D. Brewer is a Professor of Organization and Management and of Forestry and Environmental Studies at Yale University. Trained in economics, mathematics, public administration, and political science, he has in the last 5 years become increasingly concerned with the management of fisheries around the world. He has served as an advisor to the U.N. Food and Agriculture Organization and to the Woods Hole Oceanographic Institution and is at present engaged in a collaborative research project at Yale on the general topic of food, population, and natural resources. He serves on the editorial boards of five policy- and management-related journals and is the author or coauthor of seven books and over 50 scientific articles on a variety of policy and management subjects.

tion has been overlooked or treated piecemeal, or when efforts have been made, the recommendations offered have been stymied (FAO, 1980).[1] Moreover, where any movement toward implementation can be discerned at all it usually looks very much like steps to rationalize preexisting conditions, securing claims of specific interests to a nation's fishery resources (Hoole *et al.*, 1981).

The problem is complicated; however, one essential fact stands out despite many particulars that divert attention away from it. "The use of common property resource stocks inevitably involves conflicts of interest, which in the absence of appropriate institutional constraints, almost invariably lead to overexploitation and depletion of the resources" (Clark, 1981, p. 232). For anyone interested in the prudent management of the world's fish one interpretation of this fact is inescapable: the best scientific efforts to understand the extraordinary complexities of the ocean's biological systems and processes may well come to naught in the absence of their integration with political, institutional, and management systems capable of understanding and using the fruits of these efforts.

The world enclosure movement, with its creation of 200-mile Exclusive Economic Zones (EEZs), represents a profound decision, but one made with little forethought about its consequences. Among those, the virtual absence of insitutional structures capable of carrying out the decision is perhaps as indicative as it is troublesome. There is not now, nor is there likely to be, a comprehensive authority able to perform the management tasks demanded by the new regime of the sea. Magical forces, whether they be "invisible hands" in the marketplace or fairy godmothers granting fantasy wishes, are not going to do the job either in the absence of an extraordinary amount of highly unusual cooperation and coordination among nations of the world (Bergsten, 1974).

It may even be more basic than this. Institutions exist in response to policy directives and guidance; they are the practical means by which broad, strategic objectives are sought. However, even in the developed nations of the world, appropriate policies concerning the use of the ocean are scarce, and for many developing nations there are no policies at all.

> For the United States, as recently as 1975, the investigative branch of the U.S. Congress flatly stated, "the United States had no comprehensive ocean program" (U.S. General Accounting Office, 1975, p. 1).
>
> And, "In a large number of cases, developing states must begin the process of forming an ocean policy from scratch" (Friedheim, 1981, p. 286).

One may look at this situation in two very different ways: as a once in a lifetime opportunity or as an unfolding horror. The optimists, while acknowledging the difficulties, would be moved to call attention to the "window of time" in which many substantial and fundamental changes must occur (FAO, 1980, p. 1). The

[1] The analysis reported in FAO (1980) represents the considerable efforts of several concerned scientists and managers to think their ways through the many implications of the new regime of the sea. Despite its publication and its positive reception by many in appropriate positions of authority to begin taking the initial steps to implementation, nothing tangible can be discovered stemming from the work.

pessimists would counter with gloomy forecasts of mounting catastrophes as one resource after another finally gives in to gross overexploitation and collapses. As this level of the analysis, the political and management challenges are formidable, to say the least (Ophuls, 1977).

Moreover, even in those circumstances where policies and institutions do exist, they "All too often . . . are devised on the basis of ridiculously simplistic views of the resource system" (Clark, 1981, p. 237). Ordinary ideas about the problem and the kinds of "solutions" customarily advanced may actually misinform policy choices and inhibit constructive management so urgently needed in the coming years. A very different way of thinking about the problem seems to be called for (FAO, 1980; Rothschild *et al.*, 1982).

As representatives of science, many in the biological disciplines interested in fishery issues have had success in framing and answering questions posed by the disciplines.[2] However, for many other problems that fall outside the narrow bounds so defined, the rational bias, tight discipline, and assorted procedures supporting the disciplinary edifices provide little help. There is much to the matter, but it usually gets down to differences between theory and practice and the basic aims sought from intellectual work.

Science seeks the development of "theory," by generating and testing hypotheses that confirm, refine, and enlarge common understandings of events (Coleman, 1972). "Truth" as a concept and objective of science is contained in such theories. With them, small infusions of evidence from specific settings can be organized and manipulated to yield predictions about the future. However, science's success owes in part to the care its practitioners take in selecting their problems. It also results from the consistency of the "client" for the work: the disciplines, as represented by their adherents.[3]

For practical problems that demand policy responses, such as the management of specific fisheries, the scientific rules are not as applicable, There is no theory capable of predicting the social, economic, and political consequences of a natural event such as El Niño off the Peruvian coast that resulted in the collapse of one of the largest fisheries in the world. There is even confusion and dispute

---

[2]There is obviously much more to cite here than there is space to do so. Science is important, and the general theme and tone of this chapter are not intended to slight its considerable achievements in the fishery area. Neither should one conclude that improvements in other information aspects pertaining to fisheries ought to come at the expense of basic inquiry.

[3]Ecological studies hold promise for improving general understanding of marine systems, in which fisheries are one important ingredient. However, such work is not common, nor is there particular scientific reward for putting together the interdisciplinary groups needed to engage in this difficult work for long periods of time. Problems such as "target switching," by-catches and discards, multispecies and multinational predation on different kinds of fish, economic interdependencies having implications of fishing not obviously and directly connected (e.g., fluctuations in alternative forms of protein prices regionally and worldwide), "chaotic" ocsillations in population abundance, and a host of other ecologically derived topics and questions readily come to mind. However, not so obvious are efforts, anywhere, to confront these questions (Clark, 1981, p. 236).

among scientists about the underlying mechanisms of the event itself, whether the anchoveta will ever return to abundance, when this may happen, or whether anyone ought to be planning or counting on anything connected with the fishery in the future.[4] However, these are precisely the kinds of questions that managers deal with. Managers likewise gain little comfort or guidance from elegant economic theories that "solve" their problems by advocating efficiency through open markets with centralized ownership of the world's fish by a nonexistent international authority (Cooper, 1974). Despite the scientific rigor here, the suggested solution is simply not politically or practically feasible. Moreover, any recommendation that is not, is no recommendation at all.[5]

Science and the disciplined, rational principles on which it is based are unable to explain these events, account for practical realities, or even to begin sorting out or foretelling the various consequences that hold the manager's attention. The prevalent analytic paradigms used to view fishery problems are inappropriate and far more limited than most realize. Moreover, the new regime of the sea is making these weaknesses all the more obvious, particularly their disregard for major, core realities. Conflict and the passions that sustain and generate it are among these overlooked aspects (Hanna *et al.*, 1980). In other words, the "fishery problem" is not only biological and economic, it is political and institutional as well. One of the greatest challenges of the 1980s is learning to accept this as a basis of understanding so that the urgently needed changes in prevalent management practice may finally begin.

For instance, one must learn to accept the fact that politics is only partly a rational process, at least according to scientific specifications. Decisions and the analyses done in their support are always imperfect because the events dealt with are literally "unreal" by narrow rational standards. Our problems are intricate and changeable, and they often mean very different things to those involved. Neither problems, settings in which they occur, nor possible solutions stand still—they evolve naturally and in reaction to our efforts to understand and master them. Hence, one learns that a decision to regulate a certain stock in accordance with a rational abstraction, such as maximum sustainable yield (MSY), is far from the end of the matter and may have little or nothing to do with the nonstop bargaining and compromising that always attend to regulation.[6] Furthermore, despite the considerable temptations of the "optimal" or "best" solutions the

---

[4]From a 1970 reported high of 12 million metric tons (MMT), the Peruvian anchovetá fishery plummeted to less than 0.5 MMT in 1978 (Glantz, 1979).

[5]The realities of national control, increasing demand in excess of apparent supply, and many other practical matters get in the way of these kinds of proposals and analyses. So, too, do the institutional limitations already noted. Certainly for the decade of the 1980s the odds of there ever being a single international authority in control of the world's fish are slim to none, and proposals that assume its existence have nothing to say to inform the realistic problems at hand.

[6]Regulation is a political process, a truism that tends to annoy those of a rational cast of mind. "Regulatory agencies have substantial discretionary power concerning the interpretation and application of their rule-making and enforcement powers. As a consequence, regulation/enforcement becomes essentially a political process entailing bargaining between parties of unequal power" (Freeman and Haveman, 1972, p. 57).

disciplines seek and seem to provide, the information they treat is highly selective and concerns only the what has been, not the what will be that most concerns policy and decision makers. Descriptive knowledge cannot encompass the future, but it is to the future that policy and political acts taken in its behalf are always oriented.

Those honestly seeking to help with practical management problems are at a severe comparative disadvantage to their purer scientific counterparts (Brewer, 1981b). Among the many problems and handicaps, the complexity of the systems involved, human perception and value, and profound uncertainties about the future all figure prominently and cannot be ignored or assumed away. Through all of this, the pursuit of a single "optimal" or "right" solution is thwarted at virtually every turn.

Nevertheless, considerable time is spent pursuing projects narrowly defined by scientists to fit the interests and capabilities of their disciplines. In comparison, the sums so far expended for marine policy formulation, analysis, and institution building are small—far less than the considerable needs. The actual management of fisheries has been left, as a consequence, to a variety of not very well-financed and not particularly well-served individuals around the world. The following illustrative comments by Apollonio (Rothschild *et al.*, 1982, pp. 5-6)[7] suggest a bit about what the problem looks like from the manager's perspective.

On precision: "Even grossly qualitative forecasts may be most useful management tools and serve to reestablish belief in the management process." Likewise, "Our knowledge of the nature, degree, or causes of natural variability is generally anecdotal and hardly to be considered as a basis of management action when compared with the quantitative information routinely offered on the level of fishing mortality and its impact."

On problem identification: "The recent emphasis on quantitative assessments as the basic management tool has not been very helpful to managers because assessments contain little predictive content." Likewise, "A manager is not primarily concerned with an accurate estimate of the abundance of a stock. He is more likely concerned with the probable trend of relative abundance."

On relevant dimensions of the problem: "If a manager, concerned with stocks, people, economics, and regulation, is to meet his responsibilities, he must have advice on how the system works."

However, advice on "how the system works" that fails to take explicit account of the manager's and many other perspectives may be worse than no advice at all. Many individuals, having quite different views on fishery problems, must be considered when trying to figure out how the system works. A large part of the policy problem—including its definition, analysis, and resolution—centers on

[7]Spencer Apollonio, Commissioner, Department of Marine Resources, State House, Augusta, Maine. Managers have much to tell us about the way their worlds look to them. Analysts have an unrealized obligation to listen carefully here to begin getting cues and guidance about their own work. Figuring out "what's the problem?" is to a large extent a question of where one sits and how the view is from that place.

identifying specific interests and settings. Policy problems managers confront nearly always involve disparate interests, and nothing much is accomplished by assuming that everyone appreciates the situation the same way. No two individuals see, comprehend, or value identical events in identical ways. So, for example, when a manager such as Spencer Apollonio says that he wants "grossly qualitative forecasts to do his job better," he is placing a different value on measurement precision than that held by disciplinary specialists. He is saying, in effect, that he would rather be approximately right than precisely wrong—for his purposes and from his perspective. Likewise, when he says that he is "more likely concerned with the probable trend of relative abundance" in a stock, he is telling us that he cares more about the future than the past, where the data are with which scientists work. When an economist talks about efficient markets and assumes that institutions are readily created or easily adapted to fit economic visions of the problem, the equally important equity, distributional, and feasibility goals others value and seek are being slighted—and so it proceeds on through a long list of illustrative examples.

So whose perception or grasp on the problem is better? No one's and everyone's, literally. All relevant perspectives matter, each illuminates the problem differently in angle and intensity. When taken together in composite, moreover, the various views begin to give one a better sense of the complexity of the whole. I speak here not only of the biological complexity that most are already well acquainted with, but of social complexity that has until now been overlooked (Brunner and Brewer, 1971; La Porte, 1975; Simon, 1969). Impermanence is a common feature of both sorts.

Everything is changing, sometimes dramatically. Carefully analyzed data, painstakingly collected during a fixed period of time, are to fishery management what taxidermy is to a three-ring circus: It is interesting to get close to a stuffed lion or tiger, but it is not quite the same as trying to make a live one do its paces while the crowd is screaming and a bunch of clowns are disgorging themselves from a tiny little car.

Managers fulfill a somewhat similar role. They must be sensitive to a variety of interests at the same time as they strive for workable and flexible solutions to real problems. And, as with lion tamers, the best of them can expect to get chewed up from time to time.

The manager's view also emphasizes specific contexts; they have every right to question theoretical or general wisdom offered them with replies such as, "In my area of the world (my area of responsibility), it just doesn't work that way." General solutions to their problems are far less likely than highly specific, realistic, and changing ones. There is no reason to believe that what works in one place will necessarily work in others, or that what worked here last year will continue to work today or tomorrow. Theory, in the terms used previously, is both incomplete and general. Heavy reliance on it for policy and management purposes diverts attention away from important aspects of particular cases it does not and/or cannot cover. The theoretical "solution" to operate at MSY says nothing about most of management's most critical features. From the manager's perspective, there are no general solutions, just oceans of nasty particular details.

The basic themes of this chapter thus stand out. The rules of the game have changed; fisheries have always been enmeshed in ecological webs, but with continuing and increasing pressure from man, many of the interconnections are being stretched to the breaking point;[8] scientists are able to supply some of the information needed to harmonize these systems with the new rules, but they are by no means equipped or disposed to supply all that is required; and the manager's view and task are different and more difficult than many have understood until now. I next turn to a sketch of "fishery management problems," from the broadest perspective, and then move on to discuss the steps or process by which management decisions are reached and carried out. The chapter concludes with some strong opinions and unusual suggestions about the management challenges of the 1980s.

The World Most Likely

Conflict is the name of the game. The history and practice of fishery management have been mixed to terrible. Economic and biological overexploitation and ridiculous conflicts have been the dominant modes of operation. There is no reason to expect much improvement here. There is every reason to expect matters to get much, much worse.

The Northern Atlantic—both east and west—has been and continues to be fished with far more effort and far fewer results than make any sense. Overfishing there, according to one person's reckoning, is producing "needless costs of $4 to $5 billion a year" (Crutchfield, 1980, p. 45). It is hard to see how this could be changed, except for the worse. Worldwide, the situation is not terribly much better. A variety of sources report that the total take of fish has leveled off despite ever-increasing efforts to boost total production. More than 90% of the weight and value of the world's fish is taken within 200 miles of someone's coast. Human per capita fish consumption has not changed much in the same period, but demand for fish has increased, at least as much as additions to the world's population over the time. "On a worldwide basis, then, the aggregate demand for fish is expected to grow fairly steadily, at a rate which is substantially greater than our short-run capacity to expand production" (Crutchfield, 1980, p. 44).

These seem to be some of the major constraints on activity in the coming decade: Steadily increasing demand, better prices, but no apparent means to increase production. Placed against the national property rights of the new regime of the sea, these factors lead to forecasts of continuing overexploitation and increasing chances of conflict, a consistent and ancient human tendency in situations of scarce and valuable resources. More demands are going to be placed on

[8] I would go so far as to state that the social components of the ecological system dominate the biological ones. In other terms, social constraints or parameters are more potent for overall system outcomes than most biological ones. Interestingly enough, many more of the former are also subject to purposive manipulation and control.

the world's available stock of fish than ever before, but the stock is already approaching some natural limits. However, because the demand growth will be gradual, driven as it is by population increases, the destruction of the resource will not occur simultaneously or spectacularly and hence may progress quite a ways before sufficient attention can be focused to enable reasonable and constructive steps to be made. Moreover, because fish have been "nationalized," conflicts between nations will surely result as stocks continue being depleted.

Pressure to continue exploiting these resources beyond sensible limits is currently high and will only become more so. Fish are not only valuable commodities, they are political ones as well. Lacking relevant science and other missing types of information to guide intelligent management, however, decisions about the use of these resources will increasingly become simply political ones taken to insure employment and a supply of food in many hungry parts of the world. As with most political decisions, the time horizon used will be short and consideration of the longer run risks involved from overexploitation will be slight to negligible. Fish don't vote.

The most pessimistic forecast envisions collapses occurring gradually at first and then more frequently toward the end of the decade as increasingly overcapitalized fleets of the coastal states begin taking "their" fish. Developing countries, whose populations are expanding fastest and hence whose needs for food and employment are also growing more quickly, are now sensitized to the existence of an exploitable resource within 200 miles of home. It is going to be extremely difficult to channel the coastal states, some of which are developing countries, into not overbuilding fishing fleets and related processing infrastructure to take those resources. The political and economic pressures to do so as quickly as possible far outweigh arguments about the potential, longer term consequences of overexploitation and ruin.

The current trends toward overexploitation (largely through improved technologies) and diversification to take less than optimal species, already noticeable in the developed and more efficient fleets of the world, will only continue and continue to make matters worse. At some point, probably within the coming decade, a few major stock collapses and certainly numerous smallish setbacks and disasters are foreseeable. Of course, conflicts about who owns which fish will only add to the travail. Fish don't carry passports.

The future, if one's understanding of the past and sense of the present provide any insights at all, is not rosy, but what to do? One thing might be to get a better idea about realistic steps and actions to be taken to prevent the worst of the scenario from coming true and to limit the damage otherwise when it does.

Preventive, Active, and Reactive Management

Policy problems generally pass through various stages, beginning with their earliest recognition as such and analysis about what to do, through the give and take of decision making, onto execution of possible solutions, and then to actions for determining results and whether to continue, amend, or halt what is being done to resolve the problem.[9] In managerial terms consistent with the previous discussion, one can think about preventive, active, and reactive parts of the process.

Preventive management involves the identification of emerging problems, in the spirit of the previous forecast, but with more attention to important contextual details. For example, if a fishery is being overcapitalized, despite contrary advice, what are the likely consequences, when can they be expected to start showing up, and what can be done to head off or reduce their unwanted effects? Planning staffs (public and corporate) and some in universities and dedicated research units could in principle contribute to this class of preventive activities—although very few such efforts are currently visible. The positive side also exists, to seize unrealized opportunities, e.g., developing a recreational or sport fishery having high potential draw for tourist dollars. The basic ideas involved in preventive management are to imagine and figure out how potential problems (or opportunities) might be anticipated and what the probable costs and benefits will be for various alternative uses of resources.

Active management includes decision making, usually a time where "politics" stands out, and then carrying out whatever decisions are reached. Attention and energy have been concentrated on active management for most fisheries around the world. In the normal conception of "manager," most envision specific people carrying out decisions about regulation, quota enforcement, inspection, and other activities of controlling a fishery. My conception of "manager" is inclusive of all these common activities but goes beyond them. There is something gained by identifying everyone having a stake and/or influence in a fishery's intelligent use as being "managers." How I (and others like me) behave and treat the fish stock in the Long Island Sound have implications far beyond the 1 to 2 tons of fish I take there each year as a recreational fisherman. To the extent I care, am sensitive, and am willing to adapt my behavior or to work for changes in the environment, I am truly performing "managerial" functions. This personal aside is meant seriously. Others need to be made more aware of their responsibility and dependence on specific fisheries in which they participate. Active management is everyone's job. Certain official authorities are readily recognized, but many others also play a role. The "social ecology" of most fishery settings is intricate and far more complex than one first imagines (FAO, 1980). All individuals, interests, and groups who are actively involved in specific settings have the potential for affecting the well-being and future status of the resources. Their collective behavior, responsible or not, makes a difference.

Reactive management involves determinations of whether prevalent practice is beneficial or is attaining expected results. It also concerns steps needed to correct inadequate performance, to make amends, or limit damage when matters get out of hand. So, for instance, when a fishery such as the California sardine,

---

[9]This basic idea and general framework for thinking about and acting on policy problems is presented at length in Brewer and deLeon (1983), with respect to American ocean policy and problems in Brewer (1981a), and in an international fishery management setting in FAO (1980). Only the barest essentials are repeated here, and the interested reader is encouraged to consult these other sources for details.

North Sea herring, or Peruvian anchoveta "crashes," what steps are necessary to reallocate the fishing resources previously deployed? What kinds of insurance, retraining, compensation, relocation, and hundreds of other demands the disasters create can be supplied and used? What can be learned to help stave off similar future tragedies in other settings? The list of questions is long and, given the prospects for stock collapses around the world, has special appeal.

In the following section, some challenges and opportunities of the coming decade are presented from the standpoints of these three different kinds of management.

Some Possibilities: A More Hopeful View

The policy and institutional requirements for all three types of management are numerous and wide open. One common thread runs throughout—a need for education at all levels and for all participants in the process.

Take preventive management, for example. If overcapitalization is indeed the root cause driving overexploitation, as it seems to be, a preventive policy and strategy may be to educate those in banks and lending institutions to make them more wary of sinking funds into projects that individually and in the short run "add up," but which have larger collective and longer term risks. Driving this simple message home to a handful of the right financial people may be more valuable and productive than all the enforcement efforts spent around the world for a number of years—if reduction in overexploitation is the goal sought. Fishermen and politicians are motivated to exploit and hence are more resistant targets than the bankers may be.[10] Ideally, government agencies responsible for fisheries and natural resources could be invited to supply expert estimates of the state and level of effort that target stocks could sustain while investment plans and decisions are being formulated. The idea is akin to the "environmental impact statement" required in American law before large-scale capital decisions can be made that have impacts on the environment. Less ideally, and particularly in cases where official agencies have neither the talent nor the inclination to buck strong political forces, this function could be performed by independent, university, or regionally based groups.

A longer term strategy is also conceivable. Groups such as the U.N. Food and Agriculture Organization or universities having natural resource management and business administration training programs could develop and deploy fishery management cases in teaching curricula. The idea here is to provide specific case details for those in training as natural resource decision makers and managers. A less direct approach than the fishery impact statement, concentrating on teaching programs may have numerous long-term benefits as more individuals become sensitive and appreciate the problem's realistic details and implications.

---

[10]In making such a case to the investment people, for instance, analogies to the current status of overbuilding in the airline industry could help make the point. How the argument is made is important, to be sure, but making the argument is even more so.

Popular, mass education is a third general area that warrants consideration. The environmental movement in the developed countries took off in interesting ways with the publication of a few eloquent books that forcefully pointed out the consequences of continuing to dispoil the environment. An opportunity exists for comparable efforts to gain public attention about the perils of over-exploitation of fisheries. Films, books, and other media could be used for this purpose.

At a more detailed technical level, universities and other research groups have a role in creating long-range planning and policy units to concentrate on specific and regional fisheries around the world. Exactly what stocks are at risk? Which ones are poorly used? Are there new stocks that could be exploited with safety and a reasonable hope for success? Are there circumstances where less efficient harvesting techniques (e.g., "most appropriate technologies") might be acceptable as means to increase employment even though being economically "inefficient"? Additional policy and research topics might center on alternative uses of currently exploited stocks. For instance, it is well known that striped bass in the mid-Atlantic and southern New England regions of America are in serious difficulty. It is also clear that their commercial exploitation is far less "valuable" than the recreational aspects of it (a debatable matter, depending on one's perspective). To reach this conclusion one must be creative in devising the accounts to incorporate costs and benefits beyond simple landings and market prices for the fish. An enormous indirect industry in boats, beach and waterfront property, tackle and bait, and other ancillary activities exists because of the striped bass, and all of these are threatened by its current depletion and feared collapse. Work trying to understand the scientific bases of the stock's decline is underway, but I am unaware of related analyses to determine the full social costs and benefits of its demise or rehabilitation. Perhaps such would end up suggesting total prohibition of commercial fishing so as to preserve the vast recreational fishery to the greatest extent possible—or it may result in recommendations somewhat less extreme. Likewise, other strategies may emerge from a truly sophisticated analysis of the fishery, which is to say, an analysis in dimensions other than simple enumerations and prices of the fish in the market.

So who would be the "consumer" for this kind of policy analysis? In many regions of the world the question is moot because clear lines of responsibility for fisheries and institutions to guide their use do not exist. Here, thinking and defining the problem and opportunities involve institution building, e.g., prevention through policy development could mean recommending organizations, roles, and authorities that do not exist. In other parts of the world, "clients" for the research may have to be determined for the specific location and with respect to authorities that already exist but with quite different nominal responsibilities. In a developing country, as a hypothetical case, it might result in involving a tourism or development authority directly, or other plausible but unusual connections.

In the absence of an overarching world authority to own and control fish, the creation of policy and planning boards and research units for regions is attractive. The foundations for these already exist in some areas but not others; a good example is the committee on East Central Atlantic Fisheries composed of the coastal states of Africa from Morocco to Zaire. Active management of such assets common to coastal states logically requires international cooperation and, by extension, so does planning and policy research about longer term uses and implications (Gulland, 1972, p. 296). At the simplest level, regional cooperative arrangements to pool scientific research capabilities for common benefit makes far better sense than each nation's having to "go it alone." Furthermore, bases for cooperation between developed and underdeveloped countries and regions, specifically with respect to survey and assessment capabilities, could easily be created and carried out to avoid either not having as much science on hand as possible or trying to duplicate expensive investments that may be underutilized elsewhere anyway.

Technical and procedural means to monitor, coordinate, and analyze complex policy problems exist in other settings and could be adapted and put into use for preventive fishery management. Such means are necessarily multidisciplinary and site specific and are needed to keep fisheries in as broad a perspective as possible (Brewer, 1975).

Active management has its own potentials for development, and these mainly revolve around the lack of a central authority to govern world fisheries. The burden for intelligent use falls on the nations claiming their resources through EEZs. This greatly complicates matters by making any hope for general solutions just that, a hope, and a very dim one indeed. Rather, we are already beginning to see a patchwork of specific policies and programs emerging, with large variations from country to country and even within a single country (Young, 1981, Chap. 4). This trend to diversification seems hard to deny and will probably intensify in the coming decade.

Whatever the specific details, however, a fairly common pattern of resource use is discernible:

> First, there is a phase of expansion of fishing capacity and increases in annual catches. The need to control catches in order to prevent depletion eventually becomes apparent (possibly only after catches have already begun to decline), and institutions are established to assess the resource and to recommend catch limits and other regulations. But fishing capacity continues to expand—even though such expansion is clearly unnecessary—with the result that the fishing season must be progressively shortened to prevent overfishing (Clark, 1981, p. 233).

Active management that mainly restricts total catches, the usual means, treats symptoms and usually does not work to limit the progressive and unwanted overexploitation of the common-property resource. Hence, licensing, taxing, unified ownership and other normal schemes all fail to address the fundamental problem of "the commons," or they are politically infeasible. However, of the conventional tools used by active managers, allocated quotas appear to have some promise in the nationalized setting (Pearse, 1979; Anderson, 1977), as do unusual measures to discount near-term benefits by the introduction of conservative safety factors that allow for assessment errors and natural variability. We build bridges and airplanes with safety factors sometimes as much as 10 times the maximum, worst-case stresses expected, but we operate our fishing fleets from a premise diametrically opposed.

Most of these procedural matters are at least known, and unique combinations and varieties of them are to be expected in different fisheries around the world. One would expect specific patterns to be rather changeable, in light of their relative success or failure.

Not as changeable will be the common requirement to enforce whatever management decisions are reached. The role of the world's coast guards and navies has been changed indirectly by the creation of EEZs, but even in the developed world, many of these have not been restructured or prepared to take on the new responsibilities inherent in the new regime of the sea. An inkling of the problem is discernible in the following comments about the U.S. Coast Guard's role in the enforcement of the Fishery Conservation and Management Act in America:

> There appears to be no lack of will to enforce fisheries regulations among Coast Guard Officers. But the training of these officers in the realm of law enforcement is seriously deficient. In fact, most of them lack even the most rudimentary understanding of the nature and role of law enforcement (Young, 1981, p. 156).

There also seem to be implementation and coordination problems between the Coast Guard and the National Marine Fisheries Service that have yet to be ironed out (Young, 1981, Chap. 4).

Here again, an educational mission exists that is presently not appreciated or taken very seriously throughout the world. By the declaration that fisheries are to be national resources, national security forces have been drawn into their active management, whether they realize it or are willing to be or not. Discharging this educational responsibility is probably beyond the scope or ability of most fishery authorities, although they certainly should play a key role. The educational requirement is large and has further implications–for the selection of officers, their preparation and training, and the purchase of ships and equipment. No one seems particularly aware or concerned by all of this.

The realm of reactive management is developed hardly at all. It is a serious shortcoming, if one accepts the prospect of stock depletion and crashes. Now is the time when policies and institutions ought to be considered for making realistic determinations of the actual and future status of specific stocks and in preparing fisheries for the likely consequences of decline.[11]

Contingency and fall-back plans need to be thought of beforehand, not after disaster strikes. For instance, what might happen if insurance pools were created by taking a percentage fee from landings in good years to be used during the bad? (It could be assessed according to each vessel's haul or uniformly across all vessels within a fleet.) What might happen if nations required a "decommissioning

[11]As a general policy matter, termination is poorly understood (Brewer and deLeon, 1983, Part VI).

fee" to accompany any new investment in harvesting capacity–an amount to vary with the projected status of the stock? That is, the more threatened the stock, the greater the decommissioning fee. When a fishery collapses, the expenses involved are borne largely by the nation itself; hence, it seems reasonable to collect a contingency fee against likely future demands on the treasury for unemployment, retraining, relocation, and other similar expenses–this in addition to normal programs and benefits covering other sectors of the society.

Nevertheless, presume the worst does happen, and a crash occurs. Whose responsibility is it to serve as salvage specialists or receivers to reallocate investments in ruined or declining fisheries? Who ought to be worrying about and paying the price of stock rehabilitation, if this is feasible? What cooperative arrangements need to be considered between nations that share stocks so that the rehabilitation efforts of one are not expropriated without penalty by another? Who ought to be concerned with retraining and relocation of those in the fishing business when "bad years" persist and become chronic or permanent? The list of pertinent questions is as long as it is indicative of the paltry attention reactive management has received around the world (FAO, 1980). Moreover, to return to an earlier theme, not one of these questions could be considered "appropriate" under the current rules and practices of the sciences upon which fishery management has been based.

A Summing Up and Broad Topics for Consideration

This chapter could not possibly exhaust the topic of management challenges in the coming decade. My purpose has been to pick out a few glaring defects and to prick the conscience of those who are most responsible. An enormous amount of effort is required. There is little time to spend worrying about minute details, considering the magnitude and impact of decisions already made. No progress or movement, in this case, is retrogression. It poses serious threats.

The current stock of scientific information about world fisheries is grossly inadequate. Such inadequacies continue to be used as an excuse to base decisions on considerations having less to do with fish than with other matters. This must change, and the scientists must take charge in seeing improvements made.

Economic and biological overexploitation prevails around the world, and many existing management tools can be shown only to contribute to this problem. More effective controls and means to carry them out must be devised, tried out, evaluated, and then brought to more general use if appropriate. Fishery managers everywhere are responsible here.

A clearer sense of the social complexity of the management setting needs to be developed and promoted to insure that unlikely participants, having profound impacts on the resources, are not excluded in management plans and decisions. Continuing to ignore many key participants has had the effect of misspecifying the problem, which has had many needlessly harmful consequences. Any future discussions of "fishery ecology" or the "management setting of fisheries" that fail to incorporate the social-human dimensions of the problem should be avoided.

Many more specialities than just biological ones need to be enlisted, and interdisciplinary work related to specific fisheries around the world is to be encouraged whenever possible.[12]

Mainly action is needed. The current "do nothing" condition contributes to the problem and will, in time, only make its remediation more painful and difficult. An intriguing aspect of fisheries is that experience with their management could set patterns for use of other shared resources, particularly those in the sea. If the experience continues on its present course, it probably bodes ill for other resources, too.

My intent throughout this chapter has not been to advocate particular solutions, for these must be determined uniquely in hundreds of localities around the world by those who best know relevant details. Neither have I meant to single out specific individuals or institutions for fault or blame. Time is too short and precious for recrimination. Rather, I have taken a sharp, even extreme, position about the prospects to gain attention and to encourage those bearing responsibility to begin thinking and acting while there is still time to do so. If the reader is put off or angry, fine. Such reactions may lead to movement—a far better condition than the paralysis attending the new regime of the sea.

References

Anderson, L. G. (ed.) 1977. Economic Impacts of Extended Fisheries Jurisdictions. Wiley Interscience, New York. 428 pp.

Bergsten, C. F. 1974. Commodity shortages and the oceans. *In:* R. E. Osgood (ed.), Perspectives on Ocean Policy. Johns Hopkins Univ. Press, Baltimore, pp. 167-178.

Brewer, G. D. 1975. Dealing with complex social problems: The potential of the "decision seminar." *In*: G. D. Brewer and R. D. Brunner (eds.), Political Development and Change. The Free Press, New York, pp. 439-461.

Brewer, G. D. 1981a. The decision-making process and the formulation of marine policies. *In:* F. W. Hoole, T. L. Friedheim, and T. M. Hennessey (eds.), Making Ocean Policy. Westview Press, Boulder, CO, Chap. 6.

Brewer, G. D. 1981b. Where the twain meet: Reconciling science and politics in analysis. Policy Sci. 13(3):269-279.

Brewer, G. D., and P. deLeon. 1983. The Foundations of Policy Analysis. Dorsey Press, Homewood, Ill. 468 pp.

Brunner, R. D., and G. D. Brewer. 1971. Organized Complexity. The Free Press, New York. 190 pp.

Clark, C. W. 1981. Bioeconomics of the ocean. BioScience 31(3):231-237.

Coleman, J. S. 1972. Policy Research in the Social Sciences. General Learning Press, Morristown, N. J. 23 pp.

[12]"Interdisciplinary" means much more than having an oceanographer talk to a fishery biologist or to a biological statistician. We are talking about policy, and that means the policy-relevant disciplines need to be integrated and given their due, e.g., law, economics, sociology, political science, anthropology, psychology, and many others.

Cooper, R. N. 1974. An economist's view of the ocean, *In:* R. E. Osgood (ed.), Perspectives on Ocean Policy. Johns Hopkins Univ. Press, Baltimore, pp. 145-165.

Crutchfield, J. A. 1980. Marine Resources. *In:* S. Hanna, K. H. Im, and L. O. Rogers (eds.), Exploring Conflicts in the Use of the Ocean's Resources. Sea Grant ORESU-U-79-001, Oregon State University, Corvallis, OR. Republished as J. A. Crutchfield, Marine resources. Amer. Econ. Rev. 69(2):266-271.

FAO. 1980. The Scientific Basis of Determining Management Measures. J. A. Gulland (ed). Food and Agriculture Organization. UN FAO Fish. Rep. No. 236.

Freeman, A. M. III, and R. H. Haveman. 1972. Clean rhetoric and dirty water. The Public Interest 28 (Spring).

Friedheim, R. L. 1981. Providing direction to the national ocean policy research effort. *In:* R. W. Hoole, R. L. Friedheim, and T. M. Hennessey (eds.), Making Ocean Policy. Westview Press, Boulder, CO, Chap. 14.

Glantz, M. H. 1979. Science, Politics, and the Economics of the Peruvian Anchoveta Fishery. Mar. Policy 3(September):201-210.

Gulland, J. A. 1972. Population Dynamics of World Fisheries. Sea Grant, University of Washington, Seattle, WA. 336 pp.

Hanna, S., K. H. Im, and L. O. Rogers (eds.). 1980. Exploring Conflicts in the Use of the Ocean's Resources. Sea Grant ORESU-U-79-001, Corvalis, OR. Oregon State University, Corvallis, OR.

Hoole, F. W., F. L. Friedheim, and T. M. Hennessey (eds.) 1981. Making Ocean Policy. Westview Press, Boulder CO. 300 pp.

La Porte, T. R. (ed.). 1975. Organized Social Complexity: Challenge to Politics and Policy. Princeton Univ. Press, Princeton, NJ. 360 pp.

Ophuls, W. 1977. Ecology and the Politics of Scarcity. W. H. Freeman, San Francisco. 303 pp.

Pearse, P. H. (ed.). 1979. Symposium on policies for economic rationalization of commercial fisheries. J. of Fish. Res. Bd. Can. 36:711-866.

Ross, D. A., and E. Miles. 1978. The importance of marine affairs. Science 201: 305.

Rothschild, B. J., C. Clark, R. Hennemith, R. Lasker, M. Sissenwine, W. Wooster, and J. Steele. 1982. Report of the Fisheries Ecology Meeting June 8-11, 1981. Woods Hole Oceanog. Inst. Tech. Rept. WHOI-82-28.

Simon, H. A. 1969. The Sciences of the Artificial. The MIT Press, Cambridge, MA.

U.S. General Accounting Office. 1975. The Need for a National Ocean Program. B-145099, U.S. Government Printing Office, Washington, D.C.

Young, O.R. 1981. Natural Resources and the State. Univ. California Press, Berkeley. 227 pp.

10
Economics and the Fisheries Management Development Process

Lee G. Anderson

Introduction

Unregulated exploitation of fish stocks can potentially lead to the economically wasteful use of the stocks and the inputs used to exploit them. Therefore regulation to change the size, composition, and timing of landings can be justified on economic grounds if such regulation can prevent or reduce such wastes at a cost that is less than the resultant savings. To guarantee that maximum savings are achieved, the types of regulation used must encourage efficiency in effort production so that the proper amount of catch can be achieved at the lowest possible cost.

The above summarizes the most important general conclusions of fisheries economics and they are critical to the establishment of rational fisheries management. In order to use them, however, it is necessary to understand the relevant social, biological, and institutional aspects of the fisheries involved. However, the use of economics is not limited to these broad policy prescriptions. The purpose of this chapter is to show that basic economic principles can be useful at

---

Lee G. Anderson is Associate Professor of Economics and Marine Studies in the College of Marine Studies at the University of Delaware. He has been actively engaged in theoretical and applied fisheries management for over 10 years. He has written or edited four books and many scientific papers on the subject and has acted in an advisory capacity to National Marine Fisheries Service, Mid-Atlantic Fisheries Management Council, National Academy of Sciences, Food and Agricultural Organization of the United Nations, and the World Bank with respect to fisheries management and development. He has also received grants from the National Science Foundation, Sea Grant, National Marine Fisheries Service, and Food and Agricultural Organization to study Fisheries management and regulation.

every stage in the process of formulating and implementing fishery management policy. Special emphasis is given to the economic rationale for the process as a whole and to those issues which have economic content or are amenable to economic analysis.

The importance of the word "process" should be emphasized. *Webster's New World Dictionary* defines *process* as "a continuing development involving many changes" and as "a particular method of doing something, generally involving a number of steps or operations." The development of fishery management policy is definitely a process according to both definitions, and policy development becomes more tractable when viewed as such (FAO, 1980). The process must have the appropriate number and type of steps which are flexible enough to allow for required changes. It is with respect to this that the application of basic economic principles can be especially useful.

Fundamentally, the management process consists of choosing management objectives and selecting the appropriate combination of total landings, catch composition, and product quality, and the particular set of control measures used to achieve that combination which most nearly achieves the objectives. The next section describes the basics of the policy development process in detail, giving special emphasis to its economic aspects.

Following this, a step-by-step policy development process which addresses the necessary conceptual issues with due consideration for practical application is described. The institutional structure used by the development and implementation policy must be viewed as part of the development process because the motivations of the various agencies and of the individuals who comprise them determine how the process is carried out. The role of the institutional structure is discussed in the final section.

The Fishery Policy Development Process

Management Objectives

Management objectives are the stated goal or goals to be achieved by fisheries policy. Objectives are a necessary first step in the process because they provide a means of comparing different regulatory options. All else equal, when making such choices the optimum policy is the one that most closely achieves the stated objectives.

Characteristics of Operational Management Objectives. If objectives are to provide a means of comparing regulatory options they must be operational. That is, it must be possible to apply them in an unambiguous manner. There are several important characteristics of operational objectives. First, they must be stated such that criteria for success or failure can be quantified. The objective of improving fisheries may sound good, but it offers little help in comparing various policy alternatives. On the other hand, the goal of maximizing net revenues from the fishery is operational because earnings and costs can be measured and hence can provide an explicit means of comparing policy options. While some important aspects of fisheries, such as equity, industry structure, etc., are difficult to quantify, they can still be appropriate considerations for management objectives. However, unless the objective is stated such that there is a viable method of determining that it has been met, comparisons of regulatory techniques are not possible.

Related to the problem of quantification is the assigning of relative values to conflicting objectives. It is possible that there will be more than one management objective, and when this is so, there will be conflicts between them.

That is, if objectives are distinguishable from a management point of view, they will conflict. In cases where accomplishing one objective will also accomplish the others, there is really only one operational objective. It does not matter which one is selected; success in achieving one will guarantee success in the others. On the other hand, if there are separate objectives, a policy which best achieves one will not achieve the other. To make the objectives operational, it is necessary to specify the relative value of the various objectives so as to provide a means of comparing the positive and negative aspects of various regulatory options (Marglin, 1967).

Consider a fishery where maximizing the net revenue of the entire fleet is an important goal, but due to income distributional considerations, it is desired to give special importance to income in a economically depressed port. If productivity in the depressed port is lower than elsewhere, the two objectives will conflict, because if the less efficient individuals from the depressed port are given access to the stock in preference to others, their income will increase but the net revenue for the fishery as a whole will fall. It will not be possible to rationally use these objectives to select a regulatory scheme unless there is some way to directly address this conflict. While it is sometimes quite difficult to do so, it is a crucial step in the policy development process.

One way to place relative weights on these conflicting objectives would be to maximize net revenues such that a dollar in the depressed port is given a larger value, say $1.10, than a dollar in other ports. In this way, policies which favor the depressed port would have higher weighted net revenues.

Constraints are another way of dealing with conflicting objectives. For example, in the previous case, if the management authority is not prepared to place relative weights on the net income to the various ports, they may be willing to set a minimum acceptable level of income for the depressed port. The appropriate objective would then be to maximize net revenue for the fishery such that net revenue in the less developed port is greater than or equal to a specified amount. Conceptually, determining the constraint is no different from assigning relative weights, but in some instances it may be an easier way to approach the problem.

A third important characteristic of operational objectives is the time frame within which they are to be achieved. An objective to increase catch per unit of effort can be achieved comparatively more rapidly, all else equal, if harvest is stopped completely than if it is merely curtailed. The former will obviously have more arduous short-term effects on the profitability of the fleet, however. The tradeoff between gains from management and the costs associated with the rate at which objectives are achieved is an issue that must be directly faced when stating management objectives. Mueller and Wang (1981) and Clark (1976) present a more detailed discussion of intertemporal aspects of policy.

Also important in making management objectives operational is the range of issues or the accounting stance used in setting fisheries management policy. For example, a narrow accounting stance would focus only on the existing harvesting sector in a particular fishery. A slightly broader stance would include potential entrants to the fleet, while a more general stance would consider the processing sector and perhaps various aspects of other related fisheries. The most general accounting stance would also include related industries, such as boat building and net manufacturing.

Specific fisheries management policies can have significant effects on the harvesting and processing sectors and, depending on the technicalities involved, can even cause repercussions throughout the economy and society as a whole. Adopting a narrow accounting stance will mean that only those effects within the particular fishery will be considered. This makes policy development and implementation much simpler because it is only necessary to specify objectives for those aspects of the fishery within the accounting stance. Therefore, there are likely to be fewer objectives and hence less potential for conflicts. Of course, it also increases the chances that some important variables will be ignored and the policy selected will be inappropriate. The reverse arguments hold for adopting a broader accounting stance. There are more issues that must be built into the objectives which makes policy analysis more difficult and, as a result, more expensive. However, it will lower the probability of selecting a suboptimal policy. The proper balance is to select an accounting stance which includes most of the relevant issues and where greater detail is likely to increase policy analysis costs by more than the value of the extra information.

Types of Management Objectives. There is a wide range of relevant fisheries management objectives but there is no one combination that will be best for all fisheries at all times. There are, however, certain points that are useful in selecting objectives. First, for the most part, specific biological objectives are not necessary and may, in fact, cloud the real issues behind management. For example, an objective to achieve a standing stock of a specific size may be clear and concise, but it begs the important question of why such a stock is beneficial. Does this prevent extinction, keep catch per unit of effort high, etc.? These things are important for the continued commercial success of the fishery, and management objectives should address them directly. This does not mean that the biology of the stock is not important, but achievment of properly stated economic and social objectives will, of necessity, properly maintain the stock.

Second, an objective that is often overlooked is economic efficiency. The efficiency of fishery operation is important because the general scarcity of resources require that they be utilized effectively. It is vitally important to consider the value of the harvest and the cost of the resources used to obtain it. Managing such that the fishery achieves the maximum difference between value and cost of harvest will insure that resources are not used in the fishery if they could be used more beneficially elsewhere in the economy. Given resource scarcity, economic efficiency should be at least one of the objectives used when selecting proper management strategies. Granted, other issues may be important, and some of them may directly conflict with economic efficiency. However, the range of objectives selected and the relative weights assigned to each should be such that the cost in terms of lost economic efficiency for achieving other goals is directly considered. For example, while increased employment may be a valid objective, the gains from increased labor force participation must be weighed against the economic efficiency costs of improper use of other resources due to overcapitalization. Further, while increased employment which comes at the cost of reductions in total output may be acceptable, it is likely that after some point the cost of continued increases in employment may become too high.

Another important economic aspect of management objectives is that, directly or indirectly, almost all will have income distributional implications. For example, the objective to increase total harvest may be obtained only by allowing individual fish to grow. Because of the size distribution of the stock and the makeup of the fleet, however, this may mean that only larger boats which work further from shore are allowed to harvest. This may cause a redistribution of income from inshore to offshore vessels. Obviously such a distribution will be an important aspect of the management policy, especially for the individuals involved. Management should therefore take these effects directly into account, and this can only be accomplished if distributional considerations are specified in an operational management objective.

Determining Fisheries Management Policy

As indicated above, the goal of fisheries management should be to achieve the stated objectives. If this is to be accomplished, the relationship between various control measures and the objectives of management must be well understood and fully utilized in policy development. The control measures are important, because in addition to causing the fishery to obtain a particular type of harvest, they can further affect the degree to which management objectives are met due to their influence on other important attributes of the fishery (Fig. 1).

The Management Problem. In the final analysis, regulation can affect three attributes of harvest: total landings, composition of catch, and quality. These may be called the final control variables. The combination of final control variables at any point in time is determined by the nature of the control measures placed on participants in the fishery. Since fisheries management must be viewed as an ongoing process, the composition of the three control variables over time and the effect they have on the fish stock over time are of critical importance. Such considerations are pertinent to the achievement of the management objectives in both the long and short runs.

The possible types of control measures are briefly described in the center section of the figure. Total landings, composition, and quality can be affected by

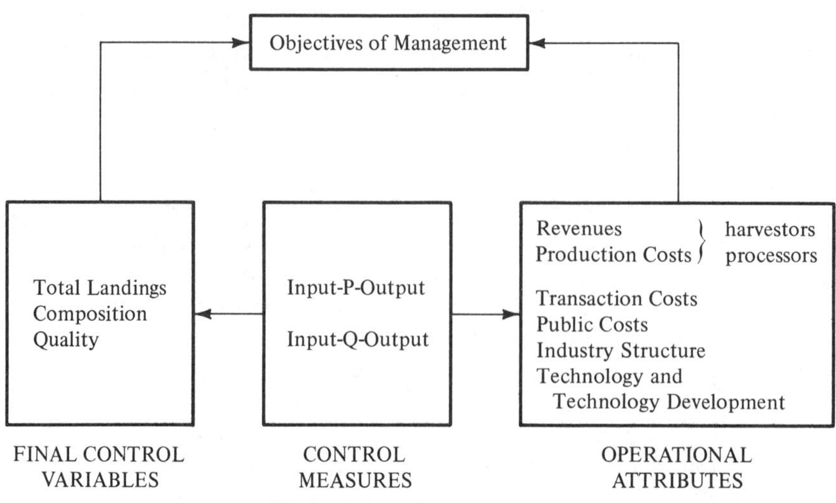

Figure 10.1. Global fisheries.

controls on either price or quantity. Prices can be affected by taxes on output or on the inputs used to produce it. Analogously, quantity can be affected by direct controls on either outputs or inputs. Figure 2, which is described in more detail in the next section, contains a more thorough description of control measures. At this stage of the discussion, however, it is only important to note that there is a wide range of alternative control measures.

The basic problem of fisheries management is to select the particular control measures which best achieve the objectives of management. This is not a simple problem, however, because each control measure can affect the final control variables differently, and in addition, each can have different effects on other important attributes of the fishery. Some of these important operational attributes are listed in the right hand side of Figure 1. Revenues and production costs of harvesters and processors are obviously important. Also significant, however, are the transaction costs of both. Following Pearse (1980) transactions costs include the internal cost of organizing individual fishing or processing enterprises, as well as the external costs of negotiating and bargaining among themselves to the extent that it is necessary to do. It also includes their cost of compliance with regulations and of dealing with regulatory authorities. These costs are to be distinguished from production costs because, while both can vary depending on the specific control measure used, they sometimes change in different directions.

The public cost of operating the regulatory program will depend upon the measures used and the willingness of participants to cooperate and comply with them. To achieve any specific combination of final control variables, public costs will obviously be higher if fishermen disagree with the main objectives of management, oppose the particular control measure, or can easily get around the regulations. Therefore the public costs of regulation will differ according to the control measures used even if similar ends are achieved.

The structure of the industry is important because it determines the efficiency with which the fish are actually produced. In addition, there is normally a different income distribution associated with each particular industry composition. For example, an industry with many small firms owned by individuals will have a different income distribution than one with a few large firms owned by absentee stockholders.

Technology and technological development can be directly affected by the particular type of regulation chosen. Certain control measures place direct limitations on the use of technology, but in one way or another all affect technology and the impetus for technological change. For example, limits on the number or size of boats focus attention on those technologies which can be used with fewer or smaller boats even though a more efficient technology could be developed that utilizes more or larger boats.

To summarize the discussion thus far, the objectives of management can be achieved by changing total landings, catch composition, and quality of catch from what would result with no regulation. However, the manner by which these changes are accomplished also affects the degree to which the management objectives are achieved.

Bearing this in mind, the solution to the management problem can be found by answering two interrelated questions. (1) What is the combination of total landings, composition, and quality which will achieve the stated management objectives? (2) Which control measures should be used, and exactly how should they be implemented so as to achieve the specified values of the final control variables? These are not easy questions, and the interrelationship between them makes the task even more difficult. To repeat, they are interrelated because both the combination of the final control variables and the control measures used to achieve them affect the degree to which the management objectives are fulfilled. Both affect aspects of the fishery which are important to the private participants, the management agencies, and consumers of fish. The problem is to find the right combination of the two.

Because of the interrelations between final control variables, control measures, and the achievement of objectives, the management problem is very complex. Moreover, if these complexities are ignored, it is unlikely that the optimal policy will be selected. Unfortunately, this broad prospective is seldom used in the actual framing of fishing management policy anywhere in the world today. In most instances, the level of total landings is selected independently on the basis of biological surpluses or commercial necessities, and then control measures to achieve it are implemented. This approach may be necessary second best alternative, especially in view of the imperfect state of biological and economic information and the limited budgets of management agencies, but it limits the potential for improvement by allowing decision makers to ignore many important aspects of management. Using the above two questions as a starting point insures that all aspects of the management problem are considered. It provides the possibility that, to the extent possible given information and budget constraints, a fishery management policy can be selected which considers both the nature of the final control variables and the control measures used to achieve them.

Control Measures. In order to more fully understand the effect various control measures can have on fisheries, and their relative efficacy of achieving fishery management objectives, it will prove useful to discuss them in more detail. The discussion will be somewhat limited but the interested reader may refer to any of the following sources for more detail: Anderson (1977, 1980), Clark (1980), Crutchfield (1961), Pearse (1979), Rettig and Ginter (1980), Stokes (1979), and Turvey and Wiseman (1957).

As mentioned above, the final control variables of a fishery can be changed through controls on price or quantity. A more detailed picture of these controls is depicted in Figure 2. Like all businessmen, fishermen try to maximize their profits and, to the extent that they can, they will change their operations so as to reduce or modify those activities or practices which are taxed. Therefore, total landings can be reduced by a landings tax while size distribution can be affected by a tax scheme which penalizes the harvest of those sizes which are to be avoided. Similarly, different taxes for each species create the motivation to adjust the composition of catch away from the higher taxed species, and finally a tax on low quality can encourage quality improvements. Practical enforcement problems, however, will more than likely limit the usefulness of all of these except simple landings taxes.

Taxes can also be used on the input side of the market. If effective fishing effort is related to a variable that can easily be measured (such as days fished by a representative vessel) then a tax on that variable will be a tax on fishing effort. In most instances, however, the measurement of fishing effort is very complex and so as a next best alternative, components of effort, such as horsepower, gear type, and vessel size can be taxed. Either option will tend to reduce effort and hence total landings. Taxes on those inputs which are particularly productive in landing certain species will reduce the motivation to use those inputs and hence will affect the composition of catch. Similarly it is potentially possible to tax mesh size to control for size selectivity.

If one were to judge controls which affect prices on the basis of economic efficiency, it is most likely that to control total landings or species composition, output taxes would be preferred to input taxes. Input taxes are less useful because of the difficulty of defining effort and also because a tax on the components of effort causes them to be combined such that production costs are increased. On the other hand, however, the preferred tax to control size selectivity is probably on the input side. A tax on mesh size is much easier to enforce than a tax on relative fish size. While it may be fairly easy to measure landings per se (it will certainly be easier than measuring effort), it would be quite difficult to determine the size composition of catch in a manner that will allow for a tax program. With respect to product quality, it is not clear whether an input or an output tax would be preferred.

On the other side of the coin, the final control variables can also be determined by control measures which focus on quantity. Rather than affect harvest by changing the relative prices of inputs and outputs, it is possible to place limits on either. On the output side, landings can be determined by specifying a total quota. Separate quotas for different species can affect catch composition if the

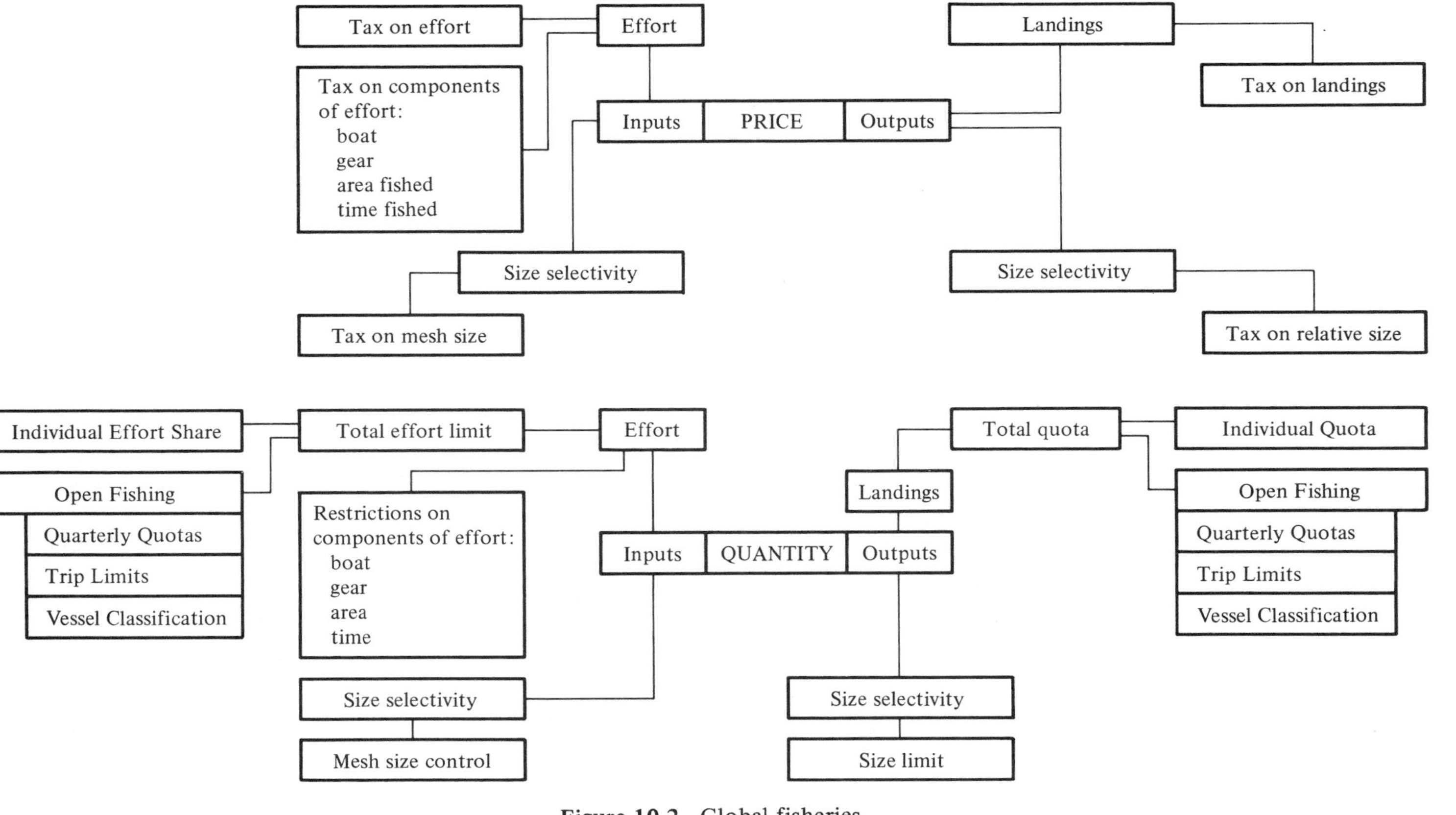

Figure 10.2. Global fisheries.

fish are not harvested simultaneously. The most common way to institute a total quota is to set a maximum catch level and then allow open fishing until the quota is harvested. However, if the time period over which fishing takes place is important, the quota can be divided into quarterly allocations. Also trip limits can be used to help distribute the catch among the fleet and to increase the length of the fishing season. In addition, in order to make even more definite distribution commitments, the quota can be further divided into vessel category shares, where various segments of the fleet are assigned a specified share of total landings. A total quota is biologically effective to the degree that it can be enforced because only the specified harvest is allowed, but it will cause inefficiencies in the production of effort (see discussion of limited entry below). In addition, the above modification to a simple total quota exacerbate the inefficiencies.

A useful way to modify a total quota program is to allocate shares of the allowable catch such that each boat or fisherman has the right to harvest a certain specified amount or a certain percentage of the total quota each year. If such a program is properly set up, it provides the potential to achieve economic efficiency in harvest because each fisherman will try to minimize the cost of harvesting his allotted share.

With respect to size composition of catch, the relevant output control is regulation on the size at which fish can be first harvested.

There are analogous quantity control restrictions on total landings, composition and quality from the input side. Total effort limits restrict the total amount of fishing activity rather than limiting catch. They are possible, however, only if effort can be defined in a meaningful way. There can be open fishing until the effort limit is reached, or various types of time, trip, or vessel classification constraints can be applied. Quantity input controls are different than quantity output controls, however, because in addition to setting up a total effort quota, it is possible to place restrictions on the various components of effort.

The input control for size composition of catch would be to regulate mesh size or size of hooks.

The above comparison of the various types of control measures is by no means complete. Most of the discussion has been in terms of how well they can achieve various harvest policies and very little has been said about how they affect the other important attributes of the fishery. Space limitations do not allow such a complete analysis here but it should be obvious that a full understanding of these effects is necessary in order to develop proper fisheries management policy.

Control measures can be classified according to the possibility of achieving efficiency. Those which permit efficiency in harvest (sometimes called limited-entry measures) are taxes on landings, taxes on effort, individual effort quotas, and individual catch quotas. All others will, directly or indirectly, cause inefficiencies (Pearse, 1979; Rettig and Ginter, 1980). Because of the problems of defining effort, taxes on landings or individual catch quotas are probably the only viable limited-entry alternatives in most fisheries.

It is interesting to note, however, that while limited-entry control measures are the ones that are capable of achieving economic efficiency, with a few possible exceptions they are not used anywhere in the world. The control measures that have been used are total quotas and restrictions on the components of effort, both of which cause effort to be produced in an economically inefficient manner. Total quotas encourage larger boats because each fisherman wants to be able to get a bigger share of the quota. On an aggregate basis, all this does is increase harvest costs. Restrictions on the components of effort also directly or indirectly increase the harvest costs because fishermen are no longer free to use the least cost combination of inputs to produce effort and, for the most part, these restrictions are not biologically effective in the long run. They lose biological effectiveness because fisherman will be motivated to increase effort by using more of those components which are not controlled.

Looking at this wide range of control measures from the perspective of Figure 1, the relevant question is which ones should be used to accomplish the final control variables such that the objectives of management are achieved. If economic efficiency is a management objective, then those control measures which cause economic inefficiency in the production of effort should be shunned. However, controls which cause inefficiency may be a rational choice if the loss in efficiency is worth the benefits from the achievement of other objectives. Another important consideration is the implementation and enforcement costs of the various control measures. While limited-entry controls may encourage production efficiency, the public costs of implementation may exceed the benefits.

The Policy Development Process in Practice

The previous discussion centered on the fundamentals of the process of developing management policy. How these fundamentals are applied is very critical to the success of fisheries policy. By using a systems approach, they will be appropriately applied by the very nature of the process (Rothschild, 1971; FAO, 1980). The purpose of this section is to provide a suggested step by step description of such a systems approach.

Step 1. Evaluate the Fishery from a Biological and Economic Point of View

As pointed out in the Introduction, economic theory indicates that an unregulated fishery results in the economically wasteful use of the fish stock and the inputs used to harvest it. The first step in the management process, therefore, is to determine the current state of the fishery because, more than likely, the stock, the fleet, and the harvest will be nonoptimal in size and composition. Some of the questions that need to be answered to ascertain the state of the fishery are: what is stock size, annual recruitment, and stock growth rate? What are the relationships between these variables and the level and composition of total landings and how do they vary with the behavior of other exploited and unexploited fish stocks? What is the size of the relevant fleet and what determines its growth and annual utilization? What are the production technicalities of harvest, processing, and marketing and how are they affected by changes in biological and economic conditions and by the various control measures? What is the income from the fishery and how is it distributed?

Once the current situation in the fishery is described and the basic nature of the interrelationships between the fleet and the stock are understood, it is possible to consider the question of how the fishery should be changed and by what means.

Depending upon the circumstances of the particular fishery, however, the ease of obtaining this information can vary quite significantly. Therefore, the policy development process must often start with that which is available. However, programs to collect more and better information should be built in to the implementation procedure, so that process can begin again if necessary as the fishery is better understood.

Before continuing, it will prove useful to discuss the two important issues at this juncture. The first is the role of information in fisheries management. While biological, sociological, and economic information is necessary to describe the current state of the fishery, and, as will be explained below, to determine whether management is achieving the stated objectives, the question of what types of information are important and how accurate does it need to be is quite relevant.

Improvements in information gathering, storage, retrieval, or analysis can be justified if they improve the ability of management policy to achieve the stated objectives. Further, they must do so such that the value of the increased success is greater than the cost of improving the information system. While information is important, it is just one part of developing management policy, and improvements can only be judged in the context of the net addition they make to the process.

The second issue is the role of models. In order to obtain the information described above, it is necessary to produce a behavioral model of the operation of the fishery. A fishery is a complex interrelationship between the fleet and the stock it exploits. The operation of the fleet depends upon certain market conditions and biological parameters of the stock. At the same time, the rate and the manner with which the stock changes through time depends upon the ecological conditions and the technological characteristics of the fleet. In other words, the stock and the fleet influence each other in rather fundamental ways but both are also affected by other unrelated items. In order to comprehend a fishery, it is necessary to understand these relationships. A fishery model is an organized way of looking at these interrelationships and it can range from a conceptual model in the manager's mind to a complex computer simulation. The former is a result of the manager's experience with the particular fishery and how individuals and the stock react to particular situations. A computer simulation model is similar, except the behavior patterns of the various agents are described by a system of equations. However, what type of model is most appropriate? The answer, of course, will vary from fishery to fishery depending on how it operates and the objectives to be achieved. Increases in model complexity can be justified to the extent that they improve the potential to understand the operation of the fishery and to better predict the results of various policies.

Since formal models are sometimes mistrusted, it may prove beneficial to discuss their use in more detail. It is not possible for simple "back of the envelope" models to provide a complete understanding of the complex ecological and economic interactions of a fishery. More rigorous models are often necessary to even theoretically describe them. However, to the extent that the true nature of these interrelationships is not completely understood, it is impossible to build a perfect model. However, it is possible to develop first approximations that provide useful insights. These models should be used with care, however, when developing and implementing policy. First, it is important not to expect too much. Given existing knowledge it is not possible to predict every variable of a fishery with precision. Therefore, a model should not be expected to be able to solve the management problem by specifying exactly how many boats should fish, and where and when they should operate.

On the other hand, since many existing models can describe general trends, their results should not be dismissed too easily. Some managers and fishermen refuse to accept them at all. This lack of confidence may have developed because the models show that accepted management practices are not proper. If the models are completely incorrect, the reluctance to use them is appropriate. It is also possible, however, that the models are correct and the individuals who dismiss them are doing so because they cannot, or will not, change their long held beliefs. If the process of developing fishery management policy is to be continually improved, new information and new insights will have to be utilized regardless of their origin. Even with their limited capabilities, formalized fisheries models may be very useful in understanding how a fishery operates. Their results should be used but they must be interpreted with care.

Step 2. State the Management Objectives as Clearly as Possible Based on the Relevant Accounting Stance and the Needs of the Fishery as Determined in Step 1

Upon successful completion of step 1 it will be possible to give a more or less detailed description of the present state of both the stock and the industry. If no other possible state is preferred to the existing one, and if it appears that no changes are likely, there is obviously no need for management. However, if other states are desired, the next step is to select management objectives such that their achievement will cause the fishery to be in the most preferred state. An interesting approach to specifying objectives when there are conflicting interests in the fishery is discussed in Bishop *et al.* (1981).

Step 3. Select the Appropriate Combination of Final Control Variables and the Management Control Package Deemed Appropriate to Achieve the Stated Objectives

Step 3 is perhaps the most important part of the process. The most suitable fishery model available must be used to undertake a policy analysis of various combinations of control variables and of control measure packages to see which joint solution best achieves the management objectives. It will be necessary to consider how both the final control variables and the control measures used to achieve them affect the degree to which management objectives are met. If no policy can be found which, after considering all effects, results in an improvement in the fishery according to the objectives, regulation is not warranted.

Step 4. Select the Implementation Procedures

Determining the proper policy is only half the battle. It is also necessary to determine the exact methods and institutions by which it will be implemented. For example, if a total quota is selected, it will be necessary to specify precisely how it will be enforced, how the catch will be measured, which agency will inspect vessels, which agency will be charged with preventing further fishing once the quota is filled, and what sanctions they can apply to violators. The institutional aspects of management are discussed in a separate section.

Step 5. Monitor the Fishery under the Regulatory Program

If the objectives have been stated properly, it will be possible to show in a quantifiable manner the degree to which they have been achieved. The purpose of this step is to monitor those critical elements of the fishery to see how they are changing as a result of the regulatory program.

Step 6. Determine if the Goals or Objectives of Management Are Being Achieved as Predicted by the Policy Analysis in Step 3

The purpose of step 6 is to determine whether the current policy as it has been implemented is achieving the objectives of management. If the answer is yes, then the management process reduces to a continuation of current policy and continued monitoring to insure that the policy remains successful. If the answer is no, there are two possible steps. First, change the implementation procedures for the current program (see Institutional Structure and Policy Development section). Second, select an entirely new harvest program and a new control option package. If the second option is deemed necessary the process should revert to step 3 and begin again.

Step 7. Periodically Reevaluate the Fishery and the Objectives of Management

If the needs of the fishery change over time, due to better perceptions of them from improved information collected during the management process or to changes in the biological or economic conditions, or because of the success of the regulation program, it may be necessary to restate the objectives of management and begin the policy development process again. To see if this is the case it is necessary to repeat step 1 occasionally.

Institutional Structure and Policy Development

Implementation is a very important part of the management process, however, it is a topic that is rarely treated in any detail in the literature. Perhaps this is why implementation is often the cause of management failure. Choosing the proper combination of harvest variables and the control methods such that the management objectives are met may well be in vain if the policy cannot be effectively implemented. Although a complete treatment of implementation is not possible here, two important issues, the nature of the institutions to implement them and the construction of actual implementation rules, will be discussed. A more detailed discussion on implementation occurs in Roemer and Warwick (1978).

The operational procedures of management institutions must be such that the individual bureaucrats are motivated to use the process in an optimal manner. The type of management achieved will be different when officials are rewarded for merely getting management plans through the system with a minimum of resistance from industry, the courts, etc., than when they are rewarded for producing plans that actually achieve the stated objectives. The diligence with which management officials ask difficult questions concerning the efficiency and distributive effects of different regulations and the extent to which they are willing to make correct (in terms of satisfying management objectives), but potentially unpopular decisions is a function of how well they are rewarded for such behavior. Similarly if enforcement officers are motivated to cite many minor violations because their performance is evaluated by the number of citations issued, the overall effectiveness of the regulatory program will be different than if officers are motivated to take and over-all view and only cite major violations.

Similarly, the lines of communication and chains of command between various agencies must be amenable to proper management. Often, fisheries management responsibility is shared by several agencies. For example, a marine resource ministry may determine the regulations, but enforcement is the responsibility of another agency. The United States, for example, has a very complex management system with many overlapping agency jurisdictions (Anderson, 1982). When there is more than one agency involved, they will often compete for supremacy. This can often cause counterproductive activities which are detrimental to the operation of the management process. To avoid this, the chains of command should firmly establish the responsibilities of each.

Another problem is that agencies are sometimes part of the fishery management system even though they have other nonfishery responsibilities. The U.S. Coast Guard, which deals with smuggling, marine safety, etc., as well as fishery regulation, is a good example. Under certain circumstances these types of agencies may not be motivated to properly handle fisheries-related problems because the reward structure is such that they receive more credit for other activities which interfere with their fisheries management responsibilities. In these cases the reward structure must be modified such that all agencies feel they are getting rewarded appropriately for their input into the system.

Regulations must be such that their purpose is obvious to fishermen and at the same time the amount and difficulties of work they entail should be kept to a minimum and should be compatible with the hectic activity on a fishing boat. With respect to the first point, fishermen are more likely to comply with a regulation if it is likely to help the fishery in general and themselves in particular. Also they are more prone to answer questions concerning catch and effort in a regular and consistent manner if the number of questions is kept to a minimum and the information needed to answer them is nonproprietary. This relates back to the issue of the optimal amount of information. While it may be marginally beneficial to get detailed information from each fisherman on each trip, such a program will not be well supported and the amount of reliable information may actually be reduced.

Since regulatory forms usually have to be filled out at sea, and while it may be useful to get detailed information on type, location, timing, and composition of catch, etc., the captain of a boat does not have time to do this during the actual fishing activity. Therefore, to insure that the information is interpreted as being as accurate as it actually is, it may be necessary to use broad-scale information. Another issue which may sound trivial is that the report has to be such that it can withstand being tossed around the cabin of a working boat. The best systems are those that ask for the minimum amount of information necessary to obtain proper regulation and to guarantee that the regulations are being observed because the fisherman is then more likely to comply in a regular and systematic way.

The ease with which regulations can be enforced is another important part of the implementation. Fishermen, like other individuals, are motivated to act in their own best interest. Therefore, they will avoid complying with restrictive regulations, especially if there are no individual advantages from compliance and if the possibility of being caught and punished is low. A catch limit for a particular area is an example of a regulation that is very difficult to enforce. Adequate enforcement would entail relatively expensive High Sea activity. All else equal, the best regulations are those which can be enforced on land, such as total quotas per boat.

The ease with which industry can make its views known is also important. A useful way to insure that a management program is supported by the industry is to provide for their input into the selection of the optimal management package and the manner in which it is implemented. This will improve the chances that the program will be something they can live with and will also increase fishermen's motivation to comply. On the other hand, opening up the door for comment can also provide the opportunity for domination by industry or its more vocal components. Such domination can shift the focus of the program to benefit those with influence at the expense of other sectors or of the nation as a whole. Providing for industry input without allowing for industry domination is a delicate balance, but it is necessary for a properly working fishery management policy development process.

Conclusions

The development and implementation of proper fisheries management policy is a very difficult task, which can be completed in a satisfactory manner only if it is approached in a systematic manner. While the process involves the use of a number of disciplines, economic analysis and the application of basic economic principles are important parts. Economics is the study of choice, and fisheries management is essentially a problem of choice. Fisheries resources are scarce and, just as important, so are the inputs which are used to harvest the stock and to regulate the activities of industry participants. The basic problem of management involves two choices. (1) What is to be done with the stock and the related economic inputs? (2) What type of regulation should be used to achieve it? The fishery management policy development process should be viewed as an organized way of making those choices such that all relevant issues are considered. As the discussion in this chapter demonstrates, economics can be very useful in all steps of this choice process.

References

Anderson, L. G. 1977. The Economics of Fisheries Management. Johns Hopkins Univ. Press, Baltimore.

Anderson, L. G. 1980. A comparison of limited entry fisheries management schemes. *In*: Report of the ACMRR Working Party on the Scientific Basis of Determining Management Measures. FAO Fisheries Report No. 236, FAO, Rome.

Anderson, L. G. 1982. Marine fisheries management policy. *In*: P. R. Portney (ed.), Current Issues in Natural Resource Policy, Johns Hopkins Univ. Press, Baltimore.

Bishop, R. C., D. W. Bromley, and S. Langdon. 1981. Implementing multiobjective management of commercial fisheries: A strategy for policy-relevant research. *In*: L. G. Anderson (ed.), Economic Analysis for Fishery Management Plans. Ann Arbor Science Publishers, Inc., Ann Arbor, MI.

Clark, C. W. 1976. Mathematical Bioeconomics. John Wiley and Sons, New York.

Clark, C. W. 1980. Fishery management and fishing rights. *In*: Report of the ACMRR Working Party on the Scientific Basis of Determining Management Measures. FAO Fisheries Report No. 236, FAO, Rome.

Crutchfield, J. A. 1961. An economic evaluation of alternative methods of fisheries regulation. J. Law Econ. 4 (October):131-143.

FAO. 1980. Report of the ACMRR Working Party on the Scientific Basis of Determining Management Measures. FAO Fisheries Report No. 236, FAO, Rome.

Marglin, S. A. 1967. Public Investment Criteria. The MIT Press, Cambridge, MA.

Mueller, J. J., and D. H. Wang. 1981. Intertemporal issues and economic analysis in fisheries management under the Fishery Conservation and Management Act of 1976. *In*: L. G. Anderson (ed.), Economic Analysis for Fisheries Management Plans. Ann Arbor Science Publishers, Inc., Ann Arbor, MI.

Pearse, P. H. (ed.). 1979. Symposium on policies for economic rationalization of commercial fisheries. J. Fish. Res. Bd. of Can. 36(7):711-866.

Pearse, P. H. 1980. Property rights and regulation of commercial fisheries. J. Bus. Admin. 11:185-210.

Rettig, R. B., and J. J. C. Ginter (eds.). 1980. Limited Entry as a Fishery Management Tool. Univ. Washington Press, Seattle.

Roemer, M., and D. P. Warwick. 1978. Implementing national fisheries plans. CIDA/FAO/CECAF Workshop on Fishery Development, Planning and Management. FAO, Rome.

Rothschild, B. J. 1971. A Systems View of Fishery Management. FAO Fisheries Tech. Paper No. 106. FAO, Rome.

Stokes, R. L. 1979. Limitation of fishing effort: An economic analysis of options. Mar. Policy (Oct.):289-301.

Turvey, R., and J. Wiseman (eds.). 1957. The Economics of Fisheries. FAO, Rome.

11

How Much Is Enough? An Essay on the Structure of Fisheries Management Agencies

Peter A. Larkin

Introduction

Many things are disputed in the broad fraternity of fisheries managers and researchers, but one thing they have in common—the conviction that the agency for which they work is less than perfect in its organization and administration. Some may complain that they are hopelessly understaffed and underfunded to do the job that is expected of them. Others may be convinced that there are enough people and funds, but their agency does not use the resources effectively. Many complain about lack of communication; many others about a lack of attention to long-term research. In brief, virtually all see something wrong with the way in which their agency is doing (or not doing) its job.

To some extent these grumbles may be just grumbles, reflecting the universal human drive for greater achievement and restlessness about the *status quo*, whatever that may be. Nevertheless, there seem to be some generalizations to be

---

Peter A. Larkin is a professor in the Institute of Animal Resource Ecology and the Department of Zoology at the University of British Columbia where he also fills the administrative positions of Associate Vice President (Research), and Dean, Faculty of Graduate Studies. He also has been Head, Department of Zoology, and Director, Institute of Fisheries, University of British Columbia, and Director, Pacific Biological Station, Nanaimo, B.C., and has served in several capacities on the Science Council of Canada, the Fisheries Research Board of Canada, and various other scientific groups. He is presently a member of the National Research Council of Canada, the International Center for Living Aquatic Resources (ICLARM) in Manila, and the Board of Directors of B. C. Packers Limited, a fish processing company. Dr. Larkin has published widely in the field of fisheries biology with specific reference to some of the salmonid fisheries of Canada.

made; grumbles seem to come in genera that older hands recognize as signs of particular kinds of institutional ailments. Surprisingly, there does not seem to be much of a literature on the subject, except as is generated by professional administrators whose prime concern is how to jolly everyone along so that they have the feeling they are doing a good job even if they are not. What is needed, it seems to me, is a technically oriented review of the subject of fisheries management that is addressed to the questions "how much management is needed" and "how should it be organized." With these more or less strategic questions answered, the day-to-day nuts and bolts of operation can be seen for what they should be, as mechanisms in a systematic approach to management and not as the *raison d'etre* for a fisheries agency.

It is to this theme that this chapter is addressed. From a technical, professional point of view, it deals with such things as how much effort is needed to accomplish a primary broad-brush level of fisheries management and research; how much to operate at higher levels of investment; the advantages and disadvantages of more complex administrative structures; and how much centralization is desirable. What follows, then, is an attempt to sketch some of the generalizations that have occurred to me as I have listened to grumbles over the last 30 years. It is an arbitrary and opinionated account, and I hope it will prompt more thoughtful and thorough analysis by others.

Primary-Level Management

There is an old saying to the effect that one man can do a lot of things, two can do a lot more, and three can do almost anything. Given only a minimal budget, it certainly should be possible nowadays for a three-man unit to manage almost any regional fishery, albeit in a crude way. Virtually any species of fish, especially one of commercial value, can be recognized with confidence without recourse to research in systematics. In tropical parts of the world it may be necessary to group species, particularly those that are uncommon, and to be satisfied with a common name such as snapper to cover all in a group. Some idea of the species composition of the catch, where it was caught, crude estimates of catch per unit effort, and of changes in gear efficiency can be obtained by soliciting the help of a few fishermen. Any dramatic rise or fall in the abundance of a common commercial species or group of species can be detected by the crudest kind of sampling. For example, the dramatic changes in the species composition and catch per unit effort in the Gulf of Thailand (Pauly, 1978) would have been evident to any competent observer who spent only a few hours a month at the dockside. Similarly, the famous collapse of the sardine fishery in California, or its recent converse in Peru, were matters of common knowledge to all participants.

With only a little more information, it is possible to foresee such dramatic changes. A substantial decline in average length of a species is usually a sound indicator of the effects of an exploitation which, taken to much greater extreme, invites disaster. To guess which species may replace one which is depleted is not completely mysterious, for sufficient is now known about the life histories and diets of many species to make some fairly educated deductions. Moreover, something similar has usually happened somewhere else and the pattern, once recognized, may be familiar.

Maintaining a crude appreciation of the biology of a fishery is probably only a job for one energetic man. As Harden Taylor (1951) pointed out many years ago, economic factors can be counted on to exert some sort of crude control on relative rates of exploitation, and one need not be too hasty to regulate every fishery. Nevertheless, a primitive level of management almost certainly requires an assistant or two to communicate with the fishermen and the processors, and to report on the success (or difficulty) of enforcing regulations (if any). Time should also be budgeted for occasional consultations with comparable teams in neighboring countries and for consultations with visiting experts.

For a fishery on a moderately homogeneous coastline, whether it involves a few or several species, a three-man team should be able to grasp what is happening and to give some useful guidelines for management. It is doubtful that much could be done to regulate the fishery unless there was an unusual level of understanding and agreement among the fishermen and the processors. More to the point, even if regulations were followed scrupulously, they would be difficult to justify except in more extreme circumstances. To assess which regulations should be applied to control a fishery in a desired way, and to adequately appreciate the social and economic contexts of fisheries, requires a more sophisticated level of management.

Secondary-Level Management

Even the simplest models of fishery assessment have relatively demanding requirements of information. The stocks of each species must be identified, and for each there must be relatively precise information on catch and effort; and preferably an estimate of growth, which implies a technique of aging. Additionally, although the ontogeny of regional fisheries biology need not recapitulate the phylogeny of the science (because there is so much already known), it is nevertheless necessary to do a certain amount of preliminary local investigation to set the stage for even a modest level of fisheries management.

For the background biology of many of the species much can be assembled from the literature, but for local estimates of growth it is usually necessary to undertake one or more of the common kinds of studies in which readings of scales, otoliths, or other bony parts are calibrated to length frequency distributions and perhaps confirmed by tagging studies. There is also much value in regional oceanographic surveys and for natural history types of studies that detail the habits of various stocks of the most important commercial species. In short, there is immediate need for a research group charged with the job of providing the broad base of local knowledge that underlies all successful resource management.

There is also immediate need for the sampling of catches and effort that provides the statistical base for stock assessment. Sampling of catches and monitoring of effort is labor intensive. To be adequate it requires trained observers who can discriminate among species and who can appreciate that random sampling is not easy. The compilation and primary analysis of data require similar discernment. The manpower requirements depend on such things as the size and diversity of the fleet, the number and distribution of landing sites, the number of species of fish, the length of the fishing season, and the cooperation of the fishermen. Even though random sampling assures us of relatively great precision from relatively small sample sizes, a fishery may be so complex as to imply the need for a great deal of stratification before the landing and effort statistics can be put together with confidence.

It is also imperative that there be a capability for digesting the information that is collected and extracting from it whatever management morals it may contain. This is by no means a trivial task, even though most professionals may have copies of the handbooks of Gulland and Ricker to follow and a computer close at hand. Most fisheries data allow for a variety of interpretations, and the choice of interpretation is often critically dependent on a close familiarity with the fishery, the scheme of sampling, and the biology of the species of fish. A close relation between the research group, the sampling team, and the assessment analyst is an important element of success.

There should be added to these biologically oriented components of the management team, a group of economists and sociologists to collect and analyze the basic data concerned with the livelihoods of fishermen and the infrastructure of the fishing business. Economists and sociologists might put this requirement ahead of the need for biologists, for in human affairs one should first pay attention to the human objectives. My bias is otherwise: where a living natural resource is to be exploited, it is first necessary to understand its potentials and fragilities.

The logical sequel to the development of understanding of a fishery is to recommend certain regulations and controls, which implies the need for public education and enforcement. To some extent some of these activities can be coupled with sampling programs, but the consequence may be that neither job is done properly. Ideally, enforcement and sampling are separate activities and each has a component of public education.

Adding up these various requirements for research, sampling analysis, enforcement, and education leads to the obvious conclusion that though three men may be sufficient to comprehend the broad nature of a fishery, at least 30 may be needed to tackle its management. Where the region is large, the fishery diverse and complex, there may well be need for many more pairs of hands for sampling and enforcement.

At this level of management there are also to be considered the ancillary services. Such simple matters as accounting, purchasing, equipment maintenance, and handling of personnel all require manpower. The astute administrator will also see the need for some expert advice on subjects in which he or his senior officers lack experience or expertise. For instance, if the administration is strong in biology, it might wish advice on economics; if strong in economics, then perhaps it is weak in planning skills. One way or another, the administrative team must have balance if the fishery is to be seen as it should be—as a whole.

The question can now be fairly posed: how should the effort be divided among the various activities? My own experience would suggest that administration and ancillary services should be no more than 10% of the manpower, research no less than 10%, sampling about 30%, stock assessment 10%, enforcement about 40%. This kind of division of effort does not allow for rapid progress in understanding, and may be insufficient for seizing opportunities for development. However, the tasks of stock assessment, and the wherewithal to recommend management policies that can be followed through with public education and enforcement, can be done effectively with a fairly small and well-balanced team.

The total cost of such a team may well be as much as 10% of the landed value of the catch. Since a staff of 30 may be considered as virtually minimal, it follows that if the cost is not to exceed 10% of the value of the catch, then for many small-scale fisheries it may not be rewarding to attempt secondary-level management on a continuing basis. For the fisheries of many lakes, for example, it may be preferable to undertake only primary kinds of activities, interspersed perhaps with infrequent intensive studies only if there seem to be troubles.

A better way of looking at total expenditures on management is to think of its cost as an investment on which there should be a return. As long as the increment of management effort is associated with commensurate increment of social or economic value, then further investment is appropriate. On this basis one might expend much more than 10% of the landed value of the catch on management. This kind of consideration leads to the next level of sophistication.

Tertiary-Level Research and Management

There is virtually no limit to how much effort can be proposed for research and management of a fishery. As Watt (1956) pointed out many years ago, increasing sophistication of conceptual models inevitably involves more diversified and more intensive data requirements. A fishery managed on a Schaefer model requires only catch-and-effort statistics. A Beverton-Holt model requires growth and mortality data, and in a self-regenerating model there is the added need for a stock and recruitment relationship. If a very complex model for multispecies interactions is to be used, it may be necessary to have basic stock assessment data for all species as well as estimates of coefficients that relate the various species to each other (through food preferences, for example). For the most elaborate conceptual models, such as those of Andersen and Ursin (1977) and Laevestu and Favorite (1977), the data requirements may be immense, involving not only the parameters of the fish populations and their interactions but, as well, a host of parameters that determine the rates and channels of primary and secondary productivity. It is one thing to conceptualize how nature works, quite another to prove it is so, and still another to collect the information that is needed to control nature if the model is correct.

For most fisheries it is highly doubtful that their worth is sufficient to manage them in accord with the most complex models. Although the models may be realistic or, more likely, gross simplifications of nature, they are nevertheless impractical for the purposes of management. They should be valued for their potentials for creating awareness, not for their utility as models for management (Larkin and Gazey, 1982).

The key question for the fisheries manager becomes which conceptual model to use, bearing in mind how rough it may be as a tool and what it will cost in relation to the value of the fishery. As the intensity of a fishery increases the value of the catch rises, and it subsequently falls depending on the effects of overfishing and the elasticity of the price of fish. The uncertain fisheries manager will cause the fishery to lose what might otherwise be gained, the losses accelerating when the fishery becomes so intense that the fish population collapses.

To avoid this possibility, recourse is taken to managing on a more complex conceptual model which, ostensibly at least, reduces uncertainty. The most complex models, being perhaps more prone to being figments of more fertile imaginations, may in fact confuse rather than illuminate and may actually increase uncertainty. As models increase in the number of their parameters, they become progressively more demanding of data if they are to be tested for validity. It is an expensive business to develop and validate a model, especially if it may be necessary to manipulate the fishery to gain knowledge. Moreover, the more complex model may well suggest a more highly refined system of regulation, entraining greater costs in public education and enforcement. Thus, with an increase in the sophistication of the guiding paradigm of management, there is a greater cost of research and application.

It follows from these considerations that if the increase in the intensity of the fishery is coupled with the development and application of more complex models, it may be expected that the potential losses from uncertainty will be progressively avoided and may be called gains arising from better management. When a fishery is at a relatively low intensity, the gains may be small, but they will approach a peak as the value of the fishery approaches or exceeds its maximum. Whether there will be a gain thereafter from increasing complexity is debatable, for the greater precision of management may cost more than what is lost from more primitive management approaches.

The flaw in the foregoing chain of argument is that knowledge, once gained, is gained forever; a fishery may collapse and only recover after decades. There is thus considerable logic in the view that managers should invest in developing a capability for a much more sophisticated level of management than they may expect to be able to implement, if only as a hedge against catastrophe. There is also the point that it is only by knowing how to manage better that one is able to judge what might be lost by reverting to simpler and ostensibly more primitive forms of management.

It is against this kind of background that most fisheries agencies of an advanced type see the need for a research-oriented group that is charged with the responsibility for developing more sophisticated understanding of the basis for managing fisheries. Some groups of this kind stress biological oceanography as the route to enlightenment; others put most emphasis on slicker and quicker numerical analysis; and those that are more old-fashioned are determined to better understand the fish. Whatever route is favored there is never, in the minds of the researchers, a sufficiency of resources.

A research vessel (or two) is usually seen as essential, entailing very large expenses for equipment and operation. A research laboratory, also at very large expense for facilities (including a library and a computer), is usually seen as mandatory. In almost as little time as it takes to blink, the research group becomes the world center of something or other and there are strong pressures from within for the resources necessary to achieve lasting distinction. In a very real sense, a tertiary-level fisheries agency has all of the necessary dimensions for expansion beyond the needs of the resource it was established to serve.

Insofar as research that is immediately related to management of natural resources is concerned, the foregoing is perhaps a reasonable summary. There may be, however, other considerations. For example, if there is concern for protection of the resource, understanding the effects of pollutants on fish (and aquatic communities) may alone be a large field of endeavor. For fisheries based on anadromous species such as salmon, concern for protection may motivate a substantial amount of the research and may occupy much of the time of management.

Additionally, there may be opportunities foreseen for developing the resource. Salmon again come first to mind, but more generally most fishery managers can generate daydreams of increasing production by one means or another. The contemporary enthusiasms for aquaculture of a wide range of species, both fresh water and marine, are typical.

Investments in research and management for protection on the one hand, and development on the other, are perhaps the most difficult to rationalize, for in both cases there are many considerations both within and external to the fisheries context. For example, all estuaries or inshore marine areas (or lakes and streams) do not have equal significance for fisheries, immediately suggesting that some may be written off in the interest of some other use, or at least that some may be given higher priority than others for intensive study. Moreover, there is an almost inexhaustible list of potential contaminants or environmental modifications for which some judgment must be made about possible effects. A profound knowledge of the physiology, behavior, and ecological relations of every major commercial species would seem to be required to adequately anticipate the potential impact of the various activities that may impinge on the fish or their environment.

In large measure, most relatively well-developed fisheries agencies undertake the necessary investigations of local situations in which major fisheries interests may be threatened. When it comes to the broad base of scientific understanding, most are content to maintain a few specialists who can draw on the combined and accumulated knowledge of similar small groups in all parts of the world.

Much the same sort of pattern is characteristic of research and management of development potential. The nuts and bolts of relatively routine activities such as stocking lakes or expanding oyster production are handled on a priority basis giving most attention to what shows most promise. A small group of specialists may undertake some experimental ventures relying heavily on rapport with similar research specialists elsewhere. However, there is always a perception that with more people and more funds, more could be done faster and would pay off in benefits sooner. For specialists in development, prosperity is just around the corner. The century of experience with salmon hatcheries (Atlantic and Pacific) is a silent testimony to the optimism that is necessary before the method finally bears fruit (maybe).

With the need for research arms in fisheries management, protection, and development, it is inevitable to suggest that the various research arms be combined into a single organization, set aside somewhat from the day-to-day hurly-burly of managing, protecting, and developing the fisheries. With such a grouping a more rarefied atmosphere can be created for research, and there are potential savings from sharing of expensive kinds of equipment. There is also the matter of scientific publication to consider: once a research laboratory (or a string of them) has been built, it is almost inevitable that there will be one or more series of publications arising from their work. While many such series are started by scientists who do the editorial work part-time, it is to be expected that arguments will be made and heeded to hire professional staff as editors.

Meanwhile, the economists and sociologists, having something to say about all phases of the fishery and its management except perhaps their own cost effectiveness, will perforce proliferate and will also seek separate identity in the organization and an appropriate cadre of assistants and accoutrements. They will attempt market surveys, scrutinize changes in fishermen's incomes and lifestyles, and debate the merits of cooperatives, the means for reducing fishing effort, the possible savings from protection, and returns from enhancement, in general striving to place the bits and pieces, if not the whole, into a broad social and economic context.

It must be apparent from the trend this discussion has taken that the fisheries administrator may have a relatively large and diversified operation to oversee. The various research and management activities entrain their various ancillary service requirements. In addition, the administrator begins to feel the need for many more staff functions within his own headquarters office, both to coordinate the total agency operation and to insure its interfacing with other agencies, both governmental and nongovernmental.

The full-fledged tertiary-level organizations of fisheries research and management all tend to have much the same character as is suggested in the foregoing. They are large and diversified, progressively require more service and directing personnel, tend to be budgeted primarily on an incremental basis, and, unless reviewed periodically, can become lopsided in relation to the original chores they were set up to do.

In a fully mature and well-balanced fisheries management and research organization it is my view that, in contrast to the secondary-level pattern, the division of funds should reflect the greater emphasis on research activities and development activities. I estimate the ideal division would be less than 10% on administration; not less than 20% on biological, economic, and social research; 20% on development; 20% on sampling, stock assessments, and management; and 30% on enforcement and protection.

It may be difficult to judge how effort is in reality divided among the various categories of activity unless one is very familiar with the details of the ways in which budgeting is done and the ways in which activities are conducted. For example, a researcher may be called on from time to time to perform staff work, which is essentially administration. The management staff may become involved in enforcement activities and vice versa. Nevertheless, it is usually possible to form rough estimates of how effort is expended, and this should certainly be a periodic exercise for any well-run agency.

The annual cost of the whole organization should probably not exceed 20% of the landed value of the fishery, if there is to be any net economic benefit to society. Preferably the long-term average expenditure on research and management should be in the range of 10 to 15% of the long-term average landed value, there being periods of high investment in development followed by large economic returns. Again, the best guide will be whether further investment in management activities of all kinds is associated with incremental benefits that offset the costs by sufficient amount to provide a return on the investment.

Some Dynamics of Fisheries Research and Management Agencies

The foregoing is largely concerned with what might be called the statics of fisheries organizations. There are features of their dynamics which also warrant comment. For example, it is typical of fisheries organizations rapidly growing through the secondary level to invest disproportionately in research (which is understandable). Subsequently, it is common for the management aspects to be stressed and research may not grow as rapidly as it should. In many tertiary-level organizations research is a common target for administrative grumbling and may be cut back in favor of what are perceived to be more immediately rewarding activities, such as investment in prospects for development. This is especially the case when the research activities are not closely directed to timely management issues. Left to their own devices, many researchers pursue a meandering interest in what has been called "hobby shop" or "grapevine" research, progressively wandering further afield or getting lost in matters of detail. The research accordingly becomes more obtuse and much less easy to relate to its original objective. This is not to say that tertiary-level organizations should not do basic or long-term research; such kinds of undertakings may be crucial to long-term management success. However, all research of fisheries agencies needs close professional direction if it is not to levitate out of its appropriate sphere.

Another common trend is for administration and ancillary services to progressively grow out of proportion, perhaps reflecting that as the organization becomes larger there is a need for more superimposed levels of control. There are many other kinds of imbalance, especially in small organizations in which personalities can have considerable impact on the character of an agency. For example, I know of one agency that, because of the impact of one man, is doing the same things today that it did 40 years ago—still undertaking surveys that are preparatory to management some time in the future.

Larger organizations that are fully fledged for fisheries research and management are somewhat less prone to the impact of personalities and tend instead to accumulate institutional styles and temperaments that shape the characters of their staffs. At the same time, the larger they get the more likely it seems that they will have organizational problems. Just as fishery management with too complex a model may not be as effective or as cost effective as with a simpler model, so may a fisheries organization that is too complex be less effective and less cost effective than one that is simpler.

There are many reasons why this should come to be so, and the almost universal occurrence of institutional arthritis has been long since documented by Northcote Parkinson. I am indebted to Mr. F. E. A. Wood for drawing to my attention the article by Elgin and Bushnell (1977) that spells out 16 problems of large, complex systems that range from diminished comprehension of the system through increased rigidity and declining performance to decreased awareness of actual system performance. The problems of bigness are evidently ubiquitous.

On the Canadian scene in fisheries much of the difficulty seems to stem from the formalization of "job descriptions" which tend to relate pay with responsibility, and responsibility with the number of people reporting to the one responsible. There is thus a strong incentive to increase the size of one's staff, whatever they do. This Roman Army type of approach, coupled with formalized negotiation, appeal, and grievance procedures, leads not only to a much larger personnel services staff, but more fundamentally to pigeonholed thinking, feathering of nests, and birdlike defense of territories. Staff are not encouraged to venture beyond their assigned roles and may in fact be discouraged from doing so. Advancement becomes a matter of promotion on a well-defined ladder according to rules that may favor seniority over accomplishment and blamelessness over initiative. It must be pointed out, of course, that these kinds of patterns may in large part be imposed from without by an agency known by such a name as the Civil Service Commission. It seems likely, however, that even if fisheries were the sole arm of government, it would nevertheless stiffen in an analogous manner with devices of its own invention, people being what they are.

Another facet of personnel management in such large agencies is the tendency for fast turnover in key operational positions. Thus the bright young man may move quickly through the system and, unless there are special precautions, much of what he has initiated or learned goes with him. Large systems may not have good memories; in consequence, they need to take special measures to preserve continuity. Complex management may be beyond their capability because they keep forgetting what they have learned.

With bigness there is also a strong trend to isolation of components. It has become trite to speak of the needs for communication between the organization and the constituency of the public of which it is a servant. To the chagrin of most who have espoused the need of coordination and communication, it frequently transpires that by becoming aware of all they apparently need to know, many become bewildered and confused. On the subject of coordination it has been said that achievement is a positive linear function of perspiration, a positive exponential function of inspiration, and a negative exponential function of coordination. Thus, the well-coordinated public servant may become something of a drone, fully aware but unproductive.

Communication with the public, who may ask anything at any time of any civil servant regardless of whether it is reasonable to do so, requires not only awareness but a gift for making complex things sound simple. Inasmuch as large regional or national fisheries agencies are highly complex and are managed according to sophisticated concepts and policies, it is little wonder that they have difficulties in explaining themselves. Anyone who has attempted to articulate the substantive and administrative nuances of proposed regulations to a group of fishermen quickly learns that what sounded simple when said in headquarters in the office jargon is received as though it were quite unintelligible when presented in plain language at a public meeting. The questions asked or comments made are often obviously based on gross misunderstanding. Frustrated and confused by the complexity of the organization and its multifaceted goals, public groups frequently react, then, with political activism or at best (or worst?) with passive acceptance.

So much is familiar to all large fisheries agencies. What is not clear is how to cut the knots of complexity and to maintain public confidence. The answer most certainly lies in the direction of simplications of various kinds, of which the most commonly suggested is the breaking up of a large central system into small decentralized units which can assume major responsibilities. Despite their apparent inefficiency (by virtue of duplication) small groups are cohesive, and freed of the trappings of large organizations (disregarding formalized lines of command and job descriptions) they are commonly characterized by the versatility of their staffs. By virtue of having significant responsibilities, they may also gain a quick rapport for the sector of the fishery for which they are responsible, whether for all species in a region or for a single species over a wide area.

Although a number of small groups will almost certain diverge with respect to their structure and effectiveness, the sum of their several individual contributions will usually be greater than that of a single consolidated unit. Essentially, say the proponents of simplification through decentralization, there is more life in a thousand mice than in one elephant. It has to be remembered, though, that the gains in the rapport with the resource users may be offset by potential losses of rapport with the central administration. The ultimate responsibility cannot be delegated; it takes a strong and adroit minister to justify the potential foibles of many decentralized units. It is little comfort to a small group of local fishermen to be told that their resource has been managed poorly, but that on the average the resources of fishermen are being handled better.

Another avenue of simplification is the farming out of many of the chores of research and management to the private sector. This kind of approach, typified by what is called the "contracting out" policy in Canada and by the sequels to the Rothschild report in the United Kingdom, may have growing pains, several different kinds of potentials for mischief, and the prospect of being more or less irreversible. In Canada, for example, there was virtually no capacity for private sector activity in fisheries management prior to the setting of government policy that favored contracting out. In consequence, a large proportion of the contracting out was done in the first instance to "false store fronts" that did nothing but issue paychecks to employees who received their direction from civil servants. This was followed and replaced in part by the awarding of contracts to retired (or resigned) civil servants who found it more to their liking to work outside of government while still enjoying its largesse, and in part by various devices for laundering government money so that it appeared to be spent "outside," such as at universities. Eventually, there was developed and now exists an independent assortment of consulting companies that could competently undertake many of the chores and routines of research and management. Now established, these operations have, as might be expected, their own perpetuation at heart and are not slow to remind the government that they deserve a continuing piece of the action.

In addition to these developments, which provoked various indignations both within and without the civil service, there has been the added feature that from a taxpayer's perspective contracting out may be just another device for expanding government service in a less visible fashion. In the argot of Canadian government circles, contracting out is a way of disguising "man years" that would otherwise be reported as still greater expansion of the government sector.

Despite these several cynicisms there is much to be said, with detachment, for the value of simplifying fisheries agencies by farming out many of their chores to consultants. Indeed, in the modern world of specialization (and rapid transportation), it is tempting to suggest that the most economical route to good fisheries research and management could be to have almost all of it done by contracting out. If I were the newly named Minister of Fisheries of a newly formed country, I would certainly consider the option of having a staff of three with a large budget for buying the best services in the world for all highly technical matters, and the services of local entrepreneurs for the routines that were best performed with local *savoir faire*. The risk in pursuing this kind of option, especially when taken to such extremes of hyperbole, is in the likely lack of continuity and the potential lack of long-term accountability on the part of consultants.

A related alternative is to involve the users of the resource in its management. Most users play a passive role in management by observance of the regulations, a role which they may augment by occasional participation in advisory boards, public hearings, and the like. It is also not to be forgotten that by virtue of their personal investments they comprise the infrastructure that transforms a natural resource into economic wealth. Their return on their investments is their long-term incentive to maintain a healthy interest in its continued sound husbandry. Pursuing this theme, it is commonly suggested nowadays that the users would have an even more likely sense of responsibility if they had a greater voice in management.

Central to the pursuit of this line of conjecture is the notion that people look after what they own. Given, then, that something akin to the principle of private property rights might be applied to fisheries, it might be expected that the rights holders would become sufficiently well informed to manage as well as, if not better than, the paternalistic government administrator of public property. Words such as these have been frequently put together in the past few years, perhaps largely in idealistic reaction to what is usually polemicized as bureaucracy.

Before launching one's boat on such an idyllic sea, it is probably useful to go back to considering the fish. Unknowing of the distinction between public and private property or of national borders, it is characteristic of many major commercial species to wander along continental coastlines from one potential private zone to another, and from one exclusive national fishing zone to another, or to go inshore and offshore from fishing zones to international waters. In such circumstances, it is simple neither to allocate rights nor to decide who might bear the burden of the wide-scale and penetrating research that might underlie such allocation. If I own the rights for Atlantic salmon, will I buy my own research to establish how much of my catch is intercepted by my national friends and my international neighbors and how much I should be compensated? To these and similar rhetorical questions the answer must certainly be that, by their nature as a resource, marine fisheries require that there be a central governmental control for resource allocation and research.

This is not to say that the private operators cannot do research; indeed, they should be encouraged to do so. Moreover there are some aspects of management, especially for some species, which can be safely placed in the private sector. For sedentary marine fisheries such as for oysters, clams, or seaweeds, is the only need for government supervisory? Should the private sector do all the necessary research and development in aquaculture? Could enforcement of commercial fishing regulations be contracted out to fishermen's groups? Are there similar aspects of protection that fishing companies and fishermen can undertake? The answer to all of these kinds of rhetorical questions is probably a guarded "yes," followed by a listing of reservations. Fishermen and fish processing companies in the modern world are much better informed than their predecessors, much more capable of accepting such responsibilities, and much more in a mood to assume them.

If this is to be the trend in the future for fisheries research and management, it may be timely to start thinking about how to structure the government fisheries agency of the future, which it seems may be characterized by a superintendential style, a great knowledgeability, a strong emphasis on education of users, and little attention to the day-to-day custody of the resource. Clever and lazy, that is the style, in which the central authority assumes the proper roles of a superintendent: to organize, deputize, supervise, and criticize.

Quaternary-Level Fisheries Research and Management

Larger countries that face on two oceans or that span many degrees of latitude or longitude are posed with special problems of organization for fisheries research and management. There may be wide differences in the nature of the fisheries. The contrasts between United States fisheries in the Bering Sea and the Gulf of

Mexico, or between Atlantic, Arctic, and Pacific fisheries of the USSR (or Canada) are good examples. There may also be wide differences in the social and economic circumstances of regional fisheries within a country, and such differences may bear heavily on management decisions. For countries in which these sorts of situations exist there is, of necessity, an added dimension to the organization of fisheries research management.

The character of large-scale fisheries organizations is largely determined by the division of powers among national and regional levels of government. In Australia and the United States, for example, fisheries are within the purview of the state governments; but insofar as international matters are concerned, or questions of interstate commerce may arise, the federal government has constitutional prerogatives. In Canada, by contrast, fisheries are a federal concern until the fish are caught, when they come under provincial regulations concerning processing. In all three countries the federal government may itself conduct research or provide grants and contracts to have it done. There are many jurisdictional details which need not be developed here. Suffice it to say that in countries such as these there is added to fisheries research and management an element of political complexity that one way or another may imply more opportunity for confusion and duplication, and the corresponding need for coordination.

How to organize in these circumstances is largely a matter of political taste, but the net result may well be much the same in terms of manpower and cost. Highly centralized structures may seem more efficient to the mind's eye but in practice may invisibly dissipate much of their energy internally with disproportionate expenditures on administration at all levels. Highly decentralized structures may be visibly duplicating each other's efforts but in so doing they may be highly efficient. The question then becomes "which style produces the most effective management?" My preference lies strongly on the side of decentralization.

In the matter of research, for example, there is little to be said for highly structured national arrangements that attempt to coordinate federal and state (provincial) research programs, or even to coordinate federal research activities. Communication among researchers can and should largely take place through the well-established and traditional routes of scientific publication. One of the consequences of an excessive amount of committee and task force activity may be that much research is not published, becoming instead part of the mystique of the civil services, and continuing to influence policy for some time before it is eventually exploded as myth. Particularly in research it should be appreciated that individual creativity and the rigors of publication are far more important than committee agreement.

There is perhaps less to be said for individual initiative in management, for the national good may be at stake in matters involving provincial (or state) cooperation or in the development of consistent international postures. On the other hand, as was suggested in my chapter in the first edition of this book, fisheries management is not a science and is much in need of experimental approaches. How better to learn than to encourage diversity in management by delegating responsibilities in highly decentralized ways?

While this may be good advice to highly complex and well-developed fisheries organizations, it may leave much to be desired for the less well developed. For these organizations, largely in developing countries, it may be preferable to first go through a phase of centralized planning and administration to insure an appropriate level of sophistication before responsibilities are delegated. Unfortunately, though, it appears to me (admittedly from a distance) that in many developing countries only a complex central administrative mode may be put in place and at the expense of an adequate infrastructure of trained fisheries researchers and managers in the regions. In consequence, national plans for expanding fisheries may be implemented without appreciation of regional realities, to the ultimate detriment of the fisheries.

To return to the developed countries, they too may have national policies of various degrees of articulation, the intent of which is to provide guidance to administrators in the achievement of long-term objectives. To the extent that the organization is decentralized, these policies are progressively more meaningless. The Canadian national policy statement, for example, reflects sufficient centralization to have some apparent bite, but on close reading many of the policy stances are either empty platitudes or are sufficiently well qualified to allow for wide political latitude in their application. The Canadian document took a long time to prepare, involved substantial consultation with the regions, and emerged as a document that indicated a painful awareness in the capital of the need to recognize the social and economic differences among the regions. As a practical guide for day-to-day administration in the regions the national policy is next to useless, and it seems unlikely that the copies in any of the regional offices are well thumbed.

The United States, to my knowledge, has no single fisheries policy statement, but in recent years the establishment of Regional Fisheries Councils to apply the concept of optimum sustained yield has been a strong centrifugal force for decentralization. The federal government retains the right to change the decisions of the Councils, but as time goes by and the Councils gain experience and confidence, one can expect the exercise of the federal prerogative to become more fraught with political tensions.

It seems to be the moral that in large and complex national settings there is a tendency to development of additional upper levels of administration which are progressively more abstract and which contribute little to productivity. Just as the production of fish from an ecological community may increase as a result of thinning out the populations at the top levels of the food chain, so may the thinning of upper administrative levels increase the cost effectiveness of regional fisheries management. There may, of course, be some loss of stability in both cases!

It perhaps only remains in such a sweeping discussion to add a remark about the ostensible fifth level of administration represented by the United Nations in general and the Fisheries Division of the Food and Agricultural Organization (FAO) in particular. Although it may seem to be a paradox, there is much to be said in support of this seemingly still higher level of organization. In fact, FAO is a very small enterprise and has a catalytic and consultative, rather than a supervisory or managerial, role. With its capability for providing expert assistance, it can help shortcut the path to development of fisheries management capacities. By virtue of its neutral status, it can undertake studies that are free from national biases. By facilitating the exchange of scientific information, it can accelerate the progress of understanding. By responding primarily to requests for its services rather than imposing its assessments and opinions, it is assured of a useful existence as long as it remains expert. In short, each national fisheries agency has good reason to encourage the expenditure of national funds on such an international good cause.

This chapter has aimed at assessing how much of what kind of management is needed to do what kind of a job in managing fisheries. It has been suggested that for the purpose of appreciating the character of a fishery a three-man team is sufficient. For management of a fishery, a 30-man team is minimal. The cost of this minimal level of management should not exceed 10% of the landed value of the catch. It follows that some fisheries may not be sufficiently valuable to warrant management. For more sophisticated management and research that may lead to development of fishery resources, it is necessary to develop a more comprehensive program that may cost as much as 20% of the landed value of the resource.

There is a tendency for large fisheries agencies to grow beyond a cost-effective size. Of the various alternatives for simplification it is suggested that decentralization may be the most useful to consider. While it is true that decentralization may not appeal because of its untidiness, it is perhaps in the long run more efficient. The structuring of higher and more abstract levels of organization at national levels in large countries accomplishes little of value. The same may be true in complex regional fisheries. By comparison, international organizations such as FAO are not seen as supernumary but as valuable free agents that should be given vigorous support by national fisheries agencies.

This kind of assessment is not meant to be unkind to middle- and upper-level administrators in large fisheries organizations. Most of them claim to be long suffering of the burdens of administration, and indeed they probably are. However, their burdens are largely imposed by a system that is controlling what need not be controlled, collating viewpoints into policies that have little relevance. In their own private assessments they often see what they are doing as largely futile. Even when the roles are filled by administrators of great skill and dedication, the effect is relatively transient because the system has an inherent forgetfulness and potential for fragmentation that in the long term is overwhelming.

Fisheries research can be as sophisticated as is demanded by nature; fisheries management can only succeed when it is matched to the understandings of the users of the resource. For this reason it is wise to associate research and management in decentralized regional units with sufficiently small mandates that coherence can be maintained albeit at the expense of overall national consistency. Diversity may be as valuable in fisheries organizations as it is in communities of fishes. The best single guide to how much fisheries management is needed is how much the resource is worth or could be worth. It follows, therefore, that increases in expenditures on management should be justified by the contribution they will make to increasing the net social and economic revenue from the resources. The inevitable moral for professionals such as ourselves is that research should be as relevant and sophisticated as is achievable, while management should be only as centralized and complicated as is necessary.

References

Andersen, K. P., and E. Ursin. 1977. A multispecies extension to the Beverton and Holt theory of fishing, with accounts of phosphorus circulation and primary production. Meddr. Danm. Fisk.-og Havunders. N.S. 7:319-435.

Elgin, D. S., and R. A. Bushnell. 1977. The limits to complexity: Are bureaucracies becoming unmanageable? The Futurist, December 1977:337-349.

Laevestu, T., and F. Favorite. 1977. Preliminary report on dynamical numerical marine ecosystem model (DYNUMES II) for eastern Bering Sea. U.S. Natl. Mar. Fish. Serv., Northwest & Alaska Fish. Center, Seattle, Washington, 81 pp. (Mimeo.)

Larkin, P. A. 1977. An epitaph for the concept of maximum sustained yield. Trans. Am. Fish. Soc. 106(11):1-11.

Larkin, P. A., and W. Gazey. 1982. Applications of ecological simulation models to management of tropical multispecies fisheries. ICLARM/CSIRO Workshop on the Theory and Management of Tropical Multispecies Stocks, Cronulla, Australia, 1981.

Pauly, D. 1979. Theory and management of tropical multispecies stocks. ICLARM Studies and Reviews, No. 1. Manila, Philippines. 35 pp.

Taylor, H. F. 1951. Survey of marine fisheries of North Carolina. Univ. North Carolina Press, Chapel Hill, N.C.

Watt, K. E. F. 1956. The choice and solution of mathematical models for predicting and maximizing the yield of a fishery. J. Fish. Res. Bd. Can. 13(5): 613-645.

12
Fisheries Research in Developing Countries

G. L. Kesteven

Introduction

The role of science and technology in economic development is a theme which holds a prominent place in the agenda of the United Nations and the specialized agencies. No less prominent is the place it commands in the minds of governments of developing countries. However, whereas the international community is most concerned with the immediate application of the results of scientific enquiry and the introduction of technological innovation in proportion that they promise economic return, the Third World must see science and technology in a different light. While Third World countries must, of course, seek the economic benefits that can come from successful application of the results of scientific research, the cultural effects of the conduct of scientific work, the political advantages of being masters of their own science, and the administrative and managerial consequences of participation in the planning and conduct of scientific research and interpretation of its results are perhaps of equal importance to them. Imported technology and secondhand science can have considerable effect (even if not always precisely the effect expected by the recipients) but such contributions maintain a recipient country in a dependent state. The conduct of one's own sci-

---

G. L. Kesteven, now retired, does consultant work (at present for FAO, UNESCO, UNU, and a fishing company), with editing *Fisheries Research* (Elsevier), and advises the Director of INIDEP, Mar del Plata, Argentina. He began work in fisheries in 1937 with the N.S.W. Department of Fisheries and since then has engaged in fisheries research and administration from positions with the Australian Government, UNRRA, FAO (in regional office, headquarters, and field projects), IOI, and other organizations.

ence is as much a part of national integrity and independence as are the taking of one's own political decisions and the preservation of one's traditions.

The foregoing observations are not to be understood to convey a conclusion that developing countries are unaware of the cultural value of science; such a conclusion would be quite silly, considering that some of the world's oldest universities are located in such countries. It is perhaps true, however, that in these countries respect for science has had a substantial elitist component, and it can scarcely be denied that development of applied sciences has been gravely retarded there. While the being, still, in developing state cannot rationally be attributed solely to deficiencies of applied sciences, there can be little doubt that a strengthening of those sciences in one of these countries will significantly increase its ability to overcome its problem. Neither can it be denied that a lack of appropriate technology sprung from, or at least supported by, indigenous applied science, can seriously disadvantage a country, especially at certain critical periods of the industrial history. Perhaps the converse of this proposition will be the more readily understandable, namely, that the opportune possession of appropriate technology and its associated applied science has at times given various countries some considerable advantages; one has only to survey the world today with regard to electronics, communications, and computers to appreciate the truth of this.

The present is a critical time for fisheries. It is critical because of changes which have followed the UN discussions of a law of the seas; because of great technological changes in the industry, with marked economic and social consequences; because of an approach to or passing beyond the limit to catch from so many stocks; and because of the numerous threats to environment. Among the challenges presented by the foregoing, that with respect to acquiring technologically advanced equipment is perhaps the least; it is relatively simple to buy boats—and to have them lie idle at the wharf, or to acquire them through joint-venture—and receive no more than some part of the catch or the profits. The other challenges are more demanding, and the consequences of erroneous decision with respect to them are generally much more serious. Whatever differences there may be between them, however, they all have this in common: that they are to be met with information; and for the developing countries, that information must be not only accurate and comprehensive but their own.

This information must be their own, drawn from their own scientific activity, and that, as argued in the first paragraph above, they must have for their very existence. However, the information must be their own also because in negotiation, whether diplomatic or commercial, they must be in possession of facts and have the appreciation of facts that can come only from having had some part in establishing them.

The argument of this chapter is not, however, solely that developing countries with valuable fishery resources need fisheries science in self-defense and positively to their own advantage. It goes further, to assert that they need a fisheries science which will be able to assist them to meet the current challenges, not merely to try to patch up, some time in the future, the consequences of erroneous decisions now; and, still further, to claim that they can organize and conduct their science according to a new paradigm, whose full realization is rapidly coming within our reach.

That new paradigm will be characterized, in the first place, by being thoroughly responsible, seeking the information required with regard to identified problems, and doing this by describing accurately the phenomena associated with the problems, searching for regularities and periodicities, and out of this work developing practical forecasting systems adequate to the situation in which the problems arise. And it will proceed deeper, looking for causes of what has been observed at the first level, in the hope of being able to forecast the behavior of those determinants, but subject to its being demonstrable that greater accuracy, or longer lead-time of forecast, will then be possible and useful. The new paradigm will next be characterized by working in real time, to provide information required for real-time decision making, to the order of accuracy required for such decisions. Its motivation then will be a desire to render service, not a pursuit of reputation.

Perhaps the most distinctive characteristic of the new paradigm will be its exigency with respect to the data required for its models. It will insist on direct measurement of the properties of its objective systems, rejecting gross averages, unchanging ratios, and invariable constants to which values have been given from theoretical assumptions. Since its models will be developed from experience it will not find itself wondering why reality does not respond to the dictates of its models.

Realization of this paradigm is becoming possible by virtue of the remarkable technological advances of the past few decades, in sensors (electroacoustic equipment, satellite navigation, and the sensors for measuring physical and chemical properties), in data-logging and transmission, and in data processing, to name major items. However, to this list must be added, with respect to the primary phase of the industry, the changes that have taken place in fishing boats, which are of social significance almost as much as they are technically important. The combination of these two sets of changes is of dramatic import. Today, thousands of fishing vessels are better equipped than boats of the past were to search for, find, catch, handle, and store fish and are even better equiped to observe oceanic phenomena and to record and transmit their observations than were research vessles of a not too distant past. Moreover, new generations of fishermen are acquiring the skills to operate the new equipment and an understanding of the systems of which they themselves are a part. This means that observations which in the past had to be made by special equipment, operated by scientists, today are routinely made and interpreted by fishermen.

Developments of these kinds are by no means reserved for the primary phase of the industry; they are to be found of equal effect in secondary and tertiary phases. For the new paradigm the meaning is this: that the conduct of fishery science must become a cooperative enterprise in which the operatives of the industry will play an indispensable part. They already provide a great proportion of the raw data and in the future their contribution will be immense. They will undoubtedly become much more active in compiling and analyzing data and making their own interpretations, and thus armed they will offer well-informed views on fishing regimes, quality control, pricing practices, and similar matters.

From these views on the need of developing countries for fisheries science and on the kind of fisheries science they might practice, the purpose of this chapter is to suggest to those countries a different approach to planning and carrying out fisheries research, by which they may take full advantage of all that modern technolgy can offer while making full use of what has been established so far in fisheries science, discarding the myths of naturalistic fisheries biology, and avoiding the pitfalls of the numbers game.

Since fisheries research is an application of science to a particular industry, it is to be organized and executed for the achievement of objectives which are of that industry or relate to its role in the economy and its place in society. This proposition might seem so simple and so obvious as to be scarcely worth stating, were it not that at various times in various places it has been neglected, subverted, and even rejected outright. That being so, it is necessary, for the purposes of this chapter, to examine the terms of this proposition and to give a clear definition of the context of fisheries research.

Fishery Industries: Their Development and Management

The central point of the above proposition is the concern for "industry," a term for which, however, different definitions are offered. In some countries the term "indsutrial fishing" signifies the activity of incorporated fishing enterprises, to the exclusion not only of artisanal and subsistence fishing but also of owner-operated vessels working alone or as members of cooperatives. Instead the term is given the connotation, in this chapter, of "any activity with economic intention or effect" directed to the exploitation of living aquatic resources and to the utilization of whatever is taken from those resources; with this definition the term embraces all commercial fishing, of all levels of organization down to artisanal fishing, subsistence fishing, and sport fishing insofar as it is organized and conducted with economic intention and effect. Perhaps the definition excludes only the small boy with a line of cotton thread and a bent pin. The definition emphasizes that the fishing with which we are concerned is engaged in for the purposes of society and has consequences for society.

The expression "economic intention and effect" covers, of course, a wide range of kinds and grades of intention and effect; moreover, it admits of recognition, especially in models of the industry, of a modulation of intention and effect in response to noneconomic objectives of individual and of society. These differences of intention, effect, and objective, within and between individuals and society, are matters for which information is required and are the root cause of the conflicts to the resolution of which the greater part of the effort of fisheries research is directed. The fundamental justification of fisheries administration is the resolution of conflict; the task of management is to choose between alternative courses of action, and is of course the most difficult when there is difference of opinion as to which is the best.

While for mankind in general the objective of fishing and of the associated activities in secondary and tertiary sectors is procurement of food for humans and domesticated animals and of materials for industrial use, most governments restrain those activities, or seek (or merely declare an intention) to do so, in favor of conservational objectives, to ensure the availability of the resources to future generations. For fishermen, however, the objective is otherwise, namely their livelihood, and in pursuit of this objective they often find themselves in conflict with society and with one another: the latter set of conflicts, between fishermen, comprises the competition between fishermen on particular grounds and the conflicts between commercial fishermen and sport fishermen.

These conflicts emerged during the last hundred years and have greatly intensified in the last 50 years as a consequence of the population explosion, the great advances in technology, the rising standards of living, the efforts to eliminate hunger and to create a new economic order, and the awakening of awareness of the threat to human environment. Moreover, conflict has been sharply intensified in particular areas as sequel to the United Nations discussions of the law of the sea.

Concretely, these conflicts are with respect to issues such as: by whom should the resources of each fishing ground be fished? What magnitude and composition should the catch have? When should fishing take place and with what gear should it be carried out? What relative weight should be given to fishing and fish culture as against other uses of land and water (for inland fisheries)? What use should be made of the catch (direct human consumption or other use, for example, as pet food)? What quality standards should be observed? What rewards for their work should fishermen and other operatives receive? At what level of efficiency should the operations be conducted? What relative weight can be given to an objective of creating employment opportunity through the creation of fisheries, or through placing restraints on efficiency? What importance should be given to the earning of foreign currency with fish and fishery products, and how much of added-value should be generated in the country of origin. All these and other issues are involved in the management of fisheries and in their development.

In the sense of the preceding paragraph, the differences between management and development of fisheries are not great and are largely of degree rather than of kind. "Development" can be taken to mean any quantitative or qualitative change of structure, organization, or procedure which results in increased benefit of some kind. (Whether the result is judged a benefit by all concerned is another question.) In fisheries, development ranges from the creation of an entirely new fishery down to small changes in the practices of catching, handling, storing, processing, transporting, and selling fish. In this sense development is part of the task of management and is taking place constantly; the management of a fishery enterprise, unless it is moribund and on the way to bankruptcy, is always on the lookout for ways of developing the enterprise.

The distinction thought to lie between "management" and development" is really a difference, even at times an opposition, between management of industry (in the sense of business management) and fisheries administration as public adminstration. Seen in this way, fisheries development is represented by projects, generally large, aimed at bringing about major changes in a country's fisheries by creating new fisheries or reorganizing them (especially in secondary and tertiary sectors) to effect a redistribution of rewards. Such projects of course offer opportunity for political propaganda, whereas developments accomplished by management are represented as being no more than what ought to be done by management and in any case undertaken only for the sake of profit. The irony of this contrast is that the propaganda value of fishery development projects especially in development of economic infrastructure, often derives from the fact that the changes to be effected are large precisely because of the failure of administrators to take decisions which were only theirs to take. To the contrary the changes in those matters that are in the charge of administration should march with those being effected by business management and the special role of government should be in guiding and complementing the changes made by business management and at times anticipating them. From this point of view development is seen to be not distinct from but part of managment, to be brought about both by business management and public administration acting in collaboration, each with its particular responsibilities.

A major consequence of this discussion is that if we can accept that fisheries development is essentially a result of a special part of the proper discharge of the management function, in business and administration then we find that we are concerned with a single, integral array of information. Briefly put, the indications of opportunity for change are to be found in analysis of existing information, and the discoveries that may lead to change today become tomorrow the commonplace of everyday action.

The Role of Research-Obtained Information in Fisheries Management

Considering the conflicts that arise in fisheries, and the numerous different issues involved in those conflicts, it is obvious that what happens in a fishery cannot be understood from an examination of resource, alone: and even less is it possible only from biological data to formulate regimes by which resources will be exploited to serve the objectives of present and future generations. Therefore "fishery" must be seen and studied for what it is, a bio-tecnico-economic-social system, and fisheries science must be developed to become capable of diagnosing the state of the whole system and of offering a prognosis of it. Anything less is scarcely worthwhile. I should add here that the term "fishery" has, in this chapter, an extended meaning, to include fish culture and aquaculture; this is in accord with the definition of the industry given above and corresponds to extension of the term "fish," by legislative definition, to include all living aquatic organisms.

Fisheries management, which means a taking of decisions with respect to biological, technical, economic, and social components, obviously, requires information with regard to each of these components. The total array of such information which can be assembled for a single country is extensive and varied. The array has a set of compartments for each of the four components, each set having divisions for inventory and for operational data; and the whole is repeated for each sector and replicated within sectors, for species, fish stocks or fishing grounds in the primary sector; for classes of products in the secondary sector; and for markets in the tertiary sector. While the items of information lodged in particular parts of this complex array vary in number from part to part, in most, if not all cases, each part will have some items which are constant and others which are variable. Finally, it is possible to specify, as FAO has done with its Species Synopsis Scheme, the items to be placed in each part of the array to meet the requirements of the current paradigm, of research, and of business management.

A very great proportion of this array is established and maintained by industry; it certainly precedes the scientific contribution and for much of the time it is sufficient for industry's purpose. Some items, however, are beyond the reach of methods employed by individuals and enterprises of industry. The reasons for this inaccessibility vary. The simplest case is that of whole-industry information; the compilation of a total inventory of a single fishery, or of a country's fisheries, with detailed record of all operations, and the subsequent processing of such data and interpretation of summarizing statistics could, theoretically, be carried out by industry itself. The reasons it is not done are familiar, but neither they nor the complexities of the required procedures make this a task for research workers, although the procedures themselves might be the subject of methodological research.

Other items are inaccessible to most fishermen because they, the fishermen, do not have the time, patience, or temperament to observe objectively over extended periods and to record carefully their observations. Nevertheless, many fishermen have done a great deal of such work and have developed a detailed and reliable account of the distribution and behavior of particular species of fish. Moreover, since many fishing skippers today have academic qualifications in biology as well as in technology and have the use of electroacoustic and other observing equipment, of data-logging equipment and calculators, and perhaps access to computers, we may expect that fisherman knowledge of resources will steadily increase and extend beyond distribution and behavior. Perhaps this process could be accelerated by providing courses in scientific method–its logic and discipline–for fishermen.

It is unnecessary to look in secondary and tertiary sectors for parallels to this development of the primary sector, since the participation in research by industry of those sectors is well advanced.

It then emerges that the items of information that are truly inaccessible to operatives of industry, as operatives, are those for which it is necessary to work full time with specialized equipment and knowledge. (However, note that this exclusion holds against individuals and small enterprises but not against large companies, for the last can and do employ research workers, directly or as consultants. Thus, when talking of information to be obtained by research we must bear in mind that at times we must make a distinction between that which is obtained by institutions of the community–generally of the government–and that which is obtained by industry.) Moreover, they are items characterized by relevance to current decision-making situations and by having an order of accuracy appropriate to those situations.

The question then is: what are these items of information that are to be obtained by research, and what is the effect in the industry of their being available? They have often been represented in the past as being answers to the questions: what is available to be fished? Where can it be caught, at different times of the year? When can it be caught, in particular places? How can it be caught? How much can be caught? To which ought to have been added—but so rarely has been—how can it best be utilized? As every fishery biologist knows, however, or ought to know, the bulk of the information in answer to the first four questions has been provided by fishermen, either in verbal and written reports or through the record of their operations. This is not to disparage the work of fishery biologists who have compiled such information from fishermen and, after supplementing it with information from survey operations and from tagging and other methods, have assembled definitive accounts of the distribution and behavior of particular species; it is, instead, to ensure that the perspectives of this work are correct.

The other two questions, however, of how much to take and how best to use it, merit much more attention here. It is fair to say that the greatest part of fisheries research of this century has been devoted to the question of how much to take; and perhaps just as fair to say that the least attention has been given by fishery scientists to the question of how best to use. From the end of last century fishery administrators have been concerned to restrain fishing (by setting size limits and declaring closed areas and closed seasons) and to limit catches, and inevitably fisheries research of the first half of this century converged upon population-dynamics models, of which the most well known is that of Beverton and Holt. With these models came the concept of maximum sustainable yield (MSY) and the practice of stock assessment; from this work fishery biologists were able to announce, as it would appear from the presentation of their results, an inherent and seemingly invariable property of the stocks they had studied, and to tell fishermen what they could expect to take from those stocks, provided they behaved themselves. Behave themselves they did, as fishermen and even, sometimes, and under duress, in observance of biological dicta, only to find (mostly to their distress, some few times to their joy) that MSY was a rather weak strut upon which to lean. It is unnecessary to go further here in an examination of the sad story of MSY; Peter Larkin has already attended the ceremony of its interment. Again it is necessary to make a reservation: these remarks are not intended to suggest that the work of stock assessment aimed at MSY was of no value, far from it. As Kuhn, and Nietsche, and others before him, have observed, a false hypothesis can be powerfully effective in promoting and enlarging knowledge and understanding; and MSY was not entirely false, the trouble came from the inadequacy of fisheries science. And that is where the question of how best to use the catch comes in, along with a host of economic and social questions. For, if fishermen fish for cash while the community wishes them to fish for food and other materials and at the same time to observe some rules of conservation, the science that is to serve the community and fishermen (and processing plant operators and others) ought to work with more than merely biological models; nor should it put its trust in the fantasies of bioeconomic models.

There is, however, another aspect of the history of fisheries research. It is the matter of direct assistance to industry. There have been cases of research having discovered a resource, and cases of research having given a better account (than existed) of the distribution of a resource, or of its magnitude, or of the fluctuations in resource availability. Technology has contributed tremendously to development of fishing gears and of fish-finding equipment, and the research of institutions such as Torry has greatly aided the secondary and tertiary sections. Nevertheless, at least in the case of the primary sector, the direct contribution to day-to-day decisions has been somewhat less than one might suppose that it might have been. While some entrepeneurs may have based their investment decisions on the results of stock assessment (with unfortunate results in some cases) almost all fishermen have been aware of research only through the effect on them of regulations based upon research results of which they know little and understand even less. It is a major part of the thesis of this chapter that if fishermen could be made aware of fisheries research in the course of their operations, be assisted in their choice of strategy and in the execution of their tactics, and become partners in the conduct of research, they would become acquainted with what lay behind the regulations and more disposed to assume responsibility for autoregulation which, surely, is the only way out of the competitive common-property dead end. The route to this condition is by way of fisheries science becoming able to deal with real problems in real time instead of offering biological diagnosis some months, or even years, after the event. For example, fishermen have as yet little understanding of exploitation rate and virtually no interest in a retrospective measure of it; what interests them is catch per hour when they are actually fishing and its variation from place to place and time to time; but their interest should be illuminated by some understanding of the short-term and long-term consequences of their own rates of exploitation.

The Situation of Developing Countries

If it is true that research-obtained information is required for the fisheries of developed countries it would seem that the need for such information in developing countries must be even greater, but this is not necessarily so. The argument cannot validly be advanced with respect to all the fisheries of a country at once, because the need for research-obtained information varies from fishery to fishery, depending on the level of development of each and of the ability of those concerned with each fishery to make use of such information. Perhaps the greatest need for research-obtained information is with regard to the most highly developed fisheries in which, undoubtedly, there ought to be the best prospect of putting such information to use. Hence the condition of being a developing country is not of itself a sufficient argument in favor of the conduct of fisheries research, so far as the needs of the fisheries are concerned; the arguments ought to and can be much more exact, with regard to the state of the fisheries and the changes that could be brought about in them if specific items of information obtainable by research could be available. But the prevalance of academic attitudes in fishery research insitutions of developing countries, and the conduct in them of fundamental rather than of applied research, hinder the application of this rule. These matters were discussed by John C. Marr at the Vancouver Conference in 1973 and were described more recently by the present writer in a paper for a Working Party of FAO's Advisory Committee on Marine Resources Research (ACMRR). Perhaps the developing countries are not much worse off in this respect than are the developed countries, but they are less able to bear the cost and to accept the delays of nondirected fisheries research.

The situation of the fisheries of developing countries is generally well known, if not well understood. It is sufficient here to name some of the major features of that situation. They are: the large part occupied by artisanal fisheries, at least in respect of numbers of fishermen and of fishing boats but also, in many cases, of catch (at least 60% of the Venezuelan catch is taken by artisanal fishermen); the powerful and sometimes malevolent role played by middlemen in artisanal fisheries; inadequate systems of storage, transport, and sale; inadequate and sometimes nonexistent economic infrastructure; the presence of foreign capital and management, through joint ventures and locally incorporated enterprises, dominating the "industrialized" phase and appropriating the added values; a weakness of administration, in many cases; the lack of national policies for fisheries and for research; and vulnerability to the pirating of the resources of the Exclusive Economic Zone (EEZ). The list could be extended, but better still it can be translated for particular countries with the incorporation of detailed local information to describe particular situation.

Many of these features are developing-country versions of conditions which exist in developed countries, more especially insofar as they involve questions concerning the rewards to labor and to capital. However, for present purposes, they cannot be said to be, of themselves, problems calling for scientific research. They become problems when someone wants to do something about them and can move forces to that effect. At that time a special opportunity is offered for the application of research methods, to establish the facts with regard to the feature (or features) said to present problems to weigh evidence with regard to those "problems," and to propose a plan of action which, appropriately to the conclusions drawn from the evidence, indicates remedial measures and/or further fact finding. Whether the fact finding should be by way of scientific research or not and, if research should be indicated what should be its kind and at what level and with what intesnity it should be conducted, are matters which can best be decided from the evidence assembled in the primary study.

The same procedure can be followed, and the same principles can be observed, subsequently in the reshaping of programs. If a primary diagnosis has proved to be correct and the information yielded by research has been found to be effective, then the situation has changed and it ought not to be assumed that in the new situation what is required is simply more of the same research, or even that further research is required. It may be the case that the procedure by which the effective information has been obtained can be transmuted into routine procedures which perhaps industry itself can apply. This is a sequence which has often taken place in secondary sector. On the other hand, it may be the case that the procedures are not yet well enough advanced, which might mean "more of the same" or a change of direction of research; in either case a decision will best be made from a second diagnostic analysis. More particularly, however, this procedure allows, even induces, the studied system (the whole bio-tecnico-socio-economic system) to direct the course of research by its exposure to its problems and progressive re-specification of the degree of coverage and order of accuracy of information required for efficient management.

These subsequent, program-review decisions will be strengthened by a special element of the procedures (implicit but not remarked in the preceding paragraphs), namely, that they call for a formulation of hypothesis with regard to the effect that the information to be obtained will have. Each subsequent diagnostic analysis will thus be an examination by tests of significance, of whether the hypothesis is to be accepted or not, and will point to the consequences of the result obtained. In this way the planning of research itself becomes a scientific practice, with effects which could be of considerable significance for developing countries.

These procedures have the effect, for fisheries biology of virtually turning the usual sequence of research end for end. That sequence followed the logical sequence of a text on fish biology, beginning with classification, passing to anatomy, morphology, and embryology, on to physiology and behavior, and then to distribution and ecology, arriving eventually to expoitation; racial studies, age determination (scales and otoliths), and growth curves quite easily slip into this sequence. And after all that, one turns to Ricker. Without in any way suggesting that any item of the foregoing is unimportant, the procedures proposed here suggest that there is now a more direct route to the information required for management (both private and public) and that to take this direct route is a course especially opportune for developing countries. The message of the procedures is, roughly: first, give the fishermen the results of a statistical analysis of their operations, showing, in the detail in which they work, the patterns in space and time of distribution and abundance, and the cycles and trends; the presentation of course being appropriate to them; and then, as they and the managers require more accurate information and more specific and more reliable prediction, pursue the investigations deeper and deeper in response to requirement while carefully watching the play of the law of diminishing returns to investment in information gathering for fisheries management.

One would assume that each director of a fisheries research institution wishes to organize a program of research which, within the means available to him, will yield most results and bring credit upon the institution (and himself). Therefore, one would assume him to seek matters for investigation which at the one time are within the competence of his staff, are likely to yield to research and bring the credit he seeks, and are of a kind for which his equipment is suitable and his funds adequate. Directors of fisheries research in developing countries do not have such an easy path. Their funds are generally meagre and uncertain; their staff is generally small, often only skeletal, especially in workshops and other technical support services; often the professional staff has had little training in fisheries science, even if they have had much experience in the elementary practices of fisheries biology; and generally the knowledge of statistics is weak. Despite these disabilities, many directors are called upon to assign their staff to ad hoc tasks relating to administration of the fisheries, even to settling disputes among fishermen. They are not immune from political interference, and in some cases have to participate in or give way to massive changes of staff and program as sequel to change of government, and these changes mean, of course, a lack of continuity of research as well as interruption (even truncation) of career prospects of the staff. Under these conditions research itself suffers directly, experience in planning research and in organizing logistic and technical support to research cannot be accumulated, and expertise for these tasks cannot be developed. In especially serious situations of this kind there is little opportunity for staff to become well acquainted and to learn to accept the exigencies of each other's kind of work; in fact, the condition favors the growth of animosity between research staff and those responsible for administration of the institute. The longer that these conditions prevail in an institute the poorer is the image of it held by industry and the public, and of course this tarnishing of its image weakens the institute's contact with industry and further aggravates the situation by lessening support to it.

Clearly a set of conditions such as the foregoing cannot be cleared up at a stroke. Each aspect calls for particular attention, in principle and with regard to local circumstances. Nevertheless, they are all closely related to program, either arising out of or causing program weakness, and this being so it is resonable to believe that closer attention to programming procedures should have a major place in a strategy aimed at strengthening fisheries research in developing countries. A scientifically planned program, formulated from an analysis of fact and oriented to testable hypothesis, can be well armed against attack; with it the preferability of retaining staff and of not dispersing their effort can be demonstrated logically. Such a program, sprung from identification of problems and having the objective of providing information of real-time significance, should attract the attention and support of industry.

Appendix: Some Suggestions to a Scientist Newly Appointed to Establish and Direct a Fisheries Research Institute in a Developing Country

Some basic precepts are to be kept in mind:

The institute is for fisheries research and even if its mandate relates only to resources its work is in fisheries biology as part of fisheries science, not in marine biology.

The program is the alpha and omega.

Contact with industry is indispensable; service to industry is a major reason for existence of the institute; industry is one of the most important sources of information upon which the institute will work.

Fisheries research can be rigorously scientific, extremely stimulating, and thoroughly respectable, or quite otherwise; which it will be in your institute will depend very much on your decisions.

First step: seek a policy statement from the atuhorities responsible for the institute; if the institute is governmental, seek a policy on fisheries and ascertain the policy with regard to research, in respect to matters such as delimitation of areas of responsibility, relations with universities and other institutions, career prospects for staff, use of common services, and, especialy, the role that institutes are expected to play in advising government and industry.

Do not be in a hurry to fix the organizational structure of the institute and in particular avoid classical labels such as "Department of Mollusca": they encourage dreams of leisured serendipitous marine biology. But do make early arrangements to establish, physically and organizationally close to your own office, an intelligence unit to take care of a monitoring system and of a compendium (see below) and in other ways to assist you. Don't call it "Programming and Planning" because those are tasks for you and the research workers to perform in collaboration.

While you have time, set yourself and your intelligence unit a program with the following elements:

a. Initiate the compilation of a compendium of information concerning those fisheries and their place in the national economy. This will consist of maps, graphs, and statistical tables; of synopses of species and other components of each fishery; and other summary statements of essential information. Keeping this up to date, with periodic discard of old material, will be a task for the intelligence unit; from this compendium assess the state of knowledge with regard to matters said to be problematic.
b. Make a diagnosis and prognosis of the fisheries with which you will be concerned.
c. Establish a system for monitoring the fisheries.
d. Promote good relations with industry, bringing its representatives into consultation in identification of problems, in planning and conduct of research, and in interpretation of results.

Formulate and make known your own policy with regard to the matters within your competence, indicating in particular the rules to be observed in the formulation of projects and emphasizing the importance of good relations with industry.

Make a first identification of problems with regard to which research might be undertaken, drawing upon your own diagnosis and consulting industry. Submit this list of problems to the scientific staff and invite them to review it, add to it, and place its items in order of priority, and to suggest the kinds of research activity which might be undertaken with regard to each item.

Establish a classification of projects such as the following:

Priority I: projects addressed to problems judged to be of first importance, and with regard to which the prospects of obtaining reliable research results on an acceptable time scale, and of the results being applied and having expected effect, are good.

Priority II: projects not so directly addressed to problems as those of priority I but required in support of them and evaluated in terms of the effect that an output from a project of this prioirity will have on one or more of the projects of first priority. These projects are of at least two kinds:

a. To provide factual support to priority I projects, e.g., through environmental research, market surveys, studies of fishing power.
b. To refine the models employed in priority I projects, with respect to either their conceptualization or the algorithms of their operation.

Priority III: projects concerned with the long-term development of models and of the research paradigm itself, insofar as this can be seen likely to become necessary for continued evolution of your program in conformity with national fisheries policy.

Priority IV: hobby projects to which scientific staff can devote some small part of their time, with their own objectives and at their own pace, each in pursuit of his own specialization and in exercise of his own imagination.

Establish a set of criteria for placing problems in an order of priority, first with regard to their own nature (present and potential value of the fishery or fisheries, losses caused by the problem or, conversely, benefits to be gained by its solution, social aspects), and second with regard to the nature of a project which would be required for it (magnitude, likely duration, prospects of success in the project itself and in securing application of its results).

Place the problems in order of priority and, for each, indicate projects of priorities I and II which might be undertaken with respect to it. To each indication of project add notes on probable duration and cost (roughly estimated) and, with regard to projects of priority II, some assessment of the degree to which realization of the project (or projects) or priority I with which they would be related would depend on the output from such projects.

The next step will be much affected by budgetary decisions made by you or those in authority above you. They will be with respect to such questions as:

To what order of accuracy and degree of coverage of problem areas should projects of priorities I and II be conducted?

Are there any problem areas which should be given precedence of others?

Is all the budget to be devoted to projects of priorities I and II? If not, what proportion of it may be reserved for projects of the other two priorities?

Can a proportion be reserved for contingencies?

Armed with these decisions the next step is essentially arithmetic, to designate a set of projects which can be executed within the means available to you.

Having made this designation, select staff to be responsible for preparing project documents. These documents will be most useful if they are constructed in three parts; the first presenting a justification of the project, the second describing the strategy and tactics to be adopted, the third presenting estimates. Justification of priority I projects will be focused on the propositions with regard to the effect expected of the information to be obtained, which will constitute a testable hypothesis. Projects of priority II will draw their justification in part from the justification of the project (or projects) of priority I to which they give support, and in part from propositions as to improvements in those priority I projects and perhaps economies in theie execution expected to follow from priority II project results. Similar arguments of justification will be made with regard to priority III projects. Justification of projects of priority IV will derive from evidence, presented by the research worker concerned, of the strength of his interest in and of his competence to undertake the work he proposes. The second part will be focussed upon a hypothesis with regard to the system to be studied; the lines of activity will converge upon a testing of the nominated hypothesis, and the volume of work on many of these lines will be determined by well-designed sampling plans. Parts 1 and 2 having been completed by research workers, with their estimates of staff time to be spent in the field and in laboratories, of equipment, services, and facilities required, the third part–a budgetary costing–will be completed by administrative staff.

From these documents a composite program budget can be prepared and decisions made as to whether one or more projects must be postponed or curtailed, or additional projects admitted. Once these decisions are made, and necessary adjustments are effected, the included projects are approved for execution.

Then follows the task of assigning responsibility for projects, of assembling project teams, and organizing support services. The last are of three kinds: scientific, technical, and administrative. Scientific services are such as age determination, plankton sorting, data storage and processing with statistics, and electroacoustics; and it may be noted that many of the priority II and III projects may be lodged in these services. The technical services are workshops, research vessel management, artists' studio, and documents reproduction. Library occupies a special place. Administration, in addition to its customary tasks of accounting, personnel records, and purchasing, has responsibility for logistic support–transport, travel arrangements, supplies, and communications.

From here the subject expands greatly, far beyond the scope intended for this chapter, to matters such as monitoring the progress of projects, regular reporting, publications, attendance at meeting, advising government, collaboration with industry, and a host of others.

13

The Outlook for Fisheries Research in the Next Ten Years

David H. Cushing

Introduction

The first stage of the industrialization of fisheries had been completed between 1879 and 1910: steam-driven purse seiners worked for menhaden off the eastern seaboard of the United States from 1870, steam trawlers operated in the North Sea from 1881 onwards, and steam longliners fished in the Pacific halibut fishery from the first decade of the present century. The second stage of industrialization took place in the 1950s and 1960s, with the introduction of factory methods aboard stern trawlers. At the beginning of this period, most stocks were unexploited (Graham 1951) but by the mid-1960s most were heavily fished; distant water fisheries expanded with freezer trawlers from Europe and the USSR, with pelagic longliners from Japan and Korea in the subtropical oceans (Rothschild and Uchida, 1968), and the four major upwelling areas (off California, Peru, North West Africa, and Namibia and South Africa) were exploited by purse seiners working for fish meal and oil. The present world catch is about 60 million tons (averaged from 1975 to 1979) and today most regions, but not all, on Graham's chart would be recorded as being overfished, that is, more fish are killed than are needed for the optimal or even the maximal yield. There were three components of demand: (1) that of the Russians and Japanese for a main source

---

David H. Cushing, a retired fisheries biologist, worked at the Fisheries Laboratory in Lowestoft on problems of marine production, exploitation of herring, acoustics in fisheries, and the dependence of recruitment on parent stock. He has provided advice on management in the International Council for the Exploration of the Sea and in the European Commission. He is a Fellow of the Royal Society of London and a Rosentiel mediallist.

of protein, (2) that of Western Europe and North America for frozen fish rather than lightly iced fish, and (3) that of the need for fish meal as an additive in animal food stuffs (as a cheap source of the essential amino acids lysine and methionine, absent in vegetable protein such as soya).

It was no accident that the need for management emerged as a consequence of industrialization. In the preindustrial times inshore fisheries were often subject to detailed regulation, but this may well have been for social and economic reasons for local markets could not absorb too much fish. Evidence that stocks might decline under the pressure of fishing did not appear until the North Sea trawl fisheries (Garstang, 1901-1903) and the Pacific halibut fishery (Thompson, 1936) had been practised for one or two decades. At that time the management of fish stocks was patently a problem between nations and scientific solutions were considered an independent basis of judgement between them. The International Council for the Exploration of the Sea (founded in 1902) and the International Halibut Commission (founded in 1922) were established for this purpose. Since the early decades of the century, fisheries laboratories and fisheries commissions have proliferated.

Throughout the world ocean there is now an overwhelming need for good management (FAO, 1980). The central problems are simple and were expressed to me by an owner on the quay in Yarmouth when the ancient East Anglian herring fishery was collapsing: he said, "Burn every bugger's boat but mine." Although his proposition was simple, execution is never so easy, and too often a stock of fish has collapsed while scientists, administrators, and politicians have sought agreement. In the past there have been failures both scientific and institutional and since the mid-1970s, the coastal states have extended fishery jurisdiction and have become responsible for the management of fish stocks in their waters. Extensions of jurisdiction were stimulated when foreign distant water fleets appeared too often off the shores of coastal states and because the international organizations for the conservation of fish stocks sometimes reacted too slowly and without results.

Because this volume is intended for a readership broader than the parish of fisheries biologists, a brief summary follows of the concepts and terms used. A fish population comprises a number of age groups to which a new recruitment is added each year as the youngest age group. The year-class passes through the ages of the stock, or population, and it is called a cohort. The central formulation in fisheries science (Ricker, 1948) is the catch equation, $C = F\bar{N}$, where C is catch in numbers, F is the instantaneous coefficient of fishing mortality, and $\bar{N}$ is the average number in the stock or in the stock at a given age. A string of catch equations by age within a cohort can be used with good estimates of F and M (the instantaneous natural mortality rate) for the oldest age group to determine F (or stock) by age within the cohort. The method is called cohort analysis. Following Murphy (1965) and Jones (1964), Gulland (1965, 1977) established a series of catch equations for each age group in a year-class; with the estimates of F and M in the oldest age group, computations proceed back in age to the youngest age group. Estimates are biased by poor estimates of M (Agger *et al.*, 1973) and those of the last three age groups in a year class are biased by the estimates of F (Pope and Garrod, 1975). As the calculation proceeds back up the year class, the catches in numbers improve the precision of estimates of F. Hence it became possible to construct a matrix of F or stock by age and year. With care, the method of cohort analysis (originally called virtual population analysis) can be used to estimate stock in the current year and hence catch, or quota, in the subsequent one. However, independent estimates of stock are needed from time to time if those from cohort analysis are not to diverge from true abundance, a potential and inherent bias.

The catch per unit of effort is the catch per unit time spent fishing and is an index of stock density. The time spent fishing, or fishing effort, f, is proportional to fishing mortality, i.e., $F = qf$, where q is the catchability coefficient. The yield in weight from the fishery is often expressed as a function of fishing effort and such a formulation provides the basis for management; its maximum is the maximum sustainable yield (MSY) and the more desirable optimum sustainable yield (OSY) is taken at a somewhat lower level of fishing effort. A stock is said to be overfished when the fishing effort is greater than that needed to take the OSY or the MSY. We may distinguish between growth overfishing and recruitment overfishing. Many fishes grow by an order of magnitude during their adult lives and if the fishing effort is too high the fish caught may be so small that the yield is less than the maximum. The solution to the problem of growth overfishing is to prevent the capture of small fish, with minimum landing sizes and mesh sizes in the cod end of the trawl regulated. For this purpose Beverton and Holt (1957) introduced the yield per recruit formulation as function of fishing mortality which provided a precise solution once fishing rate and growth pattern were determined, although the variability of recruitment was not included in their formulation. When a stock suffers from recruitment overfishing, the fishing effort is so high that the magnitude of recruitment declines, which diminishes the stock, and a vicious spiral toward collapse is generated.

Stocks were originally managed by adjusting mesh size or by closing seasons to prevent growth overfishing, but as fishing effort increased, catch quotas became necessary and they are now the rule, at least in the North Atlantic. A trawl catches many species of fish, each in different stock quantities; most fisheries are hence called mixed fisheries, which provide many opportunities for the more complex manipulation of catch quotas.

Before proceeding directly to the subject of research I shall set the stage by summarizing events in the management of fisheries (and the research which supported it). The summary, while comprehensive illuminates significant advances in fisheries research which have taken place in the last decade.

The Events that Have Dominated the Last Ten Years

By 1970, the Russian and Japanese fleets fished all over the world ocean; Russian trawlers worked the Atlantic and Pacific Coasts of North America, around the British Isles, in the Barents Sea, off many other coasts, and in the Southern Ocean. The Japanese and Koreans sailed across all three oceans equatorward of 40° latitude for tuna-like species and from the Bering Sea they landed large quantities of Alaska pollack. In the four major upwelling areas (off California, Peru, Northwest Africa, and southern Africa), stocks of sardine-like fishes were heavily exploited: these included the largest fishery in the world, that for the anchoveta off Peru which yielded twelve million tons in 1970, the year before it collapsed. In the northeast Atlantic, the herring and mackerel stocks collapsed in the late 1960s; catches of the two species dropped from an annual catch of more than 2 million tons to less than 300,000 tons. Most demersal stocks were heavily exploited from one side of the ocean to the other.

During the 1950s two solutions to the problem of overfishing were proposed by fisheries scientists. The first was to increase mesh sizes in the North Atlantic trawl fisheries to avoid growth overfishing (Beverton and Holt, 1957). At the time it was thought that simple enforcement of the proper mesh size would solve the problem of growth overfishing, but there were two difficulties. The first was that most trawl fisheries were mixed and the mesh size had to fit the smallest species. Second, if the mesh size has been fixed for such a reason fishing effort can still expand and this is what happened in the North Atlantic during the 1960s. To maintain the yield, mesh size should have been increased, but then the smallest species in the mixed fishery would be lost.

The second solution to the problem of overfishing proposed during the 1950s was developed by Schaefer (1954, 1957) with an approach analogous to that of Graham (1935); from the decrement in stock density in weight with increased fishing effort, a parabola of yield on fishing effort is derived from which the MSY is obtained. The method is simple but may be biased because stock density does not always decline as much as abundance when fishing effort increases (Ulltang, 1976 on the Atlanto-Scandian stock of herring; Garrod, 1977, on the Arcto-Norwegian cod stock); hence the estimated MSY may be too high. Ideally, the method should prevent recruitment overfishing, but the bias would have to be eliminated. The herring stocks in the northeast Atlantic collapsed because the dangers of recruitment overfishing were not sufficiently appreciated early enough by either scientists or managers. Major pelagic stocks off California, off Namibia, and off South Africa have all been reduced, probably by recruitment overfishing. The causes of collapse of the stock of Peruvian anchoveta were more complex (Kesteven *et al.*, 1977; Csirke, 1980; Cushing, 1981) because the great El Niño of 1972-1973 played a notable part in reducing two year classes independently of fishing; the lack of recovery by that stock since suggests that a condition of recruitment overfishing may have persisted after the collapse.

In the North Atlantic, many demersal stocks were overexploited in the period of expansion during the 1960s. The only ways in which the fishing could be controlled was by quota management or by limitation of effort. The control of fishing effort is in theory preferable because it affects fishing mortality directly, but the collection of effort statistics throughout the North Atlantic is of such appallingly poor quality that there was no choice but to accept quota regulation. The annual quota is set by international scientific agreement. It is then divided between nations.

Quotas for yellowfin tuna in the eastern tropical Pacific were introduced during the late 1950s and for most stocks in the northwest Atlantic in 1971 and in the northeast Atlantic in 1974. Cohort analysis devised in 1965 might have been developed for the purpose of estimating quotas. However the most important event in the 1970s was the establishment of 200-mile limits in early 1977 by many nations as a *de facto* (but not *dejure*) consequence of the Law of the Sea Conference. Management then became the overriding responsibility of the coastal state, but scientific advice was to be provided by a regional scientific organization comprising a number of coastal states. McHugh (1970) suggested that from past history within the United States, management by the coastal state might not be very effective because sovereignty was divided between federal and state administrations; in other words, successful management was the result of cooperation between nations. However, in 1976 the Canadians had demanded that all foreign fleets in the Canadian northwest Atlantic out to 200 miles reduce their effort by 40% and this demand was enforced in subsequent years. As consequence, the stocks of cod and of other species have increased considerably. Because of analogous restrictions, stocks in the United States northwest Atlantic out to 200 miles have also increased. The stock of Georges Bank haddock generated strong year classes in 1975 and 1978 following reduced effort (Anonymous, 1980), which implies that the stock had suffered previously from recruitment overfishing. Elsewhere in the world the consequences of coastal state management are less obvious because the tradition of such management is frail. In European waters, management has been poor since 1977 because the member states of the European Economic Community have not been able to agree on a Common Fisheries Policy. On the other hand, Norway, Iceland, and the Faroe Islands have been able to increase the mesh size of their trawls to the limit and have insisted that foreign fleets fishing in their waters do the same.

In the last 10 years one scientific problem has evaded any full solution, that of the dependence of recruitment upon parent stock. The problem is important because if displayed in a convincing manner recruitment overfishing could be avoided. Unhappily fisheries have collapsed because the cause of the decline in recruitment cannot be shown convincingly. Solutions have been proposed Beverton and Holt, 1957; Ricker, 1954; Shepherd, 1982; Shepherd and Cushing, 1980), but the variability of recruitment is so high that no choice of solution can be made. Because stock density may vary with abundance in a nonlinear manner, the Schaefer model can no longer be relied on to prevent recruitment overfishing, unless that bias is explicitly eliminated. Any other model requires some statement of the dependence of recruitment upon parent stock; in the early 1960s the problem was ignored by fisheries scientists working in the North Atlantic, but in the 1970s it was admitted and attempts were made to keep the biomasses of spawning stock higher than they had been. The collapse of pelagic stocks throughout the world was a consequence of scientific failure to reduce the problem to a manageable proportion, whether the decline of recruitment was due to fishing or to environmental factors.

The Next Ten Years

A forecast of the future can be made only through the windows of today. Some guesses can be made on how fisheries might develop if present trends continue, how fisheries science might grow as a discipline in response to these trends, and how managers might act to create good conservation.

The Development of Fisheries

Fish are pursued by fishermen of various skills, on boats large and small, for diverse markets. In Europe and North America the present consumption of fish remains roughly steady. In the past fish was considerably cheaper than meat, but as the price of fuel oil rises the difference in price between them diminishes.

The use of fish meal in animal foodstuffs is needed to add essential amino acids to vegetable protein. Fish is dried and converted to meat at a loss in weight in the transfer of a factor of 13 to 15. It would be desirable to use the fish in fish meal for human consumption; during the period of high catches in the Peruvian anchoveta fishery, children in Peru suffered from the protein deficiency disease, kwashiorkor. Within the next 10 years the essential amino acids may be obtained by other means, in which case the demand for fish meal may decline. Yet fisheries for fish meal have provided money and employment in the less developed countries and in any case a demand for fish oil would remain. Perhaps fisheries for fish meal will decline somewhat.

Some fishes are very highly priced (shrimp, sole, tuna) and each supplies a specialized market. Production of shrimp throughout the world has increased enormously since the early 1950s and it provides a luxury trade in the developed countries and hard currency for the less developed ones. Most of the North Sea sole is sent to northern Italy where it has augmented small, but high-valued catches from the northern Adriatic. Tuna is esteemed throughout the developed world. It is mostly canned, but in Japan it is especially valuable where it is eaten raw as sushima. The bluefin tuna has a special significance. Each of the three supplies luxury markets and the question arises whether more such markets develop in the next 10 years. If the developed world becomes more affluent (if less quickly than in the last three decades) other such outlets might emerge for herring or even cod.

Vessels will almost certainly change their character. Large distant water boats will finally vanish from waters foreign to them except where they are allowed to continue to fish by license. A small proportion of larger vessels will still be needed to work in bad weather in winter, for example, off Iceland and on the Grand Banks. The standard vessel of the coastal state will be smaller, around 80 to 100 feet in length in stormy waters and up to 80 feet in length in calmer seas. They will tend to be more highly mechanized with smaller crews. The four basic gears will still be used, line, drift net, trawl (and Danish seine), and purse seine; of these the expensive and labor-intensive drift nets will survive only in special circumstances. Lines handled mechanically may become more popular as a market for well-presented fish develops; tuna longlining will survive because the price is high but with fewer fishermen and shorter trips. Purse seines may become smaller as vessels decrease in size. Trawls will continue to be worked over the stern but with more mechanical handling.

In recent years the price of fuel oil has had a severe effect upon fisheries, particularly in those years when the price increment exceeded that of inflation. In the developed countries real attempts are being made now to conserve energy, for example with smaller cars or wood-burning stoves. There is no doubt that such trends will continue but perhaps the sharp increases in fuel oil prices will not recur. However, it is not easy to foresee how energy may be conserved in fishing vessels. Present Dutch beam trawlers in the North Sea use twice the power of distant water vessels in the 1950s. The simplest saving would be achieved by reducing the sizes of the boats, but it is to be hoped that more efficient engines and deck machines will be used. Industrialization came late to the fishing industry because, for the sailing smacks, the wind and tide were free. I do not believe that we shall see again the lovely sight of tan sail and tall mast at sea, but some vessels may seek some assistance from the wind.

Since the 1950s the west Greenland cod fishery has collapsed and that off Iceland has probably been reduced, and both declines are probably due to climatic deterioration, a cooling which has taken place between about 1945 and 1970 (Cushing, 1982a). Like other events in the future the climate in the next 10 years cannot be foreseen, but it may be asserted that change will take place. For example, it is likely that the increase in *Calanus* in the early 1950s in the northeastern North Sea, which persisted, has affected three fisheries: herring grew more quickly and consequently yield increased; the recruitment to the cod stock increased by a factor of three in 1962 and subsequently; and perhaps the great industrial fisheries for Norway pout, sprat, and sandeel owe their origin to the same cause. From a study of climate and fisheries, changes take place by decades and all we may foresee is that during the 10 years events associated with climatic change may occur. Indeed if large changes in abundance occurred in the past, we must expect them in the future.

In summary, the markets for fish may change a little during the next decade, perhaps toward essential protein for people in the less developed countries and toward a quasiluxury market in the developed countries. Most, but not all, the distant water vessels will disappear and the typical vessel will be smaller, more efficient, and manned by fewer fishermen. The price of fuel may well play a part in reducing the great excess number of vessels throughout the world.

The Development of Fisheries Science

The prominent problems which confront the fisheries biologists today are (a) the need for stock estimates independent of present methods based on numbers or weight of fish caught, (b) the need to assess stocks as a group which are mixed in the gears that capture them, (c) the need to assess predation on a fish stock at all ages in order to improve estimates of recruitment and natural mortality, and (d) the need to discover the mechanisms by which recruitment is generated.

Hence, during the next decade fisheries science may develop in three directions: (1) stock estimation independently of catches; (2) interactions between species, biological or operational; and (3) the generation of recruitment and its dependence on parent stock. Probably the most important addition to present practice is the provision of an adequate statistical base; skippers' logs should be mandatory and should report catches of species haul by haul on a daily basis. Such a system reports the time spent fishing, the fishing effort, automatically, and the catch (and catch per unit of effort) by daily position. Such systems exist already in a few places and in the next decade they may well be adopted broadly throughout the world.

Already annual cohort analyses are supported by independent estimates of stock, egg survey, groundfish survey, and acoustic survey. Recently, Lockwood *et al.*, (1981) showed that the errors of egg survey are relatively low; the technique is relatively simple but is expensive in ship time. Groundfish surveys are now made on the eastern seaboard of the United States and in the North Sea. They are also expensive in ship time, but necessary; in the course of the next decade they will be improved by better estimates of catchability, q, derived from the proper collection of statistics on fishing effort with skippers' logs. At the present time acoustic survey is not used to estimate stock quantities, although many exploratory surveys have been made. It is desirable to count fish larger than a specified size, but an integrator records biomass, which may include fishes smaller than that size; a solution may be to count and integrate in the same transmission.

With present cohort analysis, increments of weight from age to age are entered at the last stage of calculation to convert stock in numbers to stock in weight. During the next decade as the time series lengthen, studies of density dependence will proliferate and the study of growth per se will become more intense. Forney (1977) and Pope (1980) have developed a form of cohort analysis based on the gut contents of predators which describes mortality due to predation and the small consequential modifications to recruitment; this is the first step towards an analytical approach to the study of natural mortality, although Pauly (1980) has made a considerable step based on comparative growth studies.

At the present time it is fashionable to dream of multispecies models. The first and simple stage is the study of mixed fisheries (Brander, 1977; Brown *et al.*, 1975). In any trawl haul a number of species is caught and a greater number in lower latitudes. Nearly all fisheries, save those in the highest latitudes, comprise more than one species and the problem is to sum the stocks, partition their catchabilities, or analyze the interactions between them. Much of this work is well underway and by the end of the next decade most assessments will be based upon mixed fisheries. A useful discipline would be to compare the assessment of the mixed stocks with the sum of those of single species; they should be the same but not until the biological interactions, if any, have been discovered assuming that the catchabilities have been properly partitioned between stocks.

There are more ambitious multispecies models which seek to describe ecosystems. The simplest and the best is Steele's (1974) model of the North Sea ecosystem, which divides the energy output from the herbivores into three parts –invertebrate carnivores, pelagic fish, and faecal pellets; the latter are transferred to the benthos, which supports the adult demersal fish stocks. As constructed, it is a fixed engine and let us consider how it may vary. First, primary production fixed in Steele's model at 90 g carbon per square meter per year probably varies by a factor of three (Boalch *et al.,* 1978, from a decade of radiocarbon samples from the western English Channel). Bakun and Parrish (1981) have associated recruitment with primary production for three or four stocks in different parts of the upwelling system off California, northern anchovy, sardine, hake, and Dungeness crab. Yamanaka and Yamanaka (1970) associated high recruitment of several tuna species with cool water and vice versa, which is presumably a link with the offshore divergences or primary production. The failure of the two anchoveta year classes off Peru in the El Niño of 1972-1973 is associated with a decline in primary production of a factor of three. Necessarily a link between primary production and recruitment is not simple and so it may surprise us that such correlations relationships do in fact emerge. If primary production varies by a factor of three, such differences may be transmitted directly to the variability of recruitment. A second source of variation stems from the different spawning seasons of fishes in waters poleward of 40° latitude, each of which occurs at slightly different times during the spring or during the summer. In the North Sea and the northeast Atlantic the time of onset of the outburst of primary production, as indicated by the continuous plankton recorder network, has delayed by about a month since 1948 (Glover *et al.*, 1972). If species spawn at a fixed season, such a delay would lead us to expect some fish species to be replaced by others in the course of time. Such changes might account for the events in the Russell cycle (Russell, 1973). Hence as climate changes across the decades, changes in the proportions of fish stocks are to be expected. The third source of variability is the ratio of energy transferred to invertebrate carnivores and recruiting fish. Differences in recruitment are probably determined in the lives of larvae or early juveniles of both pelagic and demersal fish and the transfer of energy to the adult stock is irrelevant to recruitment. The gadoid outburst in the North Sea may have been caused by change from small copepods to *Calanus,* one of the larger copepods (Cushing, 1982b). The three sources of variation, in primary production, in the representation of fish species, and in the ratio of energy flows from the herbivores, are the most important sources of change in recent decades. Perhaps in the next decade analogous changes will occur but probably within the same group of agents.

The application of the three sources of variation to Steele's North Sea model is a multispecies structure which might describe some changes in recruitment from year to year or perhaps even from decade to decade. Such a structure does not exist and has been described merely to indicate the scale of experiment which might be undertaken. During the next decade such models will proliferate, some fairly simple and some complex, and for reasons of parsimony the simpler ones will survive.

A converse procedure suggests that the problem of the dependence of recruitment parent stock will evaporate with full information on growth and mortality of all species at each trophic level, or a hypothesis will be induced when enough data have been collected, a claim refuted by Popper (1963). Progress will be made toward the solution of this problem, if only in the addition of 10 extra data points. However, more information will be collected on the larval and post-larval lives of fishes, and a train of models of their growth and death may well be tested during the next decade. The problem must be solved because recruitment overfishing is prevented at the present time only by increasing the size of the spawning stock. What is needed is a formulation as precise as the yield per recruit.

In summary, during the next decade, fisheries science will develop in three ways: (1) in polishing procedures associated with cohort analysis, (2) in small but experimental multispecies models, and (3) in continuing the slow progress toward solution of the stock and recruitment problem. It will remain slow because the integration of the growth and mortality of larval and juvenile fishes demands expensive work at sea or in artificial enclosures.

The Development of Management

The next decade will show how well the coastal state manages the fish stocks which it often shares with a neighbor. One or two coastal states may rely primarily on imported fish, others may license their grounds to foreigners, but most will have started to manage the fish stocks under their protection. The quality of management will depend finally upon the quality of science in the regional organization. However the stocks are managed, conservation rests on the consent of fishermen. A convenient summary of present views of management is given in Anonymous (1977); FAO (1980).

Probably the most difficult task is how to manage mixed fisheries properly partly because the problem often awaits scientific solution. In the 1950s the management problem appeared simple in the trawl fisheries, to increase mesh sizes throughout the North Atlantic. We have already seen that this was not enough, but in the Gulf of Thailand now a number of additional measures would be needed. The real problem in a mixed fishery is that the component stocks are often in different states of exploitation. Given minimum mesh sizes and minimum landing sizes to secure the stocks from the excesses of growth overfishing, quotas can be used to regulate fishing mortality in the stocks. However, sometimes the quotas conflict: a restricted quota for one species might result in too low a catch of another which would infuriate the fishermen. There is a sense in which quota regulation may become, as was mesh regulation, a least regulation. Recently, Brander (1981) has shown that the skate has virtually disappeared from the Irish Sea; perhaps the only form of conservation would have been to discard these hardy animals after capture. The point is that quota management may require additional forms of regulation, closed areas, closed seasons, etc. We cannot foresee how regulations will develop in the next decade, but they may become more complex, which is perhaps undesirable.

There is a particular difficulty with pelagic fisheries exploited by purse seiners, as noted above. The purse seine allows no escape by small fish and the fisherman cannot distinguish small fish from large fish before he sets the net; if he catches small fish they are discarded and some or all die unrecorded. Pelagic stocks tend to suffer from recruitment overfishing and immature fish should not be caught. The disease might be mitigated if skippers could be persuaded to record their discarded catches, but in the end the activity of this very efficient gear will have to be restrained when immature fish are caught unnecessarily.

Perhaps the most important development in management during the next decade will be the resolution of conflicts between fishermen in the light of conservation. The industrial fishery for Norway pout in the North Sea in the early 1970s took a greater weight of very small haddock and whiting than was caught in the adult fisheries. The United Kingdom government prohibited an area off the Scottish coasts to industrial fishermen. After argument in the European Commission and work at sea by United Kingdom research vessels, a modified area was developed in which the industrial fishermen would be free to catch Norway pout in deep water regions within the prohibited area. There have been analogous conflicts between tuna fishermen off California and conservationists concerned about porpoises caught in the purse seines and between *Nephrops* fishermen in British waters and those that need the white fish discarded in the *Nephrops* fishery. The tuna/porpoise conflict was resolved by devising a seine which retains the tuna but allows the porpoise to jump out. The *Nephrops* conflict will be resolved when the fishermen realize that with larger meshes they will catch larger prawns.

Such conflicts are the result of mixed fisheries in which one part of the catch is desired and the other part is discarded dead. The resolution depends upon advanced in gear design (as in the tuna/porpoise conflict) or upon an optimal mesh size for all species. The latter depends upon a form of multispecies model in which the exploitation of the two or more species is optimized. Present conflicts have often arisen through malpractice: it is not necessary to catch *Nephrops* with small-meshed nets, particularly when the market size is greater than that retained by the best mesh size. In the next decade such practices will tend to disappear as management slowly improves, but different conflicts will arise, particularly as more complex regulations appear.

In recent years the proper objective of management has been discussed in terms of maximum sustainable yield, optimum sustainable yield, or maximum economic yield. In particular, the maximum sustainable yield has been defined in exact terms (Anonymous, 1977) and an economic yield has also been defined in terms of fishing mortality. In brief, a maximum can be defined but should never be reached; yet the present trouble is that many stocks, particularly in the northeast Atlantic, are exploited well beyond the maximum. Let us suppose that management improves in the next decade and that exploitation is less than that needed to yield the maximum. Then if regulations become more complex with perhaps more conflicts, an exact technical objective may become harder to obtain or even become somewhat obscure. Perhaps the objective will be best stated as the maximum sustainable yield in its self-evident sense as it is written into certain treaties.

The progress of management depends upon the development of fisheries science and upon the consent of fishermen. The initial step in the reduction of fishing effort is the hardest one and the rising price of fuel oil has probably played a predominant part at least in the North Atlantic. The prospect is of more complex science, an array of regulations and an increase in conflict between interests. On the other hand, if they are successful, as we must believe, the consent of the fishermen will be obtained more easily. The paradox is explained easily in that with more restraint on fishing effort in all directions stock densities will rise and so will the average catch on each vessel for all to see. Hence profits to each vessel should also rise.

Fishermen know that the highest catches and the best profits were made in the North Sea after both the world wars in this century. All desire to return to this happy state but are locked in a competitive rack of falling catches. Once they learn to spend the gains on anything other than an extra boat, the fishermen will accept the changes in regulation even if they become more complex. This is an optimistic view based on the fact that fishermen do believe in conservation and accept new regulations provided that they understand them. Perhaps the most important job in the next decade will be the long process of persuasion that accompanies any new proposal, and this job must fall on the fisheries scientists. They have learned to go to sea, to develop fairly complex models of how the populations work, and to translate the results of their researches into regulations, but they are not always good advocates and this is a trade they must learn.

Summarizing, management will become more complicated with an array of choices. Fisheries science will depend more on models which will display that array. Because such models will be somewhat complex, there will be increased need for independent estimates of stocks to make the estimates confident. The essential tool of management will be the consent of the fishermen to the advocacy of options by the fisheries scientists.

Conclusion

Despite the fact that many of the old markets for fish have disappeared as stocks have been reduced, the demand for fish in the world will remain. In the developed countries it may be quasi-luxury market that predominates rather than an essential one, fish which you like rather than protein which is cheaper. It is to be hoped that fish is fed to people in the less developed countries rather than to animals in the developed ones with a considerable loss.

Coastal state management will stand or fall by its success in the next decade. The important point is that the coastal state has the power to act. The essential step is to make a clear action the results of which fishermen can recognize; a small cut in fishing effort may only disappear in the noise and a severe and quick reduction in fishing effort can be justified scientifically. In Norway, Iceland, and the Faroe Islands, dramatic increases in mesh sizes have been made since 1977 and in Canada and the United States large reductions in fishing effort have been obtained by the exclusion of foreign fleets. Although the price of fuel oil discourages the expansion of fisheries, the sharp reduction in fishing effort by the coastal state is the prime step to be taken. Once the dramatic step has been made and the consent of the fishermen obtained, progress in the more complicated stages of management will become possible. In an area like the North Sea, one might imagine a system of annual changes, as part of more long-term objectives. To obtain the best value from the fishery various modifications to management aims will be developed.

During the next decade, fisheries science will become more confident as the data sets lengthen supported by independent stock estimates. Today the error on an estimate of stock may be low, about 10%, which could be reduced by spending more money or, as stocks will increase in numbers, reducing variance. The science always has been cooperative in an international sense and will become more so under coastal state management because most fish migrate, if no further than into the waters of a neighbor. In the past, under international control, stocks were protected by blanket regulation, but under coastal state management , contiguous states must negotiate with each other in detail. Because of the high data requirement there will be greater cooperation at sea for research vessels tend to become larger than fishing vessels and their costs might be shared.

Since the industrialization of fisheries, managers have hoped that an ideal step can be taken and that the problem of the depletion of stocks would have been solved once and for all. Now that the coastal state can in fact take that step with a sharp reduction of effort, it is becoming clear that management must continue. Indeed it probably will become more complex.

References

Agger, P., I. Boetius, and H. Lassen. 1973. Error in the virtual population analysis. The effect of uncertainties in the natural mortality coefficient. J. Cons. Int. Explor. Mer. 35:93.

Anonymous. 1977. Report of the Ad Hoc Meeting on the Provision of Advice on the Biological Basis for Fisheries Management. ICES Coop. Res. Rept. 62, 16 pp.

Anonymous. (1980). Summary of Status of the Stocks, December 1980. Northeast Fisheries Center, Lab. Rept. Doc. 80-37, National Marine Fisheries Service U.S. Dept. Commerce.

Bakun, A., and R. Parrish. 1981. Environmental Inputs to Fishery Population Models for Eastern Boundary Current Regions. Workshop Rept. 28, IOC UNESCO pp. 1-37.

Beverton, R. J. H., and S. J. Holt. 1957. On the dynamics of exploited fish populations. Fish. Invest. London 2. 19. 533 pp.

Boalch, G. T., D. S. Harbour, and E. I. Butler. 1978. Seasonal phytoplankton in the western English Channel 1964-1974. J. Mar. Biol. Assor. UK NS 58(4): 943-954.

Brander, K. J. (1977). The management of Irish Sea fisheries; A Review. Lab. Leaflet 36, Min. Agr. Fish. Food, London, 40 pp.

Brander, K. J. (1981). Disappearance of common skate *Raia batis* from the Irish Sea. Nature 290(5801):48-49.

Brown, B. E., J. A. Brenhan, M. D. Grosslein, E. G. Heyerdahl, and R. C. Hennemuth. 1975. The effect of fishing on the marine finfish biomass in the Northwest Atlantic from the Gulf of Maine to Cape Hatteras. Res. Bull. Int. Comm. North West Atl. Fish. 12:49-68.

Csirke, J. 1980. Recruitment in the Peruvian anchovy and its dependence on the adult population. Rapp. Proc. Verb. Cons. Int. Explor. Mer 177:307-313.

Cushing, D. H. 1981. The effect of El Niño upon the Peruvian anchoveta stock. *In*: F. A. Richards (ed.), Coastal Upwelling. Amer. Geophys. Union, pp. 449-457.

Cushing, D. H. 1982a. A simulacrum of the Iceland cod stock. J. Cons. Int. Explor. Mer. 40(1):27-36.

Cushing, D. H. 1982b. Sources of Variability in the North Sea Ecosystem. Meeting on the North Sea. University of Hamburg, 3-8 September, 1981.

FAO. 1980. ACMRR Working Party on the Scientific Basis of Determining Management Measures. FAO Fish. Rept. 236, 149 pp.

Forney, J. L. 1977. Reconstruction of yellow perch (*Perca flavescens*) cohorts from examination of walleye (*Stizostedion vitreum vitreum*) stomachs. J. Fish. Res. Bd. Can. 34(7):925-932.

Garrod, D. J. 1977. The North Atlantic cod. *In:* J. A. Gulland (ed.), Fish Population Dynamics. John Wiley, New York, pp. 216-242.

Garstang, W. 1901-1903. The impoverishment of the sea. J. Mar. Biol. Assoc. UK NS 6:1-70.

Glover, R. S., G. A. Robinson, and J. M. Colebrook. 1972. Plankton in the North Atlantic—an example of the problems of analyzing variability in the environment. *In:* M. Ruivo (ed.), FAO Marine Pollution and Sea Life. Fishing News (Books), West Byfleet, Surrey, and London, pp. 439-445.

Graham, G. M. 1935. Modern Theory of exploiting a fishery and application to North Sea trawling. J. Cons. Int. Explor. Mer 10(2):264-274.

Graham, G. M., Chairman. 1951. Report of discussion on developing Fishery Resources. Proc. United Nations Sci. Conf. on the Conservation and Utilization of Resources. Vol. 7, Wildlife and Fish Resources. pp. 60-66.

Gulland, J. A. 1965. Estimation of mortality rates. Annex to Arctic Fisheries Working Report. ICES CM 1965. (Mimeo).

Gulland, J. A. 1977. The analysis of data and development of models. *In:* J. A. Gulland (ed.), Fish Population Dynamics. John Wiley, New York pp. 67-95.

Jones, R. (1964). Estimating population size from commercial statistics when fishing mortality varies with age. Rapp. Proc. Verb. Cons. Int. Explor. Mer 155:210-214.

Kesteven, G. L., J. A. Gulland, R. Jones, R. Barber, and L. Boerema. 1977. Report of the consultative group convened by the Minister of Fisheries in Peru to advise him on the state of the stocks of anchoveta and other pelagic species and on the course of action taken for the management of the fishery. Lima, Peru. July 1977, 17 pp.

Lockwood, S. J., J. H. Nichols, and W. A. Dawson, 1981. The estimation of a mackerel (*Scomber scombrus* L.) spawning stock size by plankton survey. J. Plankt. Res. 3(2):217-234.

McHugh, J. L. 1970. Trends in Fishery Research. Amer. Fish. Soc. Spec. Publ. 7, pp. 25-56.

Murphy, G. I. 1965. A solution to the catch equation. J. Fish. Res. Bd. Can. 22(1):191-202.

Pauly, D. 1980. On the interrelationship between natural mortality growth parameters and mean environmental temperature in 175 fish stocks. J. Cons. Int. Explor. Mer 38(2):175-192.

Pope, J. G. 1975. The application of mixed fisheries theory to the cod and redfish stocks of Subarea 1 and Division 3K. Selected Papers Int. Comm. North West Atl. Fish. 1:163-169.

Pope, J. G. 1980. Phalanx analysis: an extension of Jones' length cohort analysis to multispecies cohort analysis. ICES CM G. 19, 6 pp. (mimeo.)

Pope, J. G., and D. J. Garrod. 1975. Sources of error in catch and effort quota regulations with particular reference to variation in the catchability coefficient. Res. Bull. Int. Comm. North West Atl. Fish. 11:17-30.

Popper, K. 1963. Conjectures and Refutations. 412 pp. Routledge and Kegan Paul, London.

Ricker, W. E. 1948. Methods of Estimating Vital Statistics of Fish Populations. Indiana Univ. Publ. Sci. Ser. 15, 101 pp.

Ricker, W. E. 1954. Stock and recruitment. J. Fish. Res. Bd. Can. 11:559-623.

Rothschild, B. J., and R. N. Uchica. 1968. The tuna resources of the oceanic regions of the Pacific Ocean. Univ. Washington Publ. Fish. NS 4, pp. 19-51.

Russell, F. S. 1973. A summary of the observations of the occurrence of planktonic stages of fish off Plymouth 1924-72. J. Mar. Biol. Assoc. UK NS 53: 347-355.

Schaefer, M. B. 1954. Some aspects of the dynamics of populations important to the management of the commercial fish populations. Bull. Inter-Amer. Trop. Tuna Comm. 1(2):27-56.

Schaefer, M. B. 1957. A study of the dynamics of fishery for yellow-fin tuna in the eastern tropical Pacific Ocean. Bull. Inter-Amer. Trop. Tuna Comm. 2: 245-285.

Shepherd, J. G. 1982. A versatile new stock and recruitment relationship for fisheries and the construction of sustainable yield curves. J. Cons. Int. Explor. Mer 40(1):67-75.

Sheperd, J. G., and D. H. Cushing. 1980. A mechanism for density dependent survival of larval fish as the basis of a stock-recruitment relationship. J. Cons. Int. Explor. Mer. 39(2):160-167.

Steele, J. H. 1974. The structure of marine ecosystems. Harvard Univ. Press, Cambridge, MA, 123 pp.

Thompson, W. F. 1936. Conservation of the Pacific Halibut. An International Experiment. Smithsonian Report for 1935, pp. 361-382.

Ulltang, O. 1976. Catch per unit of effort in the Norwegian purse seine fishery for Atlanto-Scandian (Norwegian spring spawning) herring. FAO Fish. Tech. Pap. 155, pp. 91-101.

Yamanaka, I., and H. Yamanaka. 1970. On the variation of the current pattern in the equatorial western Pacific Ocean and its relationship with the yellowfin tuna stock. Proc. 2nd CSK Symposium, Tokyo. 527-533.

Index

Springer Series on Environmental Management
Robert S. DeSanto, Series Editor

Natural Hazard Risk Assessment and Public Policy
Anticipating the Unexpected
by **William J. Petak** and **Arthur A. Atkisson**

This volume details the practical actions that public policy makers can take to lessen the adverse effects natural hazards have on people and property, guiding the reader step-by-step through all phases of natural disaster.

1982/489 pp./89 illus./cloth
ISBN 0-387-**90645**-2

Gradient Modeling
Resource and Fire Management
by **Stephen R. Kessell**

"[The] approach is both muscular enough to satisfy the applied scientist and yet elegant and deep enough to satisfy the aesthetics of the basic scientist . . . Kessell's approach . . . seems to overcome many of the frustrations inherent in land-systems classifications."

—*Ecology*

1979/432 pp./175 illus./27 tables/cloth
ISBN 0-387-**90379**-8

Disaster Planning
The Preservation of Life and Property
by **Harold D. Foster**

"This book draws on an impressively wide range of examples both of man-made and of natural disasters, organized around a framework designed to stimulate the awareness of planners to sources of potential catastrophe in their areas, and indicate what can be done in the preparation of detailed and reliable measures that will hopefully never need to be used."

—*Environment and Planning A*

1980/275 pp./48 illus./cloth
ISBN 0-387-**90498**-0

Springer Series on Environmental Management
Robert S. DeSanto, Series Editor

Air Pollution and Forests
Interactions between Air Contaminants and Forest Ecosystems
by **William H. Smith**

"A definitive book on the complex relationship between forest ecosystems and atomospheric deposition . . . long-needed . . . a thorough and objective review and analysis."

—*Journal of Forestry*

1981/379 pp./60 illus./cloth
ISBN 0-387-**90501**-4

Global Fisheries
Perspectives for the '80s
Edited by **B.J. Rothschild**

This timely, multidisciplinary overview offers guidance toward solving contemporary problems in fisheries. The past and present status of fisheries management as well as insights into the future are provided along with particular regard to the effects of the changing law of the sea.

1983/approx. 224 pp./11 illus./cloth
ISBN 0-387-**90772**-6